D1088189

THE ONCE AND FUTURE TURING: COMPUTING THE WORLD

Alan Turing (1912–1954) made seminal contributions to mathematical logic, computation, computer science, artificial intelligence, cryptography and theoretical biology.

In this volume, outstanding scientific thinkers take a fresh look at the great range of Turing's contributions, on how the subjects have developed since his time, and how they might develop still further.

These specially commissioned essays will provoke and engross the reader who wishes to understand better the lasting significance of one of the twentieth century's deepest thinkers.

Until his death in 2015, S. BARRY COOPER was Professor of Mathematical Logic at the University of Leeds. He was both a leading figure in the theory of Turing computability, and a noted advocate of multidisciplinary research, especially in connection with the theoretical and practical limits of the computable. As President of Computability in Europe, he was responsible for many international conferences. He chaired the Turing Centenary Committee, and edited the prize-winning critical edition of Turing's publications: *Alan Turing – His Work and Impact.*

ANDREW HODGES is a Senior Research Fellow at the Mathematical Institute, University of Oxford. His main research is in fundamental physics, as a colleague of Roger Penrose, but he is also the biographer of Alan Turing. His book *Alan Turing: The Enigma* (1983) has reached a wide audience and has inspired works of drama, music, art and film.

THE ONCE AND FUTURE TURING

Computing the World

S. BARRY COOPER
University of Leeds

ANDREW HODGES
University of Oxford

CAMBRIDGE
UNIVERSITY PRESS

University Printing House, Cambridge CB2 8BS, United Kingdom

Cambridge University Press is part of the University of Cambridge.

It furthers the University's mission by disseminating knowledge in the pursuit of education, learning and research at the highest international levels of excellence.

www.cambridge.org
Information on this title: www.cambridge.org/9781107010833

First published 2016

Printed in the United Kingdom by Clays, St Ives plc

A catalogue record for this publication is available from the British Library

ISBN 978-1-107-01083-3 Hardback

Contents

Contributors

Scott Aaronson, Department of Electrical Engineering and Computer Science, Massachusetts Institute of Technology, Cambridge MA 02139, USA.

aaronson@csail.mit.edu, www.scottaaronson.com

Ruth E. Baker, Wolfson Centre for Mathematical Biology, Mathematical Institute, Andrew Wiles Building, Radcliffe Observatory Quarter, Woodstock Road, Oxford OX2 6GG, UK.

baker@maths.ox.ac.uk

Andrew R. Booker, Department of Mathematics, University of Bristol, University Walk, Clifton, Bristol BS8 1TW, UK.

www.maths.bris.ac.uk/~maarb

The late S. Barry Cooper, School of Mathematics, University of Leeds, Leeds LS2 9JT, UK.

Martin Davis, 3360 Dwight Way Berkeley CA 94704–2523, USA.

www.cs.nyu.edu/faculty/davism/

Solomon Feferman, Department of Mathematics, Stanford University, Stanford CA 94305–2125, USA.

math.stanford.edu/~feferman/

Eamonn A. Gaffney, Wolfson Centre for Mathematical Biology, Mathematical Institute, Andrew Wiles Building, Radcliffe Observatory Quarter, Woodstock Road, Oxford OX2 6GG, UK.

gaffney@maths.ox.ac.uk

Richard Gordon, Embryogenesis Center, Gulf Specimen Marine Laboratory, Panacea FL 32346, USA, and C.S. Mott Center for Human Growth & Development, Department of Obstetrics & Gynecology, Wayne State University, Detroit MI 48201, USA.

http://tinyurl.com/DickGordon

Douglas Richard Hofstadter, Center for Research on Concepts and Cognition, Indiana University, 512 North Fess Avenue, Bloomington, IN 47408, USA.

www.soic.indiana.edu/people/profiles/hofstadter-douglas.shtml

Martin Hyland, DPMMS, Centre for Mathematical Sciences, Cambridge University, Wilberforce Road, Cambridge CB3 0WB, UK.
https://www.dpmms.cam.ac.uk/people/j.m.e.hyland/

Stuart Kauffman, Departments of Mathematics and Biochemistry, University of Vermont, USA.
http://en.wikipedia.org/wiki/Stuart_Kauffman

Philip K. Maini, Wolfson Centre for Mathematical Biology, Mathematical Institute, Andrew Wiles Building, Radcliffe Observatory Quarter, Woodstock Road, Oxford OX2 6GG, UK.
https://people.maths.ox.ac.uk/maini/

Kanti V. Mardia, School of Mathematics, University of Leeds, Leeds LS2 9JT, UK; and Department of Statistics, University of Oxford, Oxford OX1 3TG, UK.
http://www1.maths.leeds.ac.uk/~sta6kvm/

Ueli Maurer, Department of Computer Science, ETH Zürich, CH-8092 Zürich, Switzerland.
http://www.crypto.ethz.ch/~maurer/

Roger Penrose, Mathematical Institute, University of Oxford, Andrew Wiles Building, Radcliffe Observatory Quarter, Woodstock Road, Oxford OX2 6GG, UK.
http://www.maths.ox.ac.uk/people/profiles/roger.penrose

Christof Teuscher, Portland State University, Department of Electrical and Computer Engineering, P.O. Box 751, Portland OR 97207-0751, USA.
christof@teuscher.ch, http://www.teuscher-lab.com

Philip D. Welch, Department of Mathematics, University Walk, Clifton, Bristol BS8 1TW, UK.
http://www.maths.bris.ac.uk/people/faculty/mapdw/

Stephen Wolfram, Wolfram Research Inc., 100 Trade Center Drive, Champaign, IL 61820, USA.
http://www.stephenwolfram.com

Thomas E. Woolley, Wolfson Centre for Mathematical Biology, Mathematical Institute, Andrew Wiles Building, Radcliffe Observatory Quarter, Woodstock Road, Oxford OX2 6GG, UK.
woolley@maths.ox.ac.uk

Preface

This volume was the inspiration of the mathematical logician Barry Cooper. In 2007 he was already planning a conference to mark the centenary of Alan Turing's birth, but this was just the start of his huge and energetic dedication to a renaissance of Turing studies. In 2009, while taking on the project of a new critical edition of Turing's papers, he also raised with Cambridge University Press (and with me) an idea for a book about 'Turing and the future of computing'. After correspondence with David Tranah and Silvia Barbina of the Press, Barry and I conceived it as an opportunity for leading scientific thinkers to bring exciting and challenging aspects of Turing's legacy to a wider readership.

In 2010, we settled on the title of *The Once and Future Turing*, and extended invitations on this basis. Our enterprise rested on Barry Cooper's networking power, arising from his role as chair of *Computability in Europe* and countless conference committees. More profoundly, it was steeped in his intellectual quest to see logic interact with new advances in physical and human sciences. His very individual vision of 'computing the world' is prominent in the subtitle and introductions for the five Parts of the book, which were mainly his responsibility. My own contributions, including the overall introduction, revolved around the *Turing Once*. Barry wrote for the *Turing Future*.

Unfortunately, just as this book was in the last stages of preparation, he met a rather sudden death. It was a particular sadness for him that he was not able to see the publication of this book. He expressed heartfelt thanks to everyone involved at Cambridge University Press, a gratitude I keenly share, and, above all, to the generous work of the distinguished and ever-patient contributors. Their writing, in so many different ways, reflects the magic of time and human life and will speak to a future that neither Alan Turing nor Barry Cooper could live to see.

Andrew Hodges
January 2016

Introduction

Alan Turing's short life ran from 1912 to 1954. The inspiration for this volume lay in the centenary of his birth. But Barry Cooper and I, as editors, wanted the word 'future' in our title, as well as a reference to the past. We chose the provocative title *The Once and Future Turing*, alluding to the legend of King Arthur's tomb. We invited a range of distinguished contributors to give us snapshots of scientific work which rest upon Turing's original discoveries, and share the spirit of his thought, but which also give a glimpse of something lying beyond the present. The result is a volume of 15 papers, whose authors responded to our prompting in utterly different ways.

Turing himself was not reticent about advancing visions of the future. Famously, he did so in his classic 1950 paper 'Computing machinery and intelligence'. He was not always right. Few people would claim that his 50-year prediction for machine intelligence, cautiously phrased as it was, has been fulfilled. On the other hand, he underestimated the potential for fast, cheap, huge-scale computing. His 1948 picture of the future of computer hardware correctly identified the speed of light as the critical constraint governing computing speed. But his assumption of centimetre-scale electronic components overlooked the enormous potential for miniaturisation. Turing's foresight was more strikingly demonstrated by his 1946 observations about the power of the universal machine and the future of what would now be called the software industry; 'every known process has got to be translated into instruction table form ...'

In 1946 he could speak with the confidence of being the mastermind of the Anglo-American crypto war – with its own legacy for the future of international relations, which, to say the least, has not yet been evaluated. In 1939 he and Gordon Welchman had pulled off the feat of persuading the British authorities to make a huge investment in the untried technology of the Turing Bombe, on the conviction, correct as it turned out, that its logical brilliance would transform British fortunes. This vision was not his alone. To beat Hitler, Bletchley Park seems to have borrowed from the future, scientifically, organisationally and socially, as if the sixties had arrived before the forties. But Turing had always had a particular note of confidence in his own vision. In 1936 he chose the unmathematical but suggestive word 'invent': "*The universal computing machine*. It is possible to invent a single machine which can be used to compute any computable sequence ..." And in that same year he clearly foresaw the coming of war with Germany and the significance of cryptology to it. After the war, his confidence was not seriously dented by his inability to retain control of practical computer development. His mathematical theory of biological growth, with its hands-on computer experimentation, lacked any significant external support. Yet this isolation did not affect his enthusiasm, even though he would have had to wait until the 1970s to see his models taken seriously and until the 1990s to see the kind of computer power that they needed.

For Turing, Cambridge was the closest he had to an intellectual home but he never behaved as Cambridge expected. Turing completely ignored the distinction between 'pure' and 'applied' mathematics, which dominated the Cambridge faculty. Instead, he exemplified the prophetic capacity of mathematics, anticipating rather than following scientific observation. *Metamagical*, we might call it, coopting the word coined by Douglas Hofstadter, one of our contributors.

Before returning to this theme, which runs through this volume, here is something new about Alan Turing's life. For those already familiar with his extraordinary story, it will add a significant nugget. For those not so familiar, this vignette will give a taste of why his individual life has become such a source of fascination to a wide public, touching bases well beyond mathematics, science and technology. In 2013, long after we had chosen our title, the Bodleian Library at Oxford ran an exhibition on 'Magical Books, from the Middle Ages to Middle-earth'. At its entrance was that Arthurian inscription *Rex quondam, Rexque futurus*. But inside was featured the more recent work of Alan Garner, most famous for his novel *The Owl Service*. Here lies the magical connection: Alan Garner had been Alan Turing's training partner; they had run a thousand miles together through Cheshire country lanes in 1951–1952.

The meeting arose in 1951, as fellow athletes spotted each other on the road. Alan Garner was only 17, a classics sixth-former at Manchester Grammar School. But from the outset Garner felt himself treated as an equal, something he could appreciate because of his school's special ambience (a culture that yet another Alan has evoked in *The History Boys*). Equality also followed from matching prowess, Turing was already famous in amateur athletics for his long distance races and Garner was just about to become a seriously competitive young sprinter. Equality also was found in banter of the kind that would now be called 'no bullshit', full of word play and scurrilous humour. It came as no surprise to Garner when Turing asked him if he thought intelligent machinery was possible. After running silently for ten minutes, he said no. Turing did not argue. "Why learn classical languages?", Turing asked, and Garner said, "you have to learn to use your brain in a different way": the kind of answer that Turing would have appreciated.

Their chat kept away from the personal: it was focused on sustaining the six or seven miles of running. But once, probably late in 1951, Turing mentioned the story of Snow White. "You too!" said Garner, amazed. For he connected it immediately with a singular event from his childhood. When five years old, *Snow White and the Seven Dwarfs* had terrified him with the vision of the poisoned apple. Turing responded with immediate empathy, and their shared trauma – as Garner saw it – remained a bond. "He used to go over the scene in detail, dwelling on the ambiguity of the apple, red on one side, green on the other, one of which gave death."

The interaction overlapped with the period of Turing's trial and punishment as a gay man. Turing never spoke of what he was undergoing and somehow Garner only heard the news late in 1952, when he was warned by the police not to associate with Turing. Garner was very angry at this, and at what he learned had happened, and he never had the least sense of having been approached in any predatory way. And yet, inevitably, it ended sadly. Alan Garner painfully recalls seeing Turing for the last time in 1953, as a fellow passenger on the bus from Wilmslow to Manchester. Being with his girl-friend, Garner found it too difficult to say anything appropriate and so he pretended not to have noticed Turing's presence. This incident, so redolent of the fiction and film of final teenage years, was soon followed by Garner's departure to National Service, where he heard of Turing's death. Alan Garner revealed nothing of this until 60 years later in a column for *The Observer* newspaper. There must be many other stranger-than-fiction Turing encounters which will never be known. But perhaps there was none in which Turing found his own apple, the symbol of his death in 1954, so directly reflected.

In 2012 I heard Alan Garner tell this story as if it were yesterday, and it made direct contact with the 60-year sweep of history we are now marking. The time-shifting experience was accentuated by the towering presence of the Jodrell Bank radio telescope, adjacent to the ancient building that Alan and Griselda Garner have made their home and archaeological base. The telescope was itself an outpost of dynamic Manchester University science and, just like the computer, it was the outcome of Second World War technology turned to science after 1945. Now it is accepted and woven into the infrastructure of astronomy and of cosmology, but in 1954 it was all new. The scale and age of the cosmos was then unknown, with everything to unfold in the following decades. The scene emphasised the would-have-been that was cut off in 1954, the great scope of new discoveries that a living Turing might have made in such a creative period.

Barry Cooper, as co-editor, put the subtitle 'Computing the world' into our time-travelling title, and the suggestion of limitless horizons is entirely appropriate for

Turing, who refused to remain confined to any one area of thought. In his last notes of 1953–1954, Turing showed that he was thinking about fundamental physics, showing the influence of Dirac, with notes on spinors and some ideas about the reformulation of quantum mechanics. Some critics have probably regarded these as mere scribbles or perhaps as signs of incipient madness. But the mystical-sounding 'Light cone of the creation', in his last postcards, was correctly prophetic: the geometry of light has proved to be a key idea in modern theoretical physics, and the light cone of the Big Bang is made a precise mathematical idea in the work of another contributor to this volume, Roger Penrose. Turing's apparently weird dictum 'Particles are founts' actually referred to the idea that fundamental particles and forces are representations of symmetry groups: one of the greatest discoveries of the late twentieth century was that the sub-nuclear *quarks* are described by the symmetry group SU(3).

My interest is not unbiased: my own work in mathematics has been in developing Roger Penrose's ideas, using the 'twistor' geometry based on light cones to supersede the Feynman diagrams which form the central pillar of particle physics. Actually, Richard Feynman himself bears comparison with Alan Turing, his late-1940s work in fundamental physics, after his emergence from the atomic bomb project, being the analogue of Turing's computer. Strikingly, Feynman also wrote some of the most personal and popular books in science, with vignettes of life that say more about the culture of science than any textbook. And there was a closer connection. When Mrs Eisenhart emitted the immortal words 'Surely you're joking, Mr Feynman', she had only recently bid farewell to Alan Turing's equally awkward participation in the same suffocating Princeton Graduate College tea parties. Later, Feynman started the early thinking about quantum computing: he and Turing could have found common ground. Above all, they famously shared the playful and no-bullshit mentality, speaking their mind to authority without much heed for diplomacy.

But there were obvious differences: it was much less easy for Turing to be the honest man he longed to be. The crypto work at Bletchley Park remained far more secret than the atomic bomb, and Turing's sexuality was not just unconventional in Feynman's way: it was taboo, unmentionable, criminal and a state security issue. The freedom notionally fought for in the Second World War had not been completely forgotten, and speaking one's mind was not totally impossible. News of the early Scandinavian gay rights movement may well have inspired Turing to visit Norway in summer 1952. (And sixth-former Alan Garner proudly wrote an alpha-plus essay on classical Greek sexuality in that year.) By such small acts of resistance did change seep through. But not until the 1970s was the industrial-scale success of Bletchley Park codebreaking revealed, and only then did it become acceptable to mention the sexuality of its chief scientific figure.

The modern situation is very different, and the eager response of so many contributors to this volume reflects the serious interest now taken in everything Alan Turing was and did. It is hard to know what Turing himself would have thought of this renaissance or of his future place in science and history. He was never sure whether he was a reticent back-room boy or whether he deserved public attention as a special individual. His 1950 paper has an intense personal subtext: whilst its subject is 'intelligent machinery', Turing addressed it in a way that insisted 'I am human', and drew attention to himself as vividly as any media-savvy academic might today. In a world of 'publish or perish', Turing did a fine compromise between the two. Feynman's *Lectures on Physics* are famous, but there were no corresponding Turing *Lectures on Logic and Computing*, which could, if he had so chosen, have stamped his name on the whole new field of computer science. The 1950 paper was published and became one of the most cited in modern philosophy. The 1948 paper perished (at least until 1968) and no-one knew of the neural net models (described by Christof Teuscher in this volume).

That 1948 paper contained a Stoic picture of his place in scientific advance: "...the search for new techniques must be regarded as carried out by the human community as a whole, rather than by individuals." Turing kept remarkably quiet about his own intellectual history, referring to his 1936 universal machine when describing the nature of computers but never giving an explanation of how he arrived at it or how his 'practical universal computing machine' came into being. Official secrecy would have been a problem (he could never have explained how he came to have his knowledge of digital electronics), but he could have found a way round it. He gave no assessment of his relationship with John von Neumann, a subject that has given subsequent historians much trouble in elucidating. The nearest thing we have comes in the report on the sports page of the *Evening News*, after an athletics meeting of 26 December 1946, that he "gives credit for the donkey work on the ACE to Americans". When, in 1953, a first semi-popular book on computers, *Faster than Thought*, gave him space to set out his analysis, he gave the fundamental principle thus:

> If one can explain quite unambiguously in English, with the aid of mathematical symbols if required, how a calculation is to be done, then it is always possible to programme any digital computer to do that calculation, provided the storage capacity is adequate.

Into this one sentence are crammed both Turing's definition of computability and his conception of the computer as a universal machine. But no reader could possibly have understood its significance. (It is also, to the eye of a modern logician or philosopher, used to enforcing the subtlest distinctions in the expression of Church's Thesis, irresponsibly cavalier. Turing did not improve the precision by adding "This is not the sort of thing that admits of clear-cut proof, but amongst workers in the field it is regarded as being clear as day.") The editor of the volume, B.V. Bowden, summarised his reputation thus:

> Türing machine: In 1936 Dr Turing wrote a paper on the design and limitations of computing machines. For this reason they are sometimes known by his name. The umlaut is an unearned and undesirable addition, due, presumably, to an impression that anything so incomprehensible must be Teutonic.

After 1954 his reputation went into an even deeper Sleeping Death, partly because of this tendency to dig his own grave in his own lifetime, but also because of the blight around his trial and death. But the emergence of computer science as an engineering discipline, distinct from mathematics, did not favour him. (The establishment of the A.M. Turing Award by the Association for Computing Machinery was an odd exception.) Another factor might be that some mathematicians for too long regarded computing as beneath their dignity, even when complexity theory in the 1970s showed the mathematical depth of digital computing. All that has changed, in no small measure thanks to the power of the universal machine in transforming social communication. Turing's era now appears as a Dark Age, and in quite a literal sense, because of the dearth of historical texts. In probing the origin of the computer in the 1940s, historians find themselves arguing over second-hand stories from oral history rather than following the step-by-step documentation of intellectual process. The Arthurian legend has more relevance than might be expected to the *quondam* Turing. But now it is time to open the door to the Turing *futurus*.

Part One: Inside Our Computable World, and the Mathematics of Universality

Turing-Welchman Bombe rebuild at Bletchley Park. Reproduced with permission of Jochen Viehoff.

In 1954, Turing's last published paper, an article in *Penguin Science News*, conveyed the significance of computability in mathematics to a wide audience. But he had done little to promote his own formalism of Turing machines: just one paper applying it to a decidability question in algebra and nothing at all for emergent computer science. This left a gap in mathematical logic that was filled systematically by **Martin Davis** in his book *Computability and Unsolvability* from 1958. Now Davis emulates Turing's 1954 semi-popular work in explaining the question of Hilbert's Tenth Problem about the solvability of Diophantine equations, which Turing's definition of computability had rendered completely well defined. Martin Davis himself made a major contribution towards its beautiful resolution in 1970, and here he points to how this work refined and extended key aspects of Turing's seminal 1936 paper.

The most striking of Martin's observations relate to *universality* and *incomputability*. The results he describes as arising from the celebrated work of him and his co-solvers elegantly clarify the reach and limitations of Turing computability at the classical level. Turing's later work anticipated today's interest in 'new computational paradigms' derived from processes in Nature, and the scope of *virtual machines*. There are important and deep questions concerning the extent to which these new models of computation are subsumed within Turing's 1936 paradigm – at a practical level via the development of virtual machines. Notions of universality and reach continue to raise basic questions for today's researchers.

Martin Hyland fills in a missing piece of history in a way that evokes and explores Alan Turing's close relationship with Robin Gandy. This is a counterpart to Alan Garner's story: another hidden history, in the same time-frame. He brings to life a topic that Turing studied but never brought to fruition: the theory of types, then totally abstract but which now would be seen as vital to formal structures in computer science (a concept that in Turing's time did not exist) and, more widely, in relation to the use of language in scientific theories.

Martin's contribution focuses on little-known ideas quite fundamental to a number of contemporary scientific preoccupations. The title of Robin Gandy's early research thesis, 'Some Considerations concerning the Logical Structure underlying Fundamental Theories in Theoretical Physics', provides a clear link to subsequent convergences between science and logical frameworks. A key aspect of later developments is the interrelating interests of Turing and Gandy in Church's theory of types. Martin notes that:

> Turing's influence as Gandy's supervisor relates specifically to Type Theory, and I started this paper with the thought that Turing's interest in the area is largely forgotten.

He proclaims "It is time to say something about that interest." What follows is one of the most fascinating discussions in this whole volume, and brings out Turing's distinctive, and often prescient, linking of the abstract and the concrete in a quite unique way. Nobody else in the post-centenary period has covered this ground, and certainly not in such an authoritative and informative way.

Andrew Booker explains Turing's work in analytic number theory, a story which brings in the special machine Turing started to build in 1939 and then his programming of the Manchester computer in 1950, thus making an explicit example of how software would replace the engineering of special machines. Booker's discussion leads into the modern status of the outstanding problem of the Riemann hypothesis and the lasting significance of Turing's computational ideas.

Andrew Booker's description of the work in number theory emphasizes the extent to which Turing's interests tend to arise from more far-reaching questions. In this case there is an awareness of how the advent of the computer will inevitably impact on the very nature of proof in mathematics and a prescient anticipation of how mathematicians will adapt to the changes. A theme familiar from later writings of Turing is speculation about how computers will complement human creativity and about the evolving nature of the balance between mind and machine. As Booker notes, there may even be ethical consequences of this increasingly important and complex relationship.

Ueli Maurer takes up a parallel story, inspired by the little we know of Turing's pre-war ideas in cryptology and the emergent concept of crypto security. He sets out the terrain in the following terms:

> Computation and information are the two most fundamental concepts in computer science, much like mass, energy, time, and space are fundamental concepts in physics.

He goes on to describe the seminal and complementary roles played by Alan Turing's computational model and Claude Shannon's information theory. Beyond the practicalities of modern cryptography, his discussion relates to the status of another huge unsolved modern problem, that of the $P = ?NP$ question. Ueli Maurer writes

> One can only speculate what Turing might have been able to achieve in the field of theoretical cryptography had he spent more time on the subject.

But we do not know what he may have done after 1938 that remains secret. The Delilah speech-encipherment project reports remained secret for over 50 years, and the full description was published only in 2012. Also in 2012, two quite fundamental Turing papers were released from secrecy, explaining the basis of Bayesian analysis and its application to the Naval Enigma 'Banburismus' method. There may well be much more to follow.

The timely releases by GCHQ in honour of the Turing centenary connect us to one of Alan Turing's most original and important contributions at Bletchley Park. Essential to the bringing of intercepted German messages within the scope of the Bombe and Colossus decoding machines was the application of what were recognisably Bayesian statistical techniques. In 'Alan Turing and Enigmatic Statistics', statistician **Kanti Mardia** and logician **Barry Cooper** take a closer look at the history at Bletchley Park, and at the current significance of the statistics for today's 'big data'. The "enigmatic statistics", we read, "foreshadowed the style of what is now called Statistical Bioinformatics".

Following informative examples, and an outline of the content and significance of the released papers, Mardia and Cooper return us to the mathematics of typed information and the use of sampling techniques for 'type reduction' to data accessible to classical Turing computation. From this perspective, they describe how the cryptanalytical role of Turing's Bayesian techniques potentially take us to a better understanding of the challenge of 'big data' in a wider context.

1

Algorithms, Equations, and Logic

Martin Davis

From the beginning of recorded history, people have worked with numbers using *algorithms*. Algorithms are processes that can be carried out by following a set of instructions in a completely mechanical manner without any need for creative thought. Until the 1930s no need was felt for a precise mathematical definition or characterization of what it means for something to be algorithmic. The need then did arise, in those years, in connection with attempts to prove that for some problems no algorithmic solution is possible. In 1935 Alan Turing considered this matter in isolation at Cambridge University. Meanwhile, in Princeton, the combined efforts of Kurt Gödel and Alonzo Church with his students Stephen Kleene and J. Barkley Rosser were brought to bear on the same topic. E.L. Post had also been thinking about these things, like Turing also in isolation, since the 1920s. Although the various formulations that were developed seemed superficially to be quite different from one another, they all turned out to be equivalent. This consensus about algorithmic processes has come to be called the Church–Turing Thesis.

Turing's conception differed from all the others in that it was formulated in terms of an abstract machine. What was striking about his characterization is his analysis that showed why even very limited fundamental operations would suffice for all algorithms if supplemented by unlimited data storage capability. Moreover, he demonstrated that a single such machine, Turing's so-called 'Universal Machine', could be made to carry out any algorithmic process whatever. These insights have played a key role in the development of modern all-purpose computers. But, as will become clear later, the notion of universality spills over into mathematical domains far removed from computation.[1]

Turing began his investigation with a problem from mathematical logic, a prob-

[1] The paper Turing (1936) has been reprinted in numerous collections.

lem, whose importance was emphasized by David Hilbert, that can be described as follows:

> Find an algorithm that will determine whether a specified conclusion follows from given premises using the formal rules of logical reasoning.

By 1935 there were good reasons to believe that no such algorithm exists, and this is what Turing set out to prove. Turing succeeded by first finding an unsolvable problem in terms of his machines, specifically the problem of determining for a given such machine whether it would ever output the digit 0. Then he showed how to translate this so-called "printing" problem into the language of mathematical logic in such a way that a purported algorithm for the problem from logic could be used to obtain a corresponding algorithm solving this printing problem, something he had shown could not be done. Later Post used the unsolvability of the printing problem to prove that a previously posed mathematical problem called the *word problem for semigroups* is likewise unsolvable. Still later, Turing himself used an intricate construction to refine and extend Post's result. Turing's beautiful article (Turing, 1954) explained unsolvability to the general public in clear simple terms.

When Turing learned that work leading to conclusions similar to his own had been done in Princeton, he arranged a visit and spent two years there. A new branch of mathematics variously called *computability theory, recursive function theory* or *recursion theory* was now open to vigorous pursuit. One direction this new discipline took was to lead to unsolvability proofs for problems from various branches of mathematics. This essay will tell the story of where one such thread led.

The Plot

Our discussion will be framed in terms of the so-called natural numbers $0, 1, 2, 3, \ldots$ We will work with three different ways to specify a set of natural numbers:

(1) by an algorithm that lists all the members of the set;
(2) by an algorithm to decide membership in the set;
(3) as the parameter values for which an equation has solutions.

It will turn out that examining how these notions are related to one another will lead to surprising and far reaching conclusions.

In 1949, in my graduate student days, I conjectured that two of these three are equivalent in the sense that the sets of natural numbers specified by them are the same. Although the truth of this conjecture could be seen to have very important consequences, it was generally regarded as quite implausible. I hardly imagined then that the proof of my conjecture would require two decades of work involving one of America's most prominent philosophers, the first female mathematician to be elected to the National Academy of Science of the United States, a young Russian mathematician, and myself.

Although a mathematically rigorous treatment would require working with the technical notions developed by Turing and the others mentioned above, for the most part in this essay, algorithms will be dealt with in a loose informal manner,

An Example: The Set of Perfect Squares

A number obtained by multiplying some natural number by itself is called a *perfect square*. So the set of perfect squares is $\{0,1,4,9,16,25,\ldots\}$.

Algorithm for *listing* the perfect squares

```
Start with 0 and repeatedly add 1 to it generating the sequence of
all natural numbers. As each number is generated, multiply it by
itself and put the result in a list.
```

Note that there are infinitely many perfect squares and so the imagined computation to list them will continue 'forever'. The table below shows the list in the second row:

0	1	2	3	4	5	...
0	1	4	9	16	25	...

Definition A set of natural numbers is **listable** if there is an algorithm that lists its members (in any order with repetitions permitted).

Algorithm for *deciding* membership in the set of perfect squares

```
To decide whether a given number n is a perfect square, begin as
above listing the perfect squares in order. If one of the perfect
squares is equal to n, we can stop and we know that n is a perfect
square; if one of the perfect squares is larger than n, we can
stop and we know that n is not a perfect square.
```

Definition A set of natural numbers is **decidable** if there is an algorithm that decides membership in it.[2]

Using an equation to specify the set of perfect squares

In the equation

$$a - x^2 = 0$$

we regard a as a *parameter* and x as an *unknown*. This means that for different values of a we seek a natural number value of x that satisfies the equation. Since

[2] Other terms used for 'listable' are: recursively enumerable, computably enumerable. Other terms for 'decidable' are: computable, recursive.

this equation is obviously equivalent to $a = x^2$, the values of a for which such a solution exists are exactly the perfect squares.

Generally when we use the word "equation" we have in mind a polynomial expression set equal to 0. Polynomial expressions can involve any number of unknowns and may also include the parameter a. The expressions can be formed by combining the letters and any number of natural number constants using addition, subtraction, and multiplication. Here are some examples of polynomial expressions:

$$x^2 - 17y^2 - 2, \qquad x^3 - x^2y + 3axy^2, \qquad (16 + a^3)(ax + y^5).$$

Definition A set of natural numbers is **Diophantine** if there is a polynomial equation with parameter a that has natural number solutions for exactly those values of a that are members of the set.

Examples of Diophantine Sets

- $a - (x+2)(y+2) = 0$ specifies the set of *composite numbers*; that is, numbers other than 1 that are not primes.
- $a - (2x+3)(y+1) = 0$ specifies the set of numbers that are not powers of 2 (because they are the numbers other than 1 that have an odd divisor).
- The so-called *Pell equation*, $x^2 - a(y+1)^2 - 1 = 0$, has been well studied. It can be proved that in addition to the obvious solutions when $a = 0$ it has solutions precisely when a is **not** a perfect square.

Some Relationships

Theorem *Every decidable set is listable.*

Proof Let S be a decidable set. Generate the natural numbers in order. As each number is generated, test it to see whether it belongs to S. If it does, place it on a list. Move on to the next natural number. This algorithm will make a list of the members of S. □

The *complement* of a set S, written $\bar{S}$, is the set of all natural numbers that don't belong to S.

Theorem *The set S is decidable if and only if $S, \bar{S}$ are both listable.*

Proof If S is decidable then obviously so is $\bar{S}$, and hence both are listable.

On the other hand, if $S, \bar{S}$ are both listable then, given a number n, we can decide whether it belongs to S as follows. We use the two listing algorithms to start making lists of both S and $\bar{S}$. We wait to see in which list n will eventually show up. Then we will know whether n belongs to S. □

Unsolvability Theorem *There is a listable set K whose complement $\bar{K}$ is not listable. Therefore K is not decidable.*

Proof See Appendix.

Theorem *Every Diophantine set is listable.*

Proof Let S be a Diophantine set, specified by an equation with parameter a and unknowns $x_1,\ldots,x_k$. Set up an ordering of all $(k+1)$-tuples $\langle a,x_1,\ldots,x_k\rangle$ of natural numbers,[3] and proceed through them one by one. For each tuple, plug the numbers into the equation. Checking to see whether these numbers do satisfy the equation is just a matter of arithmetic. If the equation is satisfied by a particular tuple, place the value of a from that tuple on a list. This algorithm will make a list of the members of S. □

A Conjecture Becomes a Theorem: a Story

When I wrote my doctoral dissertation in 1950, I stuck my neck out with following:

Conjecture *Every listable set is Diophantine.*

On the face of it this was quite implausible. Why should any set that can be listed by an algorithm be specifiable by something as simple as a polynomial equation? Moreover, for reasons that will be explained later, the conjecture implies something really implausible, that there are constants m and n such that every Diophantine set can be specified by an equation of degree $\leq m$ and with a number of unknowns $\leq n$. However, the conjecture, if true, can be seen to lead to a solution of one of the famous Hilbert problems.

At an International Congress of Mathematicians in 1900, David Hilbert listed 23 problems as a challenge for the future. These have become known as the Hilbert Problems and, in addition to their intrinsic interest, these problems have commanded special attention because of the stature of their source. The tenth problem in the list can be stated as follows:

[3] One way to order the tuples is to introduce each natural number in succession, writing down each tuple that includes that number together with all the previous numbers. For the case of one unknown, the pairs ordered in that manner look like this:

$$\langle 0,0\rangle,\langle 0,1\rangle,\langle 1,0\rangle,\langle 1,1\rangle,\langle 0,2\rangle,\langle 2,0\rangle,\langle 1,2\rangle,\langle 2,1\rangle,\langle 2,2\rangle,\ldots$$

Problem *Find an algorithm to determine for any given polynomial equation with integer coefficients whether it has a natural number solution.*[4]

It isn't difficult to see that the truth of my conjecture would solve Hilbert's tenth problem in a negative way, by implying that no such algorithm exists. The reason is that my conjecture implies that the set K, from the Unsolvability Theorem above, is Diophantine. So there would be an equation that would have solutions for a given value of its parameter a just in the case where a belongs to K. Thus, if there were an algorithm such as Hilbert had asked for, it could be used to decide membership in K, contradicting the fact that K is not decidable.

Despite its apparent implausibility, I had a reason to think that my conjecture might actually be true. I was able to prove that there is a Diophantine set S whose complement $\bar{S}$ is not Diophantine. The comparison with the Unsolvability Theorem is striking. My proof was short and easy, but not *constructive*. (That means that the proof did not provide any example of such a set; it merely proved its existence.) It was easy to see that Diophantine sets as a class shared other properties with the listable sets.[5]

I tried to prove my conjecture, but the best I could do, reported in my dissertation, was to prove that for every listable set S, there is an equation with parameters a, q and k such that a number a belongs to S if and only if there is a number q_0 such that the equation has solutions for that value of a, for $q = q_0$, and for all values of $k \leq q_0$.[6] Although far from what I desired, this result turned out to play a significant role in what followed.

When I attended the International Congress of Mathematicians at Harvard University in 1950, I learned that Julia Robinson, a mathematician with whose work I was familiar, had also been working on Diophantine sets. But whereas I had been working top down, trying to get a Diophantine-like representation for listable sets, she had been working bottom up, trying for Diophantine definitions of various sets. Alfred Tarski had suggested that one should prove that the set of powers of 2, $\{1, 2, 4, 8, 16, \ldots\}$, is *not* Diophantine. Julia was attracted to this problem but, not succeeding in doing what Tarski had proposed, turned around and tried to prove that the set of powers of 2 *is* Diophantine. Of course, had Tarski been right it would have shown that my conjecture is false. Julia couldn't prove that this set is Diophantine either. But she did prove that if one could find what I like to call a Goldilocks

[4] Actually, Hilbert asked for an algorithm for arbitrary integer solutions, positive, negative or zero. However, it is not difficult to see that the two forms of the problem are equivalent.

[5] For example, the union as well as the intersection of two Diophantine sets is also Diophantine.

[6] Using logical symbolism,

$$a \in S \Leftrightarrow (\exists q)(\forall k)_{\leq q}(\exists x_1, \ldots x_n)[p(a, k, q, x_1, \ldots x_n) = 0],$$

where p is a polynomial with integer coefficients.

equation then she could prove not only that the set of powers of 2 is Diophantine but that the set of prime numbers, as well as many other sets, are Diophantine.

In the folktale, Goldilocks experiences the bears' accommodations and finds those that are too large, those that are too small, and finally those that are 'just right'. Here we are considering equations with two parameters, a and b. Such an equation is *too large* if there are values of a,b with $b > a^a$ for which the equation has solutions. It is *too small* if there is a number k, such that for all values of a,b for which the equation has solutions, $b \leq a^k$. So an equation is a *Goldilocks equation* if it is *just right*, neither too large nor too small. Julia tried to find such a Goldilocks equation, but she did not succeed.

It was at a month-long 'Institute for Logic' at Cornell University in the summer of 1957 that Hilary Putnam and I began working together. We did get some preliminary results that summer, but our breakthrough occurred two years later. Our idea was to see what would happen to my conjecture if we permitted variable exponents in the equation. Thus in addition to equations like $x^5y - 7y^3a^2z + 5 = 0$ we would also be considering equations like $x^5y^z - 7y^3a^2z + 5 = 0$. Although we did begin with my dissertation work, the introduction of variable exponents brought us squarely into Julia Robinson's territory, and we found ourselves using generalizations of some of her methods. Sets specified by equations with a parameter, where variable exponents are allowed, are called *exponential Diophantine*. We were trying to prove that

Every listable set is exponential Diophantine. $(***)$

We came close. But we had to make use of a property of prime numbers that was believed to be true but had not yet been proved:

PAP *For every number n, there is a prime number p and a number k such that the numbers $p, p+k, p+2k, \ldots, p+nk$ are all prime.*

What we were able to prove was only that:

Theorem *If PAP is true then $(***)$ is also true.*

Actually we now know that PAP is true, because it was proved in 2004, so in a sense our proof of (***) was correct. But in 1959, having no access to a time machine, we had to content ourselves with the mere implication.

The existence of a Goldilocks equation also came into the picture. It follows easily from Julia's work that if there is a Goldilocks equation then every exponential Diophantine set is also Diophantine. Hilary and I introduced the abbreviation:

JR *There exists a Goldilocks equation.*

So we had proved that if JR and PAP are both true then my conjecture is also true.

We wrote all this up for publication and, of course, we sent a copy to Julia. In her reply, which came rather quickly, she expressed pleasure in our work and told us that she had worked out a way around PAP. Her first proof of (***) was quite complicated and ingenious. But later she simplified it greatly, in effect, dividing the proof into a part where the primes weren't needed, and showing how to evade PAP in the part of the proof where they were still needed. The three of us agreed to publish a joint paper with footnotes explaining what our separate contributions had been.

So now, in order to prove my conjecture, it remained to prove JR; that is, to find a Goldilocks equation. But trying on and off for a decade, none of us were able to see how to construct a Goldilocks equation. The prevailing opinion was still that my conjecture is false, so there just couldn't be a Goldilocks equation. When I lectured on our work during those years, I used to point out that if there were no Goldilocks equation then every equation with two parameters would be either too small or too large, which would arguably be just as surprising as my conjecture. In the question period, someone was almost sure to ask for my own opinion as to the existence of such an equation. In reply, I would quip: "Oh, I think JR is true, and it will be proved by a clever young Russian." I turned out to be a prophet. That Russian was Yuri Matiysevich, who in January 1970 at the age of 22 was able to exhibit a Goldilocks equation. Yuri made use of the famous Fibonacci numbers: $1,1,2,3,5,8,13,\ldots$ (each of the numbers after the first two is the sum of the two preceding numbers). Writing F_n for the nth Fibonacci number, Yuri's equation had parameters u,v such that the equation had solutions just in the case $v = F_{2u}$. From what had long been known about the Fibonacci numbers, this implies that his equation is indeed 'just right'.

At last my conjecture had finally become a theorem, two decades after I had first proposed it. The theorem has often been called Matiyasevich's theorem because of his beautiful construction that completed its proof. It has also been called the MRDP theorem to acknowledge the role that all four of us had played. Yuri himself has generously suggested calling it DPRM:

Matiyasevich/MRDP/DPRM Theorem *If a set is listable then it is also Diophantine.*

The theorem has important and surprising consequences, some of which have already been mentioned but are well worth recapitulating. Recall that K is listable, but not decidable.

Corollary *There is a polynomial* $p_o(a,x_1,\ldots,x_\ell)$ *such that the equation*

$$p_o(a,x_1,\ldots,x_\ell)=0$$

specifies the set K. Hence, no algorithm exists to determine for a given value of a whether there exist natural numbers $x_1,\ldots,x_\ell$ *that satisfy this equation.*

And as already been remarked, this implies:

The unsolvability of Hilbert's tenth problem

Corollary *No algorithm exists to determine for a given polynomial equation with integer coefficients whether it has a solution in natural numbers.*

A Universal Equation

Using Turing's universal machine together with the Matiyasevich/MRDP/DPRM theorem, it is easy to prove the following:

Universal Equation Theorem *There is an equation* $p(n,a,x_1,\ldots,x_k)=0$ *such that for every listable set S there is a number* n_0 *such that S is equal to the set of all values of a for which* $p(n_0,a,x_1,\ldots,x_k)=0$ *has solutions.*

Since listable sets and Diophantine sets are the very same sets, we can alternatively write down the:

Universal Equation Theorem *There is an equation* $p(n,a,x_1,\ldots,x_k)=0$ *such that for every Diophantine set S there is a number* n_0 *such that S is equal to the set of all values of a for which* $p(n_0,a,x_1,\ldots,x_k)=0$ *has solutions.*

Notice that in this form of the Universal Equation Theorem there is no mention of algorithms; it is all about polynomial equations and natural number solutions. It was this consequence of my conjecture more than anything else that made the conjecture seem so implausible. In his review of the paper that Hilary and I had written jointly with Julia, Georg Kreisel mentioned the fixed bound k on the number of unknowns needed to specify any given Diophantine set as a reason to doubt that our work had any connection with Hilbert's tenth problem. Meanwhile he failed to mention our proof that the existence of a Goldilocks equation would imply the unsolvability of that problem.

Naturally it was of interest to determine the best possible value of k. Yuri and Julia working together were able to achieve the value $k=13$. Later Yuri improved it to $k=9$. In each case the corresponding degree of the polynomial was astronomically large. One can achieve $k=29$ with a polynomial of degree 16.

Prime Numbers and a Really Ugly Polynomial

Here is another way to use a polynomial to make a list of *positive* numbers: Successively plug in all natural number values of the variables and compute the value of the polynomial. If the value is a positive natural number, put it in the list. If the value is 0 or negative, just ignore it. In the table below the values of the polynomial $5x^2 - 2xy - 3y^2$ are shown for $x = 0, 1, 2, 3, 4, 5;\ y = 0, 1, 2, 3, 4, 5$. Extracting only the positive values, the corresponding list begins $\{5, 20, 13, 45, 36, 21, 80, 69, 52, 29, 125, 112, 93, 68, 37, \ldots\}$.

x	y	$5x^2 - 2xy - 3y^2$	x	y	$5x^2 - 2xy - 3y^2$
0	0	0	3	0	45
0	1	–3	3	1	36
0	2	–12	3	2	21
0	3	–27	3	3	0
0	4	–48	3	4	–27
0	5	–75	3	5	–60
1	0	5	4	0	80
1	1	0	4	1	69
1	2	–11	4	2	52
1	3	–28	4	3	29
1	4	–51	4	4	0
1	5	–80	4	5	–35
2	0	20	5	0	125
2	1	13	5	1	112
2	2	0	5	2	93
2	3	–19	5	3	68
2	4	–44	5	4	37
2	5	–75	5	5	0

Using an ingenious simple trick Hilary Putnam saw that every Diophantine set of positive natural numbers can be obtained in this way as the set of positive values of a polynomial. Here's how Hilary's trick works: suppose for some such Diophantine set that

$$p(a, x_1, \ldots, x_n) = 0$$

is a polynomial equation with parameter a that has natural number solutions for exactly those values of a that are members of the set. In this case, Hilary's polynomial would be $a[1 - (p(a, x_1, \ldots, x_n))^2]$. Except when the polynomial p evaluates to 0, this will be 0 or negative. And when p does evaluate to 0, so that the value of a belongs to the set, Hilary's polynomial will evaluate to this same value of a.

Therefore the positive values of Hilary's polynomial are exactly the members of the given Diophantine set.

The prime numbers 2,3,5,7,11,13,17, etc., i.e., numbers that have exactly two divisors, themselves and 1, have been a favorite interest of number theorists. In the days before my conjecture had become a theorem, I used to enjoy posing the following problem to number theorists: *find a polynomial whose positive values are exactly the prime numbers.* The quick response I often got was that it shouldn't be too difficult to prove that there could not be such a polynomial. Naturally they did not succeed in finding such a proof since the Matiyasevich/MRDP/DPRM theorem together with Hilary's trick implies that there must be such a polynomial. And here is a polynomial that does the job:

$$
\begin{aligned}
(k+2)\{1 &- [wz+h+j-q]^2 \\
&- [(gk+2g+k+1)(h+j)+h-z]^2 \\
&- [2n+p+q+z-e]^2 \\
&- [16(k+1)^3(k+2)(n+1)^2+1-f^2]^2 \\
&- [e^3(e+2)(a+1)^2+1-o^2]^2 \\
&- [(a^2-1)y^2+1-x^2]^2 \\
&- [16r^2y^4(a^2-1)+1-u^2]^2 \\
&- [n+\ell+v-y]^2 \\
&- [((a+u^2(u^2-a))^2-1)\,(n+4dy)^2+1-(x+cu)^2]^2 \\
&- [(a^2-1)\ell^2+1-m^2]^2 \\
&- [q+y(a-p-1)+s(2ap+2a-p^2-2p-2)-x]^2 \\
&- [z+p\ell(a-p)+t(2ap-p^2-1)-pm]^2 \\
&- [ai+k+1+\ell-i]^2 \\
&- [p+\ell(a-n-1)+b(2an+2a-n^2-2n-2)-m]^2\}
\end{aligned}
$$

In an article by the four authors J.P. Jones, D. Sato, H. Wada, and D. Wiens it is proved that the positive values of this polynomial when the variables are given natural number values are precisely the prime numbers.

I don't think anyone would find this polynomial a thing of beauty. I like to say that only its mother could love it. But I think that its very ugliness helps to explain why my conjecture was thought to be so implausible. This messy object is not the sort of thing mathematicians had in mind when they imagined possible polynomials. And indeed in terms of the much prettier examples of the kind that did come to mind, my conjecture was implausible.

Logic

A *formal logical system* provides:

- a special language in which propositions are represented by strings of symbols;
- a list of initial strings or "axioms";
- rules of inference for obtaining new strings from given strings.

The strings thus obtained are called the *theorems* of the system. From our point of view, a formal logical system provides an algorithm that makes a list of its theorems.

We will make use of the equation

$$p_o(a, x_1, \ldots, x_\ell) = 0, \tag{1.1}$$

introduced earlier, that specifies the non-decidable set K. Let $\mathscr{L}$ be a particular formal logical system that, for each $a = 0, 1, 2, \ldots$ uses a string we'll call Π_a to represent the following proposition:

Proposition *Equation* (1.1) *has no solutions in natural numbers* $x_1, \ldots, x_\ell$.

Note that this proposition is equivalent to saying that a belongs to the set $\bar{K}$.

Definition We say that $\mathscr{L}$ is *sound* if, whenever a string Π_a is a theorem of $\mathscr{L}$, the proposition that it represents is true, i.e., the number a belongs to $\bar{K}$.

Gödel Incompleteness Theorem *Let $\mathscr{L}$ be sound. Then there is a number a_0 for which equation* (1.1) *has no solutions in natural numbers although Π_{a_0} is not a theorem of $\mathscr{L}$.*

Proof Otherwise, it would be the case that $a \in \bar{K}$ if and only if Π_a is a theorem of $\mathscr{L}$. So by making a list of the theorems of $\mathscr{L}$ and placing a in a second list when Π_a turns up in the list of theorems, one could make a list of the the members of $\bar{K}$. But this is impossible because $\bar{K}$ is not listable.

Thus for every *sound* logic, there is a true proposition not provable in that logic! □

This suggests that one should try to prove more and more theorems by creating stronger and stronger formal logic systems. However, such efforts will not change the polynomial p_0; they will only change the value of a. Undoubtedly, for the strong formal logical systems adequate for proving the bulk of ordinary mathematics (like those based on axioms for set theory and ordinary predicate logic) a value of a for which truth with unprovability happens will be enormously large.

Writing about Mathematics

About 50 years ago, on our first trip to Europe, my wife and I were sitting on a park bench in Venice while our two young sons were pushing toy cars on the ground. A man joined us and initiated a conversation. It was a very difficult conversation because he spoke only Italian and at that time neither of us knew much more than "grazie" and "prego". But we did have a dictionary and somehow we managed. He showed us photos of his family and let us know that he was a gondola maker. Of course we knew about gondolas as the romantic way of getting around the canals of Venice, but had never considered how and by whom they were manufactured. The man succeeded in communicating to us that these gondolas were made from special wood and without such fasteners as nails and screws. He drew a picture of the dovetail joints holding the parts of the boat together. Finally, he pointed to me, evidently wanting to know what was my profession. With the help of the dictionary I managed: "maestro matematico". This produced a stream of Italian quite incomprehensible to us. Finally I realized that what he was trying to tell me was: "In school, mathematics was always my worst subject."

With so many people experiencing mathematics like the gondola maker, it is with some temerity that a mathematician approaches the task of communicating to the general public something of the excitement that we feel as more and more is learned about the abstractions whose properties we struggle to master. We are warned that just seeing examples of the symbolic language we use to navigate our way through the thickets of abstract complexity on the printed page will put potential readers off and that we should include as few mathematical formulas as possible. Indeed the great physicist Stephen Hawking's *A Brief History of Time* contained not a single mathematical formula and did become a best seller, although it is hard to imagine what a reader innocent of advanced mathematics would make of his references to complex analytic manifolds. The first draft of this chapter was submitted to the editors without the section on prime-representing formulas because I feared the necessary inclusion of complicated formulas would make readers uncomfortable. I only added that section at the suggestion of one of the editors. In any case, it is the readers of this essay who will have to judge whether, without causing readers undue discomfort, I have succeeded with my own story in exposing them to something of the pleasures and struggles of mathematical discovery.

Alan Turing wrote in a clear down-to-earth plain-speaking manner even in his technical articles. Here is a sample from his historic article (Turing, 1936) in a mathematical journal in which such plain English is quite rare amidst complicated mathematical formulas:

> Computing is normally done by writing certain symbols on paper. We may suppose that this paper is divided into squares ... In elementary arithmetic the two-

> dimensional character of the paper is sometimes used. But such a use is always avoidable ... I assume then that the computation is carried out on one-dimensional paper, i.e. on a tape divided into squares.

This is how Turing begins his analysis of the computation process in which he argues by a cogent series of reductions that anything computable can be computed by his very simple devices that have come to be called *Turing machines*.

Turing's beautiful 1954 article explains unsolvability to the general public in clear simple terms. Beginning with a kind of puzzle likely to be familiar to his readers, Turing moves on to explain why certain simple processes involving substitutions in strings of symbols can lead to problems that are algorithmically unsolvable. Turing's goal of introducing the general public to unsolvability is very similar to mine in this essay. But whereas I make my own struggles the centerpiece of the story, Turing is self-effacing. His essay ends with a list of six unsolvable problems. There is no way that a casual reader could guess that it was Turing himself who had proved the unsolvability of two of them.

Julia Robinson – The Movie

Yuri, Hilary, and I appear in an hour-length documentary film *Julia Robinson and Hilbert's Tenth Problem* by George Csicsery, available from Zala Films, Oakland, California. See

`http://www.zalafilms.com/films/juliarobinson.html`,

`http://www.ams.org/ams/julia.html`.

Appendix: Proof of the Unsolvability Theorem

Unsolvability Theorem *There is a listable set K whose complement $\bar{K}$ is not listable. Therefore K is not decidable.*

Proof Let us fix a particular programming language in which to write algorithms for listing the members of a set. Then all possible programs in that language can be written in a sequence:

$$\mathscr{P}_0, \mathscr{P}_1, \mathscr{P}_2, \ldots$$

For $i = 0, 1, 2, \ldots$, let S_i be the set listed by $\mathscr{P}_i$.

Now, let K consist of those numbers i such that $i \in S_i$. Then we claim:

K is listable. For each $n = 1, 2, 3, \ldots$ run each of the programs $\mathscr{P}_1, \mathscr{P}_2, \ldots \mathscr{P}_n$ for n steps. Make a list as follows. Whenever a particular program $\mathscr{P}_i$ outputs the very number i, put i on the list.

$\bar{K}$ **is not listable.** Suppose that $\bar{K}$ is listed by program $\mathscr{P}_{i_0}$. We ask: is $i_0 \in \bar{K}$? If so, it would be listed by $\mathscr{P}_{i_0}$ and hence would be in K. Contradiction. So i_0 must be in K. By definition, $i_0 \in S_{i_0}$, i.e., i_0 is listed by $\mathscr{P}_{i_0}$. But then $i_0 \in \bar{K}$. Again, a contradiction. □

References

Davis, Martin, *Computability and Unsolvability.* McGraw Hill, New York (1958). Reprinted with an additional appendix: Dover, New York (1982).

Davis, Martin, Hilbert's Tenth Problem is Unsolvable, *American Mathematical Monthly,* **80**, 233–269, (1970); reprinted as an appendix in the Dover reprint of Davis (1958).

Davis, Martin, Yuri Matiyasevich, and Julia Robinson, Hilbert's tenth problem: Diophantine equations: positive aspects of a negative solution, *Proceedings of Symposia in Pure Mathematics,* **28**, 323–378, (1976). Reprinted in Feferman (1996), pp. 269–324.

Feferman, Solomon, editor, *Collected Works of Julia Robinson.* American Mathematical Society, Providence, RI (1996).

Matiyasevich, Yuri V., *Desyataya Problema Gilberta.* Moscow, Fizmatlit (1993). English translation: *Hilbert's Tenth Problem.* MIT Press, Cambridge, MA (1993). French translation: *Le dixième problème de Hilbert.* Masson (1995).

Matiyasevich, Yuri V., My Collaboration with Julia Robinson, *Mathematical Intelligencer,* **14**, 38–35 (1992). Corrections, **15**, 75, (1993). Reprinted in *Mathematical Conversations: Selections from the Mathematical Intelligencer*, Robin Wilson and Jeremy Gray, editors, Springer, New York (2001). Also reprinted in Reid (1996). French translation in *Gazette des Mathématiciens,* **59**, 27–44, (1994). The French translation was reprinted in the French version of Matiyasevich (1993).

Matiyasevich, Yuri V., Hilbert's Tenth Problem: Diophantine equations in the twentieth century. In *Mathematical Events of the Twentieth Century*, A. A. Bolibruch, Yu. S. Osipov and Ya. G. Sinai, editors, Springer-Verlag, Berlin; PHASIS, Moscow, pp. 185–213, (2006).

Petzold, Charles, *The Annotated Turing: A Guided Tour through Alan Turing's Historic Paper on Computability and the Turing Machine.* Wiley (2008).

Reid, Constance, *Julia: A Life in Mathematics.* Mathematical Association of America (1996).

Turing, Alan, On computable numbers with an application to the Entscheidungsproblem, *Proceedings of the London Mathematical Society*, ser. 2, **42**, 230–267, (1936). Correction: *ibid*, **43**, 544–546, (1937). Reprinted in *Collected Works: Mathematical Logic*, R.O. Gandy and C.E.M. Yates, editors, North-Holland, Amsterdam (2001), pp. 18–56. Also reprinted in *The Undecidable*, Martin Davis, editor, Raven Press (1965), Dover (2004), pp. 116–154. *The Essential Turing*, Jack Copeland, editor, Oxford (2004), pp. 58–90, 94–96. See also Petzold (2008).

Turing, Alan, Solvable and unsolvable problems, *Science News*, **31**m 7–23, (1954). Reprinted in *Collected Works of A.M. Turing: Mechanical Intelligence*, D.C. Ince, editor, North Holland, pp. 187–203, (1992). Also reprinted in *The Essential Turing*, Jack Copeland, editor, Oxford, pp. 582–595, (2004).

Yandell, Benjamin H., *The Honors Class: Hilbert's Problems and Their Solvers*. A.K. Peters, Natick, MA (2002).

2

The Forgotten Turing

J.M.E. Hyland

In fond memory of Robin Oliver Gandy: 1919–1995

Introduction

Alan Turing is remembered for many things. He is widely known as code breaker and cryptographer, and as – at least in some sense – inventor of the computer. He is if anything even more famous as the father of artificial intelligence. Beyond that he was a mathematician, mathematical logician and pioneer in the study of morphogenesis.

Logicians remember Turing for his celebrated Entscheidungsproblem paper (Turing, 1937a), for work on the λ-calculus (Turing, 1937b) and, if they are cognoscenti, for his second great paper (Turing, 1939) in the *Proceedings of the London Mathematical Society*. All that work was completed by 1940. It is not widely appreciated that Turing's interest in logic continued to the end of his life. His later interest in the theory of types, a central area in the foundations of mathematics, is largely forgotten. That interest has had more influence than is evident. I am going to tell the story: it is an odd one.

The one and only student

My D.Phil. supervisor at Oxford was Robin Gandy: he was the only person to take a doctorate under Turing's supervision, and also one of his closest friends.

While Turing had only one student, Gandy had many. I was one of the the cohort from the early 1970s. Gandy rather liked the thought that his students were intellectual grandchildren of Turing. We mostly cared rather less about that, but from Gandy's reminiscences we all caught a glimpse of Turing through the eyes

Published in *The Once and Future Turing*, edited by S. Barry Cooper & Andrew Hodges. Published by of someone who had known him very well. Turing met Gandy in 1940 at a party in King's College, Cambridge. Gandy was a third year student, while Turing, who had intermitted his Fellowship of the College at the start of the war, was already engaged in his celebrated work at Bletchley Park. During the war, Gandy became a friend and then much later Turing's student. By the end of Turing's life Gandy had begun his academic career as a Lecturer in Applied Mathematics at Leicester. You can read about their unique relationship in the Turing biography by Andrew Hodges (1983). Here I shall focus on Turing's influence on Gandy as his Ph.D. supervisor.

It seems entirely fitting that Turing's student should have become a well-respected mathematician in his own right. But in fact, Gandy's intellectual development was not straightforward. I believe that without Turing's decisive intervention Gandy would never have become a serious logician. Turing's grandchildren owe him a great debt.

Memories

When I started research with Robin Gandy in 1971, I was a typical disoriented student of the early 1970s, with long curly black hair and no real sense of mathematical direction. Gandy, on the other hand, was a leading UK logician. He had made substantial contributions to the subject[1] and important results carried his name. As Reader in Mathematical Logic at Oxford he was, with Michael Dummett, running the new course in Mathematics and Philosophy. He was a Fellow of Wolfson College, where he lived happily until his retirement. There should have been an enormous distance between us but somehow there was not.

All Gandy's students experienced his extraordinary character. He was not what most people think of as an academic. He did not seem quite serious. He was amused and amusing, with great enthusiasm for life. He was most definitely loud: you always knew where he was down the pub. By the time I knew him he had put aside his famous motorbike and leathers, but he remained a remarkable and handsome figure. He was sufficiently unselfconscious that at times he lectured with shirt visibly tucked into underpants. He certainly did not aspire to glamour in the usual sense, but there was a glamour of personality about him. He enjoyed being a little outrageous: he would mischievously recall that he had once been a very pretty boy – and he was amused occasionally to be rewarded with a shocked reaction. But I think that even when he was young, personality must have outweighed looks.

I believe that with Gandy the dashing legend and extravagant personality obscured his real and unusual warmth. Speaking at a memorial meeting in 1995,

[1] His obituary in the *Bulletin of Symbolic Logic* (Moschovakis and Yates, 1996) gives an account of some high points of what was a very distinguished career.

Dummett said that Gandy liked people the way some people like cats. As with cats, people liked Gandy in return. (And yes, Gandy also liked cats.) Gandy had a special generosity of spirit, which I suppose is what attracted Turing initially.

It would be hard to do anyone justice in a quick description like this, but I have so far omitted one significant side to Gandy. It does not quite fit the rest; he was passionate about mathematical logic. Certainly he was as far from the dry-as-dust logician of common parlance as one could imagine. How did he come to be a logician?

Early years

Gandy worked on radar during the war and was posted to Hanslope Park where Turing went to work after his time at Bletchley Park. For a time the two of them shared a house together with Gandy's cat Timothy. Gandy returned to King's College from war service in 1945 for a fourth year, and in 1946 took Part III of the Mathematical Tripos, with distinction. He spent the next few years thinking about the foundations of physics and working towards a Fellowship. In 1949, after what seems like a reasonable period, he applied for an internal Research Fellowship at King's. This required him to submit a dissertation.

Gandy's dissertation was entitled 'Some Considerations concerning the Logical Structure underlying Fundamental Theories in Theoretical Physics'. It seems King's did not keep a copy, but I am grateful to Patricia McGuire, the Archivist, for locating the three expert reports which were considered by the Electors to Fellowships. Turing, who was by then working in Manchester, was one of the experts. The others were Frank Smithies of St John's College and Richard Braithwaite of King's, Knightbridge Professor of Moral Philosophy. We can tell from the reports that Gandy's dissertation was about how scientific theory is related to empirical observation. Gandy's approach was to consider the design of a machine – like a Turing machine but seemingly more complicated – to derive scientific hypotheses from data.

Turing took great care with his report. The general assessment is supplemented by three pages of detailed criticism and commentary. The final section of the assessment reads as follows.

> A less pretentious approach would have made it possible to cover much more ground. This might have been done by the method of example and analogy. Examples are given at some points and form some of the best parts of the thesis. The detailed criticisms are numerous, but their number reflects as much on the reviewer's industry as on the author's shortcomings. The majority of papers of this nature are too flimsy to stand up for criticism. I believe that in a year's time Mr. Gandy should be able to produce something worthy of a Fellowship.

Of the others, Smithies was sympathetic to Gandy's ambition but felt it unrealised, while Braithwaite was sceptical. Gandy was not elected.

Turing thought that Gandy should have been able to produce something worthy of a Fellowship in a further year. Gandy applied again in 1950 with a dissertation with the more straightforward title 'The Foundations of Physics'. Again the dissertation is lost, but Patricia McGuire has found the reports.

One again was written by Turing. It begins bluntly.

> I am very disappointed in this thesis.

It continues as follows.

> The writer has a good imagination and good ideas but he has failed to put them across because of poor technique and taking much too little time over the actual writing of the thesis. He has very rightly decided that symbolic logic is the right medium for these very general considerations, but unfortunately he does not really know enough symbolic logic to carry the programme through successfully. His ideas on the subject of 'groups of indifference' are very stimulating and I shall be most interested to hear whether anything comes of them in the end. But they are certainly not sound as expounded at present, and it does not seem possible to put it right by merely trivial alterations.

That is damning enough but the rest of Turing's report justifies his negative view by an illustrative analysis of a single half-page passage. Turing identifies a range of problems, some arising from lack of clarity of exposition and some from definite errors in logic and mathematics. The second reviewer, Max Newman, though less incisive, is no more supportive. Again Gandy's application for a Fellowship was unsuccessful.

There is an interesting contrast between Turing's two reports. The first is completely dispassionate. There is no hint that Turing even knows the author of the dissertation. The second is very different. The disappointment sounds personal and it is evidently on the basis of personal knowledge that Turing writes about the time spent and the lack of knowledge of symbolic logic. The friendship has not compromised the judgement of the dissertation, which is almost brutal; but the frustration Turing felt about his friend is clear.

Let us take stock at this point. In 1950 Gandy was already 30. He had written two dissertations but seemingly without what now would be regarded as research supervision. He may know what a Turing machine is, but he has almost no technical proficiency in logic. The trajectory of Gandy's intellectual development appears very unpromising. How did he become a logician at all, let alone the very distinguished logician of later years?

Student and supervisor

Something important happened in the next few years. At the end of 1952 Gandy completed a Ph.D. dissertation (Gandy, 1953). It is in two parts, and the first and more substantial of these, on the theory of types, constitutes Gandy's first steps in mathematical logic.

Andrew Hodges (1983) records simply that around 1950 Turing became Gandy's supervisor. It is not clear what the arrangement amounted to but it seems natural to suppose that it was instigated by Turing following the second unsuccessful Fellowship dissertation. Turing became responsible for arranging the oral examination and, as explained in Hodges (1983), had difficulties in doing so. But the oral must have taken place by the next summer as the dissertation was deposited in the University Library in July 1953.

The influence of a supervisor on a Ph.D. dissertation is seldom clear. I start with what Gandy says. He closes his introduction with the following acknowledgement.

> Finally I must try and show the extent of my debt to A.M. Turing. He first called my somewhat unwilling attention to the system of Church and the importance of the deduction theorem. Much of the work on permutations and invariance and on the form of theories was done in conjunction with him. Without his encouragement I should long ago have given way to despair; without his criticism my ideas would have remained shallow and obscure.

That is surely heartfelt and goes beyond the usual words of thanks to a supervisor; but in terms of content it is far from telling the full story.

So what is in the dissertation? Given Gandy's earlier failures, the title 'On axiomatic systems in mathematics and theories of physics' is worrying: it suggests more of the same. The second part of the dissertation is indeed a further attempt to describe the foundations of physics[2] in logical terms. It does not get far, and there are oddities. The section on the deduction theorem, to which Gandy refers, contains reflections on meaning with no obvious relation to the theorem or the rest of the thesis. But the sections on structure and theories show a good grasp of logical fundamentals and my guess is that overall the second part of the Ph.D. dissertation represents a substantial advance over the 1950 Fellowship dissertation. However, the first part on mathematical logic is at a quite different level.

I'll give a brief overview of it. By the system of Church, Gandy means Church's theory of types (Church, 1940). He presents it together with his own variant of it and proves their equivalence. There is a novelty stemming from Turing (1948): types come with a distinguished default value,[3] which informally Turing called 'nonsense elements'. The invariance under permutations, to which Gandy refers

[2] Gandy's interest in physics never left him and happily he did eventually succeed in making a serious contribution. His late paper (Gandy, 1980) on physics and mechanism is still much discussed.

[3] In the language of modern computer science this amounts to raising an exception.

in his acknowledgement, forms a substantial section. Presumably it derives from the groups of indifference of the 1950 Fellowship dissertation. There is a concrete application: the only definable individual is the nonsense element. There are technically proficient sections on what we would now call a notion of inner model and on the definition of truth for sentences of restricted complexity. There is one further section, the third, called 'virtual types'. In the context of the Ph.D. dissertation it does not stand out: nothing is done with the main construction. But its intrinsic intellectual significance and its importance for our story make it quite special.

Gandy thanks Turing for drawing his reluctant attention to Church's theory of types. There is a suggestion there that Turing's involvement was substantial, but there is no way of knowing the extent of it. All that is clear is that in a few years under Turing's supervision Gandy had become a serious logician.

Chinese translation

I would like to break away from the main story for a moment to say something about the construction of virtual types. I don't want to explain the mathematics, but I would like to give a sufficient flavour of the idea to place it within Turing's intellectual concerns.

Imagine as English speakers that we are interested in translation from English into Chinese.[4] We are given some machines that claim to perform this task, and we want to assess them. We have a co-operative Chinese speaker to help us, but we have nobody who speaks both languages. The task looks hopeless. We can't tell if the machines give true translations. But there are two things we can test – their consistency and their extensional equality.

What is consistency? Let's take one machine to begin with. We provide it with two sentences which mean the same thing, let's say, "The cat sat on the mat" and "The mat was sat on by the cat". We feed these into our machine and get two translations out. Then we give the translations to our Chinese speaker. He can't tell us whether they mean "The cat sat on the mat" but he can tell us whether they mean the same thing as each other. They might both mean "The moon is made of green cheese", but that doesn't matter. As long as the machine takes two sentences that mean the same thing as each other and gives two translations that mean the same thing as each other, it is consistent.

Now let us take two machines. We've tested to see if they're consistent. Now we want to see if they're extensionally equal. This simply means that we feed "The cat sat on the mat" into one machine and "The mat was sat on by the cat" into

[4] The echo of a famous debate is conscious, but I am not getting into all that.

the other[5] show the translations to the Chinese speaker and find that they mean the same thing.

All the basic ideas for handling virtual types appear in this fancy about translation. We look at operations on relevant data (English sentences). We look at those operations that take equivalent data (synonymous English sentences) to equivalent data (synonymous Chinese sentences). Operations are equivalent when their outputs on the same data (English sentences) are equivalent (synonymous). That's pretty much all there is to it. In technical terms I have just described the inductive extension of partial equivalence relations to function spaces or, in Gandy's terms, the definition of the function space of virtual types.

My purpose in this is to draw attention to the similarity between the considerations arising with virtual types and the issues involved in the well-known Turing test for artificial intelligence. They focus on input–output behaviour. In the Turing test, if the response of a machine is indistinguishable from that of a person, then we regard them as equivalent, each as conscious as the other. The flavour of virtual types is very much part of Turing's world.

Genesis of an idea

I'd now like to go back to the the story of Gandy's intellectual development. His first published papers (Gandy, 1956, 1959) were on the relative consistency of the axiom of extensionality. I'll just call this the consistency result. There is no sign of it in the Ph.D. dissertation (Gandy, 1953), but the dissertation does use the inductive construction from the section on virtual types. The connection is this. Most mathematicians think extensionally and, as a foundation for mathematics, Church's theory of types already has an axiom of extensionality built in. The idea of virtual types is to create new types with their appropriate extensional equality. Technically one is taking 'quotients by partial equivalence relations'. Gandy's insight was that if you started with a system without extensionality then you could use the same idea to construct new types for which the axiom of extensionality holds. That gives what is typical of logic, a metamathematical theorem: the consistency result. This concrete application of the inductive construction from Gandy (1953) is important. As I shall indicate, it can be regarded as the ancestor of many further ones.

Andrew Hodges (1983) records that Gandy visited Turing ten days before Turing's death and that they discussed type theory. Gandy's consistency result is not in the Ph.D. dissertation, and his paper (Gandy, 1956) was not received until July 1955, a year after Turing's death. However I am certain that the consistency result

[5] We are dealing with equality and we have consistency, so we could in principle feed both into "The cat sat on the mat". But the formulation I give is right in generalisations.

was known at the time of the 1954 visit, and for two reasons. The first is straightforward: in effect Gandy told me so himself.

My D.Phil. thesis was on what Stephen Kleene (1959) had called the countable functionals and Georg Kreisel (1959) the continuous functionals. These two original approaches to this higher type structure are quite different but both use the virtual types technique from Gandy (1953, 1956). In those days I was not much interested by intellectual antecedents, but Gandy did once talk with me of the connection between his early paper and the work in which I was interested. It was not one of our usual detached conversations. I recall from it that Gandy was evidently very proud of his early work, and that he wanted me to understand that Turing had been impressed by it and had praised it. I am confident that Gandy specifically mentioned the paper[6] in the conversation. What I remember most of all is that Turing's liking the paper mattered very much to Gandy.

Foresight and hindsight

When I had the conversation with Gandy, I did not see what the fuss was about. The idea of the construction[7] of virtual types seemed so simple. Could Turing really have been impressed by it? I once raised the question with Kreisel who teased me by observing elliptically that Turing was unusually talented and there was no telling what his views may have been. At the time I took this to mean that Turing might have hidden his true opinion. But we know that Turing could be painfully honest, and this reading does not seem quite right.

With hindsight I can see that my early sense that there was not much in the idea was mistaken. The idea of the paper may be simple but it has wide application. Gandy's own extension to set theory (Gandy, 1959) is delicate, as Scott (1962) observed; and Scott's analysis leads very naturally to the Scott–Solovay formulation of Boolean-valued models. In proof theory the same idea occurs, and when extended to an impredicative setting gives a method associated with Tait and Girard: that of "reducibility candidates". The method was soon adapted for other purposes in theoretical computer science where it is known as Plotkin's logical relations. There are many other appearances of the idea within abstract mathematics. The modern terminology for it is partial equivalence relations[8] or subquotients. Recently extensionality is back on the agenda as a consequence of Voevodsky's Univalence Axiom for homotopy type theory. There are hopes of adapting the construction for use in that area. One should not underestimate simple ideas; those with

6 In fact I had no knowledge of Gandy's Ph.D. dissertation until I started writing this piece.

7 I did not read Gandy's paper and did not refer to the connection in my thesis. So this is by way of setting the record straight.

8 Searching the web for partial equivalence relation generates more than two and a half million hits.

wide application have a special place in mathematics. So Turing's liking Gandy's paper seems to show great foresight. But again there is more to the story than that.

Turing and type theory

Turing's influence as Gandy's supervisor relates specifically to type theory, and I started this chapter with the thought that Turing's interest in the area is largely forgotten. It is time to say something about that interest. There are three published papers, all appearing in the *Journal of Symbolic Logic*. They have hardly had the impact of Turing's other work. But Turing took the trouble to write them and he did not, after all, write that much. What do the papers amount to?

The first (Newman and Turing, 1942) was written with Max Newman about a year after the appearance of Church's formulation of his Simple Theory of Types. This theory has a rather subtle axiom of infinity[9] the type of individuals and the paper shows how to derive from it the same formulation of infinity for all types involving the type of individuals. That is by way of being a sanity check: it is not desperately difficult if one keeps one's head. But we might just note the early appearance of an induction over the collection of types. It is the demands of an inductive construction which drives the definition of virtual types.

The second paper (Turing, 1942) appeared soon afterwards and is of quite different character. Russell and Whitehead and later both Quine and Curry, for example, used dots rather than brackets as a form of punctuation. For Turing this amounts to the use of conventions improving readability. He describes conventions extending those of Curry and gives a precise treatment of the evident issues of disambiguation.

Turing's third paper (Turing, 1948) appeared after the end of the war. The practical issue it considers is the use of type theory informally and so without strict regard for the typing rules. The idea is to get some of the benefits of set theory, and Turing considers explicitly a cumulative hierarchy based on individuals. Abstractly one can think of what he does as a kind of reverse engineering of a strictly typed system towards a more type-free system. In the looser system some expressions are interpretable and some not. Turing's paper may be forgotten, but the issue of the relationship between formal and informal mathematical practice remains very much alive.

[9] It is not important for us, but the essential idea is this. One is not given a handle on the type ι of individuals, so one deals with the type $(\iota \to \iota) \to (\iota \to \iota)$ of so-called Church numerals over it. One asks that it be infinite.

Turing's intellectual taste

Before coming to the final point of the story, I want to say something about Turing's intellectual attitude. The papers and records of talks making up his *Collected Works* are very varied. The pure mathematics and the contributions to logic are outweighed by the machine intelligence and morphogenesis. The Turing mythology is tied up with Bletchley Park. That story, alongside the later proposals for real computing machines, stresses the distinctly practical side to Turing's understanding of mechanism and computation. One might imagine that Turing had little interest in developments in pure mathematics in the post-war period, and little time generally for abstract mathematics. But he was surely aware of them.

Just before the tribute to Turing's influence which I quoted earlier, Gandy acknowledges other intellectual influences.

> The debt which I owe to Bourbaki[10] and to Philip Hall[11] for the development of abstract structure theory is obvious; what is new here is perhaps the technique of extending the usual definitions to objects of arbitrarily high type. Similarly my debt to Klein and Weyl will be apparent. From the many writers on mathematical and natural philosophy who have influenced me, I single out Poincaré, Russell and Ramsey.

The influences are by no means easy to see, and I think that the passage must reflect not simply Gandy's interest but Turing's as well. Both of them appear well aware of the importance of developments in pure mathematics.

From computability to morphogenesis, Turing had an unusual instinct for really fundamental questions; and like any good mathematician he sought definitive answers. He could find them: the Entscheidungsproblem paper (Turing, 1937a) provides such an answer to the fundamental question "what is it for a function to be computable?" But to understand the history of Turing's involvement in type theory, we need to appreciate something more. I think that it is captured at the end of the letter Gandy wrote to Max Newman after Turing's death.

> I thought you hit the nail on the head in the Guardian;[12] the mark of his particular genius was that however abstract the topic he always had absolutely concrete examples in mind; and this, of course, was why he found a lot of contemporary mathematics unsympathetic – he did not like developing abstract concepts merely for their own sake.

[10] Bourbaki is the name of the now famous group centred in France which was establishing a new vision of abstract mathematics.

[11] Philip Hall was the leading UK algebraist of the time. He was a Fellow of King's, but Gandy's mention of him is not college piety. He had wide interests in abstract mathematics. For example he owned a copy of the thesis of the French logician Herbrand.

[12] Newman had written Turing's obituary in what was then the *Manchester Guardian*.

The never written paper

I now return to Turing's second paper (Turing, 1942) on type theory. From a modern perspective, the work involved is a necessary precursor to the effective implementation of a formal language. Turing's intention seems to have been more immediate: he says that he will use the conventions in forthcoming papers. These never appeared. Presumably war work took over. But one planned title which Turing mentions is striking: 'The theory of virtual types'. When I read that I gulped; and I went back and looked more closely at Gandy's Ph.D. dissertation. I quote now from the end of the section there on virtual types.

> So far as I know the idea of introducing virtual types is due to A.M. Turing (see footnote in Newman and Turing (1)[13]). He has not published his version and I do not know to what extent the version given here is in agreement with his.

What should we make of that? First, we are better placed than was Gandy when he wrote. We now have long experience with the basic construction and its many applications. It is what mathematicians call canonical: there is only one way to proceed. Were there something else to do we would have seen it by now. I have no doubt Turing's version would have been the same as Gandy's.

So then what? Well, supervisors often have ideas which they have to a greater or lesser extent thought through and which they suggest to a student leaving it to the student to work out the details. Presumably something of the kind happened in this case. Given the tell-tale footnote, Gandy must have been aware that Turing had considered the question of introducing virtual types or (as we would not say) subquotients. Turing, with delicacy, discretion, reticence – who knows exactly what, left the matter of the agreement between the two approaches lie. Nobody wanted to pursue the matter. That happens more often than one might think.

So why had Turing never written up his own work on virtual types? It feels to me like a reflection of his intellectual taste. He understood what to do in the 1940s but I imagine that he thought it simple and not that important. He did not have a concrete example of its use to give it value. That I believe came later.

The following seems to me the likely run of events. Up until 1950, Gandy was pursuing his original interests in physics and Turing's influence was the casual interest of a friend. After Gandy's 1950 Fellowship application had failed, Turing became Gandy's effective supervisor. I imagine this to have been instigated by Turing on the grounds that Gandy needed to develop his logic to support his views on the foundations of physics. Turing encouraged Gandy to work in type theory, guiding him to produce his own system and working with him on the permutation ideas from Gandy's 1950 dissertation. My guess is that Turing's hand shows in the definite concrete application. Turing described the idea of virtual types; and was

[13] Gandy's reference is wrong: Turing's footnote is in Turing (1942).

happy that once Gandy appreciated the problem he found the fundamental induction step himself. The rest of the logic material of the Ph.D. dissertation involves adapting to type theory existing ideas; it looks like the result of routine research supervision. All this took place over a couple of years, and I suppose that from time to time Gandy and Turing also discussed ideas then current about structure and the like, with applications to physics in mind. In late 1952 Gandy wrote up, perhaps leaving too little time for the full working out of his ideas on physics, or perhaps leaving things sketchy on Turing's advice. Under Turing's guidance the emphasis is very much on logic.

There is usually a fallow period between submission and examination of a Ph.D. dissertation. Gandy no doubt continued to read about and work on logic and physics, but I believe that the next event in the story happened, after the oral, in the second half of 1953. I believe that Gandy himself discovered the application of the virtual types idea to the consistency of extensionality. That concrete application of the idea changed everything. Turing's old idea had become the definitive solution to a problem, thereby establishing its significance. And perhaps more importantly it resolved the difficulty about Turing's unwritten version: Gandy had the application (the consistency result) and legitimate ownership of the material. It is easy to imagine Turing's pleasure in saying to Gandy that he must write up the result for publication.

Earlier I said that I had two reasons for believing that the consistency result was obtained before Turing's death. The second one is just this. There is no way to make sense of Turing's approval of Gandy's work, the approval which meant so much to Gandy, otherwise. The construction as it appears in the Ph.D. dissertation was apparent to Turing in 1942. It is not psychologically realistic to suppose that Turing convincingly praised something he had thought of himself many years before. There had to be something new as the focus of approval. This must have been the new metamathematical application to the consistency result. I am convinced that this was Gandy's own idea[14] and I am equally convinced that this sign of a completely new insight is what really appealed to Turing.

It is a remarkable story. Turing's interest in type theory lasted from when he read Church (1940) until the end of his life. He never published the construction which was to prove to be his most influential idea in the area. Towards the end of his life, when his main interests were in areas other than logic, he taught the subject to his friend and student Gandy. Gandy found the crucial application which established the significance of the idea and so it entered the literature.

[14] Circumstantial evidence supports this. In all his papers, Turing, following Church, considers systems in which extensionality is an axiom. A system without extensionality is never entertained.

Turing's legacy

Turing can hardly have supposed that by the time of his centenary he would be recognised as a national war hero, but he must have known that he left an intellectual legacy. The Turing Machine and Turing Test are the familiar aspects of that. Together with other more specialised scientific contributions, recognised in their own area, this legacy was rightly celebrated during the centenary year. I hope I have shown here that the use of partial equivalence relations stemming from the idea of virtual types is also part of Turing's intellectual estate. It comes to us via his student Gandy, but it is part of the inheritance nonetheless.

I also want to draw attention to a less obvious legacy. Gandy was Turing's friend before he was Turing's student. The history of Gandy's Fellowship applications suggests that Turing had the frustrating sense that his friend had great potential likely never to be realised. Taking Gandy on as a student was a serious project of Turing's last years, and a successful one. Contrary to what would have been reasonable expectation in 1950, Turing turned Gandy into a mathematical logician. Turing was not to see logic become the love of Gandy's life; but, if my reading of the intellectual history is correct, he had the satisfaction of seeing Gandy succeed as a logician with his own independent ideas.

Turing left his mathematical books and papers to Gandy in a will dated 11 February 1954. This seems to me to a concrete sign of his pleasure in Gandy's intellectual development. But the real legacy to Gandy was his becoming a logician. That legacy has been passed on, to Gandy's students and to their students and so into the future of logic. In this story there are two legacies, the human influence along with the intellectual; and both stem from the forgotten Turing, the type theorist.

References

A. Church (1940). A formulation of the simple theory of types. *Journal of Symbolic Logic* **5**, 56–68.

R.O. Gandy (1953). *On axiomatic systems in mathematics and theories of physics.* Ph.D. dissertation, University of Cambridge.

R.O. Gandy (1956). On the axiom of extensionality – Part I. *Journal of Symbolic Logic* **21**, 36–48.

R.O. Gandy (1959). On the axiom of extensionality – Part II. *Journal of Symbolic Logic* **24**, 287–300.

R.O. Gandy (1980). Church's Thesis and principles for mechanisms. In *The Kleene Symposium*, J. Barwise, H.J. Keisler, and K. Kunen (editors), pp.123–148, North-Holland.

A.P. Hodges (1980). *Alan Turing. The Enigma of Intelligence*, Burnett Books. See also the website `http://www.turing.org.uk/book/`.

S.C. Kleene (1959). Countable functionals. In *Constructivity in Mathematics*, A. Heyting (editor). North-Holland.

G. Kreisel (1959). Interpretation of analysis by means of functionals of finite type. In *Constructivity in Mathematics*, A. Heyting (editor). North-Holland.

Y. Moschovakis and M. Yates (1996). In memoriam: Robin Oliver Gandy, 1919–1995. *Bulletin of Symbolic Logic* **2**, 367–370.

M.H.A. Newman and A.M. Turing (1942). A formal theorem in Church's theory of types. *Journal of Symbolic Logic* **7**, 28–33.

D.S. Scott (1962). More on the axiom of extensionality. In *Essays on the Foundations of Mathematics, Dedicated to A.A. Fraenkel on his Seventieth Anniversary*, Y. Bar-Hillel, E.I.J. Poznanski, M.O. Rabin and A. Robinson (editors). North-Holland, 115–131.

A.M. Turing (1937). On computable numbers, with an application to the Entscheidungsproblem. *Proceedings of the London Mathematical Society* **42**(2), 230–265; A Correction. *ibid.* **43**, 544–546 (1938).

A.M. Turing (1937a). Computability and λ-definability. *Journal of Symbolic Logic* **2**, 15-163.

A.M. Turing (1939). Systems of logic based on ordinals. *Proceedings of the London Mathematical Society* **45**(2), 161–228.

A.M. Turing (1942). The use of dots as brackets in Church's system. *Journal of Symbolic Logic* **7**, 146–156.

A.M. Turing (1948). Practical forms of type theory. *Journal of Symbolic Logic* **13**, 80–94.

3

Turing and the Primes

Andrew R. Booker

Alan Turing's exploits in code-breaking, philosophy, artificial intelligence and the foundations of computer science are by now well known to many. Less well known is that Turing was also interested in number theory, in particular the distribution of prime numbers and the Riemann hypothesis. These interests culminated in two programs that he implemented on the Manchester Mark 1 (see Figure 3.1), the first stored-program digital computer, during its 18 months of operation in 1949–1950. Turing's efforts in this area were modest,[1] and one should be careful not to overstate their influence. However, one cannot help but see in these investigations the beginning of the field of computational number theory, bearing a close resemblance to active problems in the field today despite a gap of 60 years. We can also perceive, in hindsight, some striking connections to Turing's other areas of interests, in ways that might have seemed far-fetched in his day. This chapter will attempt to explain the two problems in detail, including their early history, Turing's contributions, some developments since the 1950s, and speculation for the future.

Prime numbers

People have been interested in prime numbers since at least the ancient Greeks. Euclid recorded a proof that there are infinitely many of them around 300 BC (Narkiewicz, 2000, §1.1.2). His proof, still one of the most elegant in all mathematics, can be expressed as an algorithm:

(1) Write down some prime numbers.
(2) Multiply them together and add 1; call the result n.
(3) Find a prime factor of n.

[1] I think that Turing himself would agree with this; indeed, it seems clear from his writings at the time that he was disappointed with the results obtained from both programs.

Figure 3.1 The left-hand half of the Manchester Mark 1.

For instance, if we know that 2, 5 and 11 are all prime then, applying the algorithm with these numbers, we get $n = 2 \times 5 \times 11 + 1 = 111$, which is divisible by the prime 3. By an earlier theorem in Euclid's *Elements*, the number n computed in step (2) must have a prime factor (and in fact it can be factored uniquely into a product of primes by the Fundamental Theorem of Arithmetic), so step (3) is always possible. On the other hand, from the way that Euclid constructs the number n, the prime factor found in step (3) cannot be any prime written down in step (1). Thus, no list of primes can be complete, i.e. there are infinitely many of them.

Note that n can have more than one prime factor (e.g. the number 111 in our example is also divisible by 37), and Euclid doesn't specify which one we should take in step (3). In 1963, Albert Mullin made Euclid's proof completely constructive by starting with just the prime 2 and repeating the algorithm to add a new prime to the list, always choosing the smallest prime factor of n in step (3). Similarly, one can instead always choose the largest prime factor, and these two constructions result in the so-called *Euclid–Mullin sequences* of primes (Narkiewicz, 2000, §1.1.2), the first few terms of which are shown in Table 3.1. Mullin posed the natural question whether *every* prime number eventually occurs in each sequence. This is still unknown for the first sequence, though it has been conjectured that the answer is yes. On the other hand, it was shown recently (in 2011) that infinitely many primes are missing from the second sequence. This demonstrates how even very old topics in number theory can generate interesting research.

Euclid was followed within a century by Eratosthenes, who found an algorithm for listing the primes that is still in use today. These results typify the contrasting ways that one can study and use prime numbers: either by looking at individual

first sequence (smallest prime factor)	second sequence (largest prime factor)
2	2
3	3
7	7
43	43
13	139
53	50207
5	340999
6221671	2365347734339
38709183810571	4680225641471129
139	1368845206580129

Table 3.1 *First ten terms of the Euclid–Mullin sequences*

primes (as in the case of Eratosthenes) or by trying to understand general properties of the sequence of all primes, even those that may be well beyond our capacity to compute (as in the case of Euclid). As we will see, Turing's two programs on the Manchester Mark 1 fall squarely within these two respective camps.

Large primes

The first of these programs was an idea of Max Newman, then head of the mathematics department at Manchester University, conceived as a way of testing the capabilities of the new machine and draw publicity for the project. It was to search for a large prime number.

Mersenne primes

In much the same way that there is heaviest known element at any given point in history, there is also a largest-known prime number. At the time of writing this stands at $2^{57\,885\,161} - 1$, a number with over 17 million digits, but that record is unlikely to stand for long. This is an example of a *Mersenne prime*, those that are 1 less than a power of 2. They are named for the French monk Marin Mersenne, who in 1644 predicted that $2^n - 1$ is prime for $n = 2, 3, 5, 7, 13, 17, 19, 31, 67, 127, 257$, and for no other values of $n < 258$. All but the last four were known at the time of his prediction, and we now know that he was wrong about 67 and 257 and also omitted $n = 61$, 89 and 107, which do yield primes. The name has stuck nevertheless!

Throughout recent history (the last 150 years or so), the largest known prime number has usually been a Mersenne prime. The reason is probably that it turns out to be a bit easier to find primes among the Mersenne numbers than for more general classes of numbers, and thus they have received more attention. It is not hard to see that if $2^n - 1$ is prime then n itself must be prime, which helps weed out most of the non-primes. (Unfortunately, this test only works one way, i.e. even if n is prime, it is not guaranteed that $2^n - 1$ will be; the smallest counterexample is $2^{11} - 1 = 2047 = 23 \times 89$.) Second, there is very fast algorithm, described by Lucas in 1876 and later refined by D.H. Lehmer, for testing the primality of a given candidate $2^n - 1$ when n is a prime bigger than 2:

(1) Start with the number $x = 4$.
(2) Replace x by the remainder from dividing $x^2 - 2$ by $2^n - 1$.
(3) Repeat step (2) a total of $n - 2$ times.
(4) Then $2^n - 1$ is prime if the final value of x is 0, and not otherwise.

(The reader is invited to try this with $n = 3$ or $n = 5$. With some paper and a pocket calculator to hand, it's also fun to check that the test correctly proves that $2^{11} - 1$ is not prime.) Third, the form of the Mersenne numbers (1 less than a power of 2) makes some of the arithmetic in the Lucas–Lehmer test particularly easy for computers, since they do calculations internally in binary.

Another reason for studying Mersenne primes is their connection to the so-called *perfect numbers*, which are those numbers that equal the sum of their proper divisors. For instance, the proper divisors of 28 are 1, 2, 4, 7 and 14 and these total 28. These numbers have also been studied since antiquity, and in fact Euclid was aware that if $p = 2^n - 1$ is prime then $p(p+1)/2$ is perfect; thus, for instance, the perfect number $28 = 7 \times 8/2$ noted above is related to the Mersenne prime $7 = 2^3 - 1$, and each new Mersenne prime that is found also yields a new perfect number. (Incidentally, $p(p+1)/2$ is also the sum of the numbers from 1 to p, e.g. $28 = 1+2+3+4+5+6+7$, so it is no wonder that the ancient mathematicians regarded these numbers as having mystical properties compared with ordinary numbers, to the point of calling them "perfect".) About 2000 years later, Euler proved the converse statement that any *even* perfect number is produced by Euclid's construction, so there is a direct correspondence between Mersenne primes and even perfect numbers. It is still unknown whether there are any odd perfect numbers, though it is generally believed that none such exist.

Mersenne primes in the electronic era

By 1947, all the Mersenne numbers $2^n - 1$ for n up to 257 had been checked by hand, settling Mersenne's original claim. The largest prime among those was $2^{127} - 1$, discovered by Lucas in 1876. All Mersenne primes since then have been discovered by machines. (However, Ferrier discovered that $(2^{148} + 1)/17$ is prime in 1951 using only a mechanical desk calculator; it remains the largest prime discovered 'manually'.) The first such investigation was made by Turing, together with Newman and engineers Tom Kilburn and Geoff Tootill, in the summer of 1949. From a letter[2] that Turing wrote to D.H. Lehmer in March 1952, we know that the team verified all the known Mersenne primes and extended the search out to $n = 433$, though Turing described their efforts as unsystematic. Ultimately, the test brought the publicity for the new machine that Newman sought, but came up short since they ended the search before finding any new primes.

One may debate the scientific value of their work because of that, though even if they had found a new prime number, it would by now be just a footnote in history. In any case, it wasn't long before new record primes were discovered by computer; Miller and Wheeler found several new ones in 1951 using the EDSAC at Cambridge, and Robinson found the next five Mersenne primes in 1952 using the SWAC at the National Bureau of Standards in Los Angeles. Robinson's calculation, described in detail by Corry (2010), is particularly impressive since he wrote his program having never seen a computer before. He sent the punchcards containing the code by mail to D.H. and Emma Lehmer in Los Angeles. They first ran the program on January 30, 1952; it ran bug-free and found the next Mersenne prime, $2^{521} - 1$, on the same day. Turing said how impressed he was with Robinson's results in his letter to D.H. Lehmer.

The search for Mersenne primes has continued unabated ever since. This presumes, of course, that there are more to find; however, there is no analogue of Euclid's proof for the Mersenne primes and, despite clear heuristic and empirical evidence, we still have no proof that there are infinitely many of them. At the time of writing 48 are known, 36 of which were found by computers.[3] Since the mid-1990s, the search has been dominated by the aptly named Great Internet Mersenne Prime Search, set up by computer scientist George Woltman. Woltman's program uses spare time from the personal computers of thousands of volunteers, linked by the internet.[4] They have discovered new world-record primes at the rate of about one per year.

The distribution of prime numbers

The second problem that Turing investigated on the Manchester Mark 1 was the Riemann hypothesis (or RH), which has to do with the asymptotic distribution of

[2] Found in the Emma and D.H. Lehmer Archive at Bancroft library, UC Berkeley.

[3] This sentence raises a philosophical question: Should the computer be credited with the discovery along with the people involved in writing and running its program? A particularly interesting case occurred on April 12, 2009, when a computer proved that the number $2^{42\,643\,801} - 1$ is prime, making it the third largest known Mersenne prime; however, no human took notice of that fact until June 4 of that year. Which should be considered the date of discovery?

[4] The reader is invited to participate; visit `www.mersenne.org` for details.

prime numbers. This was a problem close to Turing's heart, and in fact he made an earlier attempt to investigate RH in 1939 with a special-purpose analog machine, using an elaborate system of gears. Turing had apparently cut most of the gears for the machine before he was interrupted by the war. By the time that he returned to the problem, in June 1950, the progress toward general-purpose digital computers during the war had made Turing's 1939 machine obsolete. (The machine was never completed, though we do have a blueprint for it drawn up by Turing's friend Donald McPhail. A project has recently been proposed to build it; ironically, the first step of that undertaking will be a computer simulation.) Indeed, it had become practical, if only barely so, to consider much more than was possible with any analog machine – testing RH algorithmically, with no human intervention. As we will see below, this aspect of the problem, often taken for granted in modern discussions of the subject, was of keen interest to Turing.

The story behind the Riemann hypothesis goes back to Gauss, who as a boy of 15 or 16 (in 1792–1793) made long lists of prime numbers in order to understand just how common they are. (It seems fair to say that Gauss was a computational number theorist before there were computers!) He came to the conjecture that, around a large number x, roughly 1 in every $\ln x$ integers is prime;[5] thus, if we want to know how many primes there are among the numbers 2, 3, 4, $\ldots$, x (without actually counting them), we can estimate this by the integral $\int_2^x \frac{1}{\ln t}\,dt$, which is usually denoted $\mathrm{Li}(x)$. One might call this a quantitative version of Euclid's qualitative result that there are infinitely many primes.

x	$\pi(x)$	nearest whole number to $\mathrm{Li}(x)$
10^3	168	177
10^6	78 498	78 627
10^{12}	37 607 912,018	37 607 950 280
10^{24}	18 435 599 767 349 200 867 866	18 435 599 767 366 347 775 143

Table 3.2 *Comparison of* $\pi(x)$ *and* $\mathrm{Li}(x)$.

Table 3.2 shows the number of primes up to x, usually denoted $\pi(x)$ (although it has nothing to do with the constant $\pi = 3.14159\ldots$), for various powers of 10. Also shown is the nearest whole number to Gauss' approximation $\mathrm{Li}(x)$. One thing that is immediately apparent is that $\mathrm{Li}(x)$ seems to give an overestimate of the true count, and in fact that is true for every value of $\pi(x)$ for x at least 8 that has ever been computed. It was thought for a while that it must always be the case that $\mathrm{Li}(x) > \pi(x)$ for large x, but Littlewood proved in 1914 that this is false for some x and moreover the inequality flips direction infinitely many times. In 1933, Skewes made Littlewood's theorem effective by showing that the first flip occurs

[5] Here ln denotes the natural logarithm to base $e = 2.71828\ldots$

before $x = 10^{10^{10^{34}}}$ if the unproven Riemann hypothesis (discussed below) is true. This is an unimaginably large number, so much so that Hardy called it "the largest number which has ever served any definite purpose in mathematics". (That might have been true in 1933, but mathematicians have since found ways to make use of much larger numbers, such as those encountered in *Ramsey theory*.)

Turing worked on an improvement to Skewes' argument, hoping to reduce the bound significantly and remove the assumption of RH (Hejhal, 2012; Hejhal–Odlyzko, 2012). He made some progress toward both goals in the summer of 1937, and returned to the problem again around 1952–1953, but never published his work. In any case, both Skewes' and Turing's approaches have since been supplanted by later work based on computational methods; the latest results are discussed below.

The Riemann zeta-function

The Riemann zeta-function is the infinite series

$$\zeta(s) = 1 + \frac{1}{2^s} + \frac{1}{3^s} + \cdots = \sum_{n=1}^{\infty} \frac{1}{n^s}. \tag{3.1}$$

As we learn in calculus class, this series converges for every $s > 1$ and diverges for every $s \leq 1$; the borderline case $s = 1$ is the well-known *harmonic series*, $\sum_{n=1}^{\infty} \frac{1}{n}$. The connection between the zeta-function and the prime numbers comes from another formula for $\zeta(s)$, derived by Euler in 1737 (Narkiewicz, 2000, §1.1.4):

$$\zeta(s) = \frac{1}{1-\frac{1}{2^s}} \times \frac{1}{1-\frac{1}{3^s}} \times \frac{1}{1-\frac{1}{5^s}} \times \cdots = \prod_{p \text{ prime}} \frac{1}{1-\frac{1}{p^s}}. \tag{3.2}$$

Here it is given as an infinite product rather than a sum, and the variable p runs through all prime numbers (2, 3, 5, 7, 11, ...).

The equivalence between (3.1) and (3.2) is a sort of analytic expression of the Fundamental Theorem of Arithmetic, mentioned above. To see this, we first expand the factor $\frac{1}{1-\frac{1}{p^s}}$ using the formula for a geometric series:

$$\frac{1}{1-\frac{1}{p^s}} = 1 + \frac{1}{p^s} + \left(\frac{1}{p^s}\right)^2 + \left(\frac{1}{p^s}\right)^3 + \cdots = 1 + \frac{1}{p^s} + \frac{1}{(p^2)^s} + \frac{1}{(p^3)^s} + \cdots$$

Next, we multiply together these geometric series for all choices of p. To do this, we have to imagine every conceivable product of the terms $\frac{1}{(p^k)^s}$ for various primes p and exponents k. For instance, one term that arises is $\frac{1}{(3^2)^s} \times \frac{1}{13^s} = \frac{1}{117^s}$, coming from the corresponding terms for $p = 3$ and $p = 13$. More generally, it is not hard to see that every product of terms will take the form $\frac{1}{n^s}$ for some positive integer n. In fact, for any given n, the term $\frac{1}{n^s}$ must eventually occur since n has some prime factorization. Finally, because the prime factorization of n is unique, $\frac{1}{n^s}$ occurs exactly once. Thus, we arrive at the original defining series (3.1).

All of these manipulations make sense and can be made completely rigorous whenever $s > 1$. Euler had the clever idea of letting s tend to 1, so that (3.1) tends to the harmonic series, which diverges.[6] Thus, it must also be the case that (3.2) gets arbitrarily large as s gets close to 1. From this[7] Euler concluded that there are infinitely many prime numbers p, since otherwise (3.2) would make sense and remain bounded even as s approached 1. This proof, while fiendishly clever, may seem much ado about nothing given that Euclid had already shown that there are infinitely many primes some 2000 years earlier. What makes Euler's proof important is that it can be generalized in ways that Euclid's cannot.

First, in 1837, Dirichlet showed how to modify Euler's proof to show that an arithmetic progression

$$a,\ a+b,\ a+2b,\ a+3b,\ \ldots,$$

contains infinitely many primes, as long as a and b have no common factor (Narkiewicz, 2000, §2). (If a and b have a common factor then it is easy to see that this progression can contain at most one prime; for instance the progression 6, 10, 14, 18, ... contains no primes since every term is even.) To do so, he introduced certain modified versions of the zeta-function, the so-called L-functions that now bear his name, and again studied their behavior as s tends to 1. Dirichlet's theorem is important in the sense that it has been used as an ingredient in countless other theorems in number theory. Moreover, it marks the beginning of what we now call *analytic number theory*, using techniques from real and complex analysis to study fundamental questions about numbers.

Second, in 1859, Riemann wrote a path-breaking paper on the zeta-function, his only paper related to number theory (Narkiewicz, 2000, §4). In it, he sketched how a detailed study of $\zeta(s)$ (notation introduced by Riemann) can be used not only to see that there are infinitely many primes but also to understand the asymptotics of their distribution, ultimately leading to a proof of Gauss' conjecture, which was completed independently by Hadamard and de la Vallée Poussin in 1896; we now call this result the *Prime Number Theorem* (Zagier, 1997). Riemann's key insight was to consider $\zeta(s)$ not just for real numbers s, but for complex s as well. In fact he showed, through the principle of *analytic continuation*, how to make sense of

[6] This is a modern interpretation of his argument; in the 18th century, the notions of limit and convergence were not yet formulated rigorously, so Euler would have more brazenly set s equal to 1 and not worried so much about the extent to which it made sense to do so. The pendulum may yet swing the other way; the subject of *non-standard analysis* allows for a rigorous formulation of Euler's more direct approach though, as its name implies, it is not yet fully accepted by all mathematicians.

[7] An alternative version, popular among algebraic number theorists, is to consider instead $s = 2$. Another theorem of Euler's says that $\zeta(2) = \pi^2/6$, and if there were only finitely many prime numbers then, by (3.2), this would be a rational number. However, Legendre proved in 1794 that π^2 (and hence also $\pi^2/6$) is irrational.

$\zeta(s)$ for all complex s apart from 1. The crucial point turns out to be understanding those values of s for which $\zeta(s) = 0$. It is known that this holds for $s = -2, -4, -6, \ldots$, and for infinitely many non-real values of s with real part between 0 and 1. Riemann computed approximations of the first few non-real zeros, which are shown in Table 3.3. (The zeros come in complex-conjugate pairs, i.e. for every zero at $x + iy$, there is another one at $x - iy$. Thus, it is enough to list those with positive imaginary part.) He then made the bold guess that all of them have real part exactly $\frac{1}{2}$.

$0.5 + i14.134725141734693790\,45\ldots$
$0.5 + i21.022039638771554992\,62\ldots$
$0.5 + i25.010857580145688763\,21\ldots$
$0.5 + i30.424876125859513210\,31\ldots$
$0.5 + i32.935061587739189690\,66\ldots$

Table 3.3 *First five zeros of Riemann's zeta-function with positive imaginary part*

Over 150 years later, we are still unsure of this guess, now called the Riemann Hypothesis (Conrey, 2003), though there is significant evidence in favor of it, and most mathematicians today believe it to be true. If true, RH implies that Gauss' estimate of the number of primes up to x is accurate to 'square root order', which, in other words, means that roughly the top half of the digits of the estimate are correct; for instance, while it is currently well beyond our technology to say exactly how many primes there are with at most 50 digits, Gauss' formula predicts the number to be about

$$\underline{876268031750784168878176}862640406870986031109950,$$

and it is very likely that the underlined digits are correct. In the absence of a proof of RH, we have had to make do with weaker results; for instance, we know for sure that the number of correct digits in Gauss' approximation increases with the number of digits of x (which is the qualitative statement of the Prime Number Theorem), but we don't yet know that it does so linearly.

Turing and the Riemann Hypothesis

One thing that makes RH a good conjecture is its falsifiability, i.e. if it does turn out to be false then that can clearly be shown by observing a counterexample. There are some philosophical reasons to believe in the truth of RH but, aside from that, our best evidence in its favor is the many numerical tests that have been performed, not one of which has shown it to be false. (On the other hand, as the $\pi(x)$ versus $\mathrm{Li}(x)$ question shows, one should not rely entirely on numerical evidence.) Curiously,

Turing was not convinced of its truth; indeed, it is clear from his paper on the subject (Turing, 1953) that he had hoped the Manchester Mark 1 would find a counterexample. In his defense, skepticism of the conjecture was not uncommon in the first half of the 20th century, and some near counterexamples found in early investigations made it seem that a true counterexample might be uncovered with just a bit more computation.

As mentioned above, the first computation of this type was made by Riemann himself[8], and probably figured in his formulation of the conjecture. By the 1930s, Titchmarsh had extended the computation out to more than 1000 zeros, which were all found to obey RH. Titchmarsh's method, which was essentially derived from Riemann's, consisted of two main steps:

(1) Find all the zeros with real part $\frac{1}{2}$ and imaginary part between 0 and some large number T. Although the values of $\zeta(\frac{1}{2}+it)$ for real numbers t are typically complex, it turns out that one can define a real-valued function $Z(t)$ with the same absolute value as that of $\zeta(\frac{1}{2}+it)$. Thus, the zeros of $Z(t)$ correspond to the zeros of Riemann's zeta-function with real part $\frac{1}{2}$, and they can be found simply by inspecting the graph of $Z(t)$ and noting where it crosses the t-axis (see Figure 3.2, top panel).

(2) Find, by an auxiliary computation, the total number, say $N(T)$, of non-real zeros of the zeta-function with imaginary part up to T. If this agrees with the count of zeros with real part $\frac{1}{2}$ found in step (1) then all zeros with imaginary part up to T obey RH.

Of these two steps, the first is relatively straightforward. In fact, Riemann had already found a formula (later published by Siegel) that could be used to evaluate $Z(t)$ very quickly, which he used for his computations. The second step is a great deal more complicated; the methods used in all investigations up to and including Titchmarsh's were ad hoc and not guaranteed to work for large values of T. That was not good enough for Turing, who wanted the machine to work as autonomously as possible. Instead, he found a criterion that could be used to decide if all the zeros had been found *using the values that had already been computed.* Thus, Turing effectively replaced the most cumbersome step in the verification by an automatic check.

Turing's method was based on a careful comparison of the observed values of $N(T)$ versus its known asymptotic formula as T increases. Riemann postulated, and it was later rigorously proven, that $N(T)$ can be approximated by the smooth

[8] This was only discovered decades after Riemann's death by examinination of his unpublished notes in the Göttingen library.

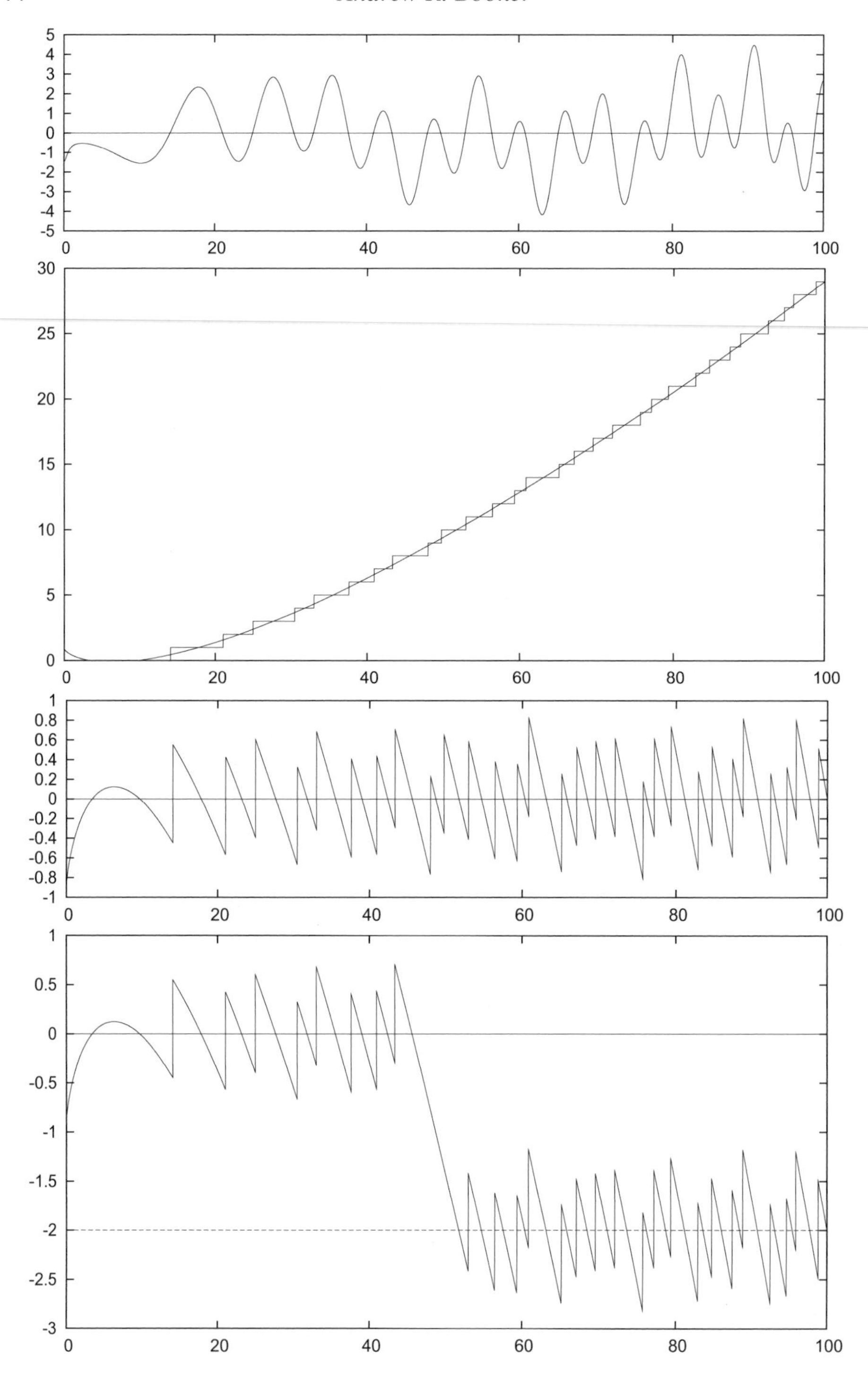

Figure 3.2 From top to bottom: $Z(t)$, $N(T)$ versus $M(T)$, $E(T)$, and $E(T)$ with a pair of 'missing' zeros

function

$$M(T) = \frac{T}{2\pi} \ln\left(\frac{T}{2\pi e}\right) + \frac{7}{8}.$$

(Figure 3.2, second panel, shows the graphs of $N(T)$ and $M(T)$ for T up to 100. Notice how every jump in the graph of $N(T)$ occurs at a zero crossing of $Z(t)$; this confirms RH up to height 100.) This is an asymptotic approximation, meaning that the percentage error in the prediction tends to 0 as T grows, but in absolute terms it can be out by a large margin for an individual value of T. In practice it has never been observed to be wrong by more than 4, though it is known theoretically that the error is typically large for very large T. In any case, this renders the formula useless when it comes to deciding whether we have found all the zeros for a given value of T.

Turing had the clever idea that one should look at the error term $E(T) = N(T) - M(T)$ for a range of values of T instead of just one value. It was shown by Littlewood that $E(T)$ has average value close to 0 when T is large, and thus it tends to oscillate around 0, as is visible in Figure 3.2, third panel. If we imagine drawing that graph using measured data, any zeros that were missed would skew the average; for instance, if two zeros were missed,[9] it would begin to oscillate around -2, as shown in the bottom pane. This can be turned into a rigorous proof that none are missing, as long as one has a version of Littlewood's theorem with explicit constants. One of the main theoretical results in Turing (1953) was a painstaking derivation of such a theorem.

Although Turing carried out his investigation in 1950, his paper was not published until 1953, just a year before his death. The project was apparently not a high priority task for the Mark 1, as the following quotation from the paper makes clear:

> The calculations had been planned some time in advance, but had in fact to be carried out in great haste. If it had not been for the fact that the computer remained in serviceable condition[10] for an unusually long period from 3pm one afternoon to 8am the following morning it is probable that the calculations would never have been done at all. As it was, the interval $2\pi.63^2 < t < 2\pi.64^2$ was investigated during that period, and very little more was accomplished.

Evidently Turing was disappointed with the results obtained. As we now know, computers have become enormously faster, less expensive, and more reliable than in 1950, and these improvements came about very quickly. However, that would

[9] Since the zeros are located by sign changes, one always misses an even number of them.

[10] The Mark 1, like all of the early electronic digital computers, employed thousands of *vacuum tubes* (or thermionic valves), a technology that evolved from the incandescent light bulb. As is the case with light bulbs, one could expect an individual tube to last for years, but when using thousands of them, it was inevitable that at least one would burn out every day. To guard against this, it was standard practice on the Mark 1 to repeat sections of code every few minutes and halt the machine when a discrepancy was discovered.

have been difficult to anticipate at the time, which might explain Turing's pessimism. D.H. Lehmer had this to say in his *Mathematical Review* of the paper:

> Although the author tends to belittle the actual results obtained in a few hours of machine time, the paper shows that a great deal of careful work has been done in preparing the calculation for the machine and this work will be of value to future computers. Since 1950 there has been a large increase in the number and reliability of large scale computers. No doubt further results on this problem will appear in due course.

Indeed, by 1956 Lehmer himself had applied Turing's method to extend the computations to ranges well beyond the reach of mechanical calculators. With modern computers and improved algorithms, they have reached extremes that would have been unfathomable in the 1950s. For instance, the first 10 trillion zeros have been found to obey RH, as has the 10^{32}th zero and hundreds of its neighbors; all such calculations continue to rely on Turing's method as a small but essential ingredient. It is unfortunate that Turing would never see any of that come to pass.

Formal proofs

There are many other interesting remarks in Turing (1953), but one in particular speaks to his mindset at the time:

> If definite rules are laid down as to how the computation is to be done one can predict bounds for the errors throughout. When the computations are done by hand there are serious practical difficulties about this. The computer will probably have his own ideas as to how certain steps should be done.[11] [...] However, if the calculations are being done by an automatic computer one can feel sure that this kind of indiscipline does not occur.

It should be noted that Turing was writing 'Computing machinery and intelligence' around the same time, and in that context the quotation is not surprising. Nevertheless, it was far ahead of its time; even two decades later, when the first proof of the *four-color theorem* was announced, there was considerable doubt over whether it could be accepted if it was not practically possible for a human to check it. Turing was declaring in 1950 that it is not only acceptable but in fact *preferable* for machines to replace humans in some settings.

The tide is slowly turning toward Turing's point of view, in that mathematicians are now more trusting of results obtained by computer, though one still frequently hears the argument that such proofs are less elegant than those obtained by "pure thought". The shift in perceptions is illustrated by another controversy, similar to that surrounding the four color theorem, concerning Thomas Hales' proof in 1998

[11] Turing's use of the word "computer" here to refer to a human reflects the common usage up until the 1940s.

of the *Kepler conjecture*; this time it was not so much about whether the *machine* could be trusted, but rather its *programmers*, since the implementation was technically very challenging.

As the use of computers in pure mathematics increases, and proofs become more complicated as a result, such controversies seem likely to become more prevalent. One response is a thriving interest in *formal proofs* (Hales, 2008), in which a computer is used to check every single step starting from the basic axioms; for instance, we now have two independent formal proofs of the Prime Number Theorem, both completed within the last decade. Following this trend, it is easy to imagine a future in which Turing's vision is the *de facto* standard, and mathematical proofs would not be fully accepted until they had undergone formal verification by a machine.

Present day and beyond

Given the economies of scale achieved in the speed, reliability and availability of computing machines noted above, it is no surprise that their use has exploded in all sorts of human endeavors. They are now ubiquitous in applied mathematics and are starting to come to prominence in pure mathematics as well, including number theory. For instance, when it comes to primality testing, it is now routine to find prime numbers with thousands of digits, and this helps to keep our online transactions secure.

For analytic number theory and RH in particular, a wealth of new understanding has come from computations of the zeta-function, much in the spirit of Turing's work in 1950. Foremost among these are computations done in the 1980s by Andrew Odlyzko, who, together with Schönhage, found an algorithm that could be used to compute many values of $Z(t)$ simultaneously and very quickly, the first theoretical improvement along these lines since the discovery of the Riemann–Siegel formula. The new algorithm enabled Odlyzko to expose a link between the zeros of the zeta-function and *random matrix theory*, a tool used by physicists to model the energy levels of heavy atoms. Figure 3.3 shows a graph produced by Odlyzko comparing the nearest-neighbor spacing distribution of zeros of the zeta-function with that of the eigenvalues of random Hermitian matrices (the so-called *GUE ensemble*). Roughly speaking, the curve gives the probability that a gap of a given size will occur between two consecutive zeros; thus, for instance, we see that the zeros are rarely close together, i.e. they tend to repel each other.

The first hint of a connection between the zeta-function and random matrix theory came from a chance meeting between mathematician Hugh Montgomery and physicist Freeman Dyson at the Institute for Advanced Study in 1972. Montgomery had conjectured a formula for one statistic, the *pair correlation*, of the zeros of the zeta-function, and Dyson immediately recognized that it was the same formula

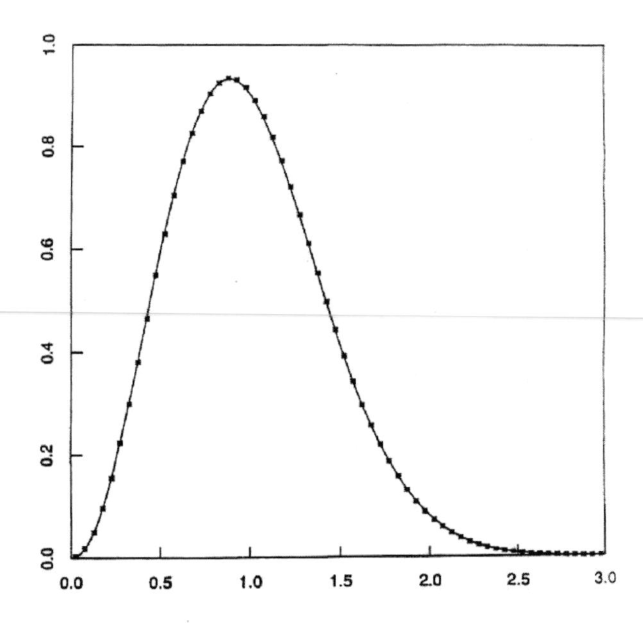

Figure 3.3 Nearest-neighbor spacing distribution of about 79 million zeros of ζ around the 10^{20}th zero (scatter plot), shown with the GUE model (smooth curve) for comparison.

as in the GUE model. However, Odlyzko's numerics, such as in Figure 3.3, were definitive in making the case for what otherwise might have seemed a curious coincidence.

What is still unclear is *why* the zeros of the zeta-function follow GUE statistics. One possibility is that there is a physical system, similar to a heavy atom, whose spectrum is exactly the set of zeros of ζ. If so, finding the system would realize an approach, now called the Hilbert–Pólya conjecture, to proving RH that was suggested by Pólya a century ago.

A more mundane possibility is that this is a universal phenomenon, in the same way that Gaussian distributions turn up a lot in nature, a view strengthened by computations involving other L-functions which have found similar results. There are now many mathematical objects that fall under the general label of L-function (Booker, 2008), the prototypes of which are the Riemann zeta-function and the original L-functions of Dirichlet. The *Langlands Program*, a major area of research in modern number theory, aims to classify the various kinds of L-functions and the relations between them, and we are just beginning to explore the full potential of computational methods in this area.[12] In part, the interest in other L-functions grew out of our inability to prove RH, since when mathematicians are stuck on one problem they often try to make progress in different directions by generalizing. Thus,

[12] See www.L-functions.org.

for every one of these more general L-functions there is an associated 'Riemann Hypothesis', though there are still no cases in which it has been proven.

Whatever the reason for the GUE phenomenon, one can argue that a thorough understanding of it is necessary before a proof of RH can be found. Although this still seems a long way off, we are just now moving past the point of verifying conjectures, as Turing and Odlyzko did, and are starting to use such computations as a replacement for RH when solving problems. Here are three examples related to prime numbers:

(1) *Counting primes.* As we saw above, Gauss' formula gives a good approximation for how many primes there are among the numbers $2, 3, \ldots, x$. But what if we want to know the exact answer for a particular x? Until the 19th century, the best known way to determine that was to find all the primes with Eratosthenes' algorithm and count them. Fortunately, mathematicians have found some cleverer ways since then. The fastest method currently known was proposed by Lagarias and Odlyzko in 1987; it works by adding various correction terms to Gauss' formula, using the computations that go into verifying RH. A version of it was used for the first time in 2010 by Buethe, Franke, Jost and Kleinjung, who computed the last entry of Table 3.2. Their method assumed RH, but that assumption has very recently been removed in an independent calculation by Platt.

(2) *The $\pi(x)$ vs. $\mathrm{Li}(x)$ problem.* Using numerical approximations of the first 22 million zeros of the Riemann zeta-function, Saouter and Demichel showed in 2010 that $\pi(x)$ exceeds $\mathrm{Li}(x)$ for some value of x below 1.3972×10^{316}, and there is reason to believe that the first such occurrence is near there. Thus, while we might never know the first occurrence exactly, the question of the best possible improvement to Skewes' bound has effectively been solved.

(3) *The Goldbach conjecture.* On June 7 1742, Christian Goldbach wrote a letter to Euler in which he conjectured that every integer greater than 5 can be written as the sum of three prime numbers.[13] Essentially no progress was made on this until the 20th century. In the 1930s, Vinogradov showed that the conjecture is true for all sufficiently large *odd* numbers, though the even case has remained elusive. (The conjecture turns out to be much harder for even numbers, since at least one of the primes must be 2 in that case, leaving only two degrees of freedom.) Here the meaning of 'sufficiently large' has been reduced over the years, but it still stands at over 10^{1300}, far too large to check all smaller numbers directly, even with modern computers. On the other hand, it is known that Goldbach's conjecture holds for all odd numbers if RH is true for the Riemann zeta-function and Dirichlet L-functions. Numerical verification of RH, as in

[13] It was still common to consider 1 a prime number in Goldbach's time, so he actually wrote 2 in place of 5.

Turing's work, promises to allow us to bridge the gap between these results in the not-too-distant future.[14]

Besides investigating existing conjectures and problems, over the last 60 years computers have proven to be extremely useful in suggesting new lines of enquiry. A good example is the *BSD conjecture*, discovered in the early 1960s by Birch and Swinnerton-Dyer using the EDSAC at Cambridge. The conjecture concerns *elliptic curves*, which are equations of the shape $y^2 = x^3 + ax + b$, where a and b are fixed integers. (Elliptic curves are the subject of another famous problem in number theory, the Shimura–Taniyama conjecture, whose proof by Wiles et al. in 1995 led finally to a complete proof of Fermat's Last Theorem after 350 years.) Given an elliptic curve, one can associate with it an L-function, and like the Riemann Hypothesis before it, the BSD conjecture is a prediction about the zeros of that function. It is now considered to be one of the most important open problems in number theory, and even partial results toward it have had striking applications. One such is Tunnell's resolution in 1983 of the 1000-year-old *congruent number problem* (Chandrasekar, 1998), which asks, for a given number n, whether there is a right triangle which has area n and sides of rational length. (Strictly speaking, Tunnell's algorithm can only be proven to work assuming the BSD conjecture, but a full proof of the conjecture is not necessary in order to apply the algorithm.) Another is Goldfeld's effective resolution of the *class number problem* (Goldfeld, 1985), posed by Gauss in 1801; for instance, this work tells us (among many other things) all the integers that can be written uniquely as a sum of three perfect squares.

Thus, on the one hand, computers have been instrumental in addressing some longstanding (sometimes ancient) problems in number theory. On the other hand, there are other questions which our current techniques seem completely inadequate to solve, leaving numerical experiments as the *only* way of attacking them at present. For instance, we have already encountered a few questions in this chapter for which our theoretical knowledge is not significantly advanced beyond what Euclid knew around 300 BC:

(1) Does every prime appear in the first Euclid–Mullin sequence?
(2) Are there any odd perfect numbers?
(3) Are there infinitely many Mersenne primes?

One should never try to place a limit on the ingenuity of human beings (or their machines), but as Gödel showed, there are questions to which the answer is simply unknowable, and it is conceivable that one of these is in that category. (In fact, finding a "natural" example of such a question was the original motivation behind

[14] Indeed, Harald Helfgott claimed a full proof of the odd Goldbach conjecture in 2013, after this chapter was written but before publication.

Mullin's construction.) In a way this is good, since it leaves open an avenue for amateur mathematicians and hobbyists, including those that may form the next generation of computational number theorists, to get involved in what is in other ways becoming a sophisticated and impenetrable subject.

The future of computers in number theory

We have arrived at a point now where nearly every mathematician has on his or her desk a tool that Gauss could only dream of. As we saw above, computers are starting to shape the outcome of research in number theory. It seems likely that this trend will continue to the point that they will become indispensable for doing research, and no one will work entirely without them. Perhaps, as a natural evolution of the current boom in formal proofs, the computers will even start to do some of the thinking themselves.

In a famous address in 1900, David Hilbert gave a list of 23 unsolved problems outlining his vision for the development of mathematics in the following century. Problem number 8 on the list was the theory of prime numbers, including both RH and the Goldbach conjecture. Sadly, we are hardly any closer to a proof of RH today than in 1900, with revelations such as the link to random matrix theory seeming to generate more questions than answers. (Hilbert might have anticipated this; he is quoted as saying "If I were to awaken after having slept for a thousand years, my first question would be: has the Riemann hypothesis been proven?") Nevertheless, significant progress has been made on most of Hilbert's problems, sometimes in unexpected ways; for instance, Gödel's work mentioned above, as well as that of Turing after him, was very much contrary to Hilbert's expectations.

With the turn of another century in 2000, several lists were proposed as replacements for Hilbert's problems. The one that has received the most attention is the list of seven Millennium Prize Problems published by the Clay Mathematics Institute, which offers $1 million for the solution to any one of them. To date, one problem, the Poincaré conjecture, has been solved. Of the remaining six, we have encountered two in this chapter while discussing Turing's work, the Riemann Hypothesis and the BSD conjecture, and one might also add the P vs. NP problem, which is discussed at length in the chapter by Ueli Maurer. That is not to say that Turing's investigations on the Manchester Mark 1 had very much direct influence on these things, but if nothing else it is testimony to his uncanny ability to recognize and get involved in problems of lasting interest.

What will the list for the 22nd century look like? Probably no one alive today can make a meaningful prediction. However, it seems a safe bet that it will include at least one problem from number theory; if so, perhaps it will be one that was discovered by a computer. Turing, who was never afraid to speak his mind, said it best in an interview following the initial press coverage of the Mark 1:

> This is only a foretaste of what is to come, and only the shadow of what is going to be. We have to have some experience with the machine before we really know its capabilities. It may take years before we settle down to the new possibilities, but I do not see why it should not enter any of the fields normally covered by the human intellect and eventually compete on equal terms.

References

Andrew R. Booker. Uncovering a new L-function. *Notices Amer. Math. Soc.*, 55(9):1088–1094, 2008.

V. Chandrasekar. The congruent number problem. *Resonance*, 3:33–45, 1998. 10.1007/BF02837344.

J. Brian Conrey. The Riemann Hypothesis. *Not. Amer. Math. Soc.*, 50(3):341–353, 2003.

Leo Corry. Hunting prime numbers – from human to electronic computers. *Rutherford J.*, 3, 2010.

Dorian Goldfeld. Gauss's class number problem for imaginary quadratic fields. *Bull. Amer. Math. Soc. (N.S.)*, 13(1):23–37, 1985.

Thomas C. Hales. Formal proof. *Not. Amer. Math. Soc.*, 55(11):1370–1380, 2008.

Dennis A. Hejhal. A few comments about Turing's method. In *Alan Turing – His Work and Impact*, S. Barry Cooper and J. van Leeuwen, editors. Elsevier Science, 2012.

Dennis A. Hejhal and Andrew M. Odlyzko. Alan Turing and the Riemann zeta function. In *Alan Turing – His Work and Impact*, S. Barry Cooper and J. van Leeuwen, editors. Elsevier Science, 2012.

Andrew Hodges. *Alan Turing: the Enigma*. A Touchstone Book. Simon & Schuster, 1983. Chapters 6 and 7 cover the period discussed here, including a detailed history of the design and development of the ACE and Manchester Mark 1 computers.

Władysław Narkiewicz. *The Development of Prime Number Theory: From Euclid to Hardy and Littlewood*. Springer Monographs in Mathematics. Springer-Verlag, 2000.

A.M. Turing. Some calculations of the Riemann zeta-function. *Proc. London Math. Soc. (3)*, 3:99–117, 1953.

D. Zagier. Newman's short proof of the prime number theorem. *Amer. Math. Monthly*, 104(8):705–708, 1997.

4

Cryptography and Computation after Turing

Ueli Maurer

Abstract

This chapter explores a topic in the intersection of two fields to which Alan Turing has made fundamental contributions: the theory of computing and cryptography.

A main goal in cryptography is to prove the security of cryptographic schemes. This means that one wants to prove that the computational problem of breaking the scheme is infeasible, i.e., its solution requires an amount of computation beyond the reach of current and even foreseeable future technology. As cryptography is a mathematical science, one needs a (mathematical) definition of computation and of the complexity of computation. In modern cryptography, and more generally in theoretical computer science, the complexity of a problem is defined via the number of steps it takes for the best program on a universal Turing machine to solve the problem.

Unfortunately, for this general model of computation, no proofs of useful lower bounds on the complexity of a computational problem are known. However, if one considers a more restricted model of computation, which captures reasonable restrictions on the power of an algorithm, then very strong lower bounds can be proved. For example, one can prove an exponential lower bound on the complexity of computing discrete logarithms in a finite cyclic group, a key problem in cryptography, if one considers only so-called generic algorithms that cannot exploit the specific properties of the representation (as bit-strings) of the group elements.

Published in *The Once and Future Turing*, edited by S. Barry Cooper & Andrew Hodges. Published by Cambridge University Press
The author is supported in part by the Swiss National Science Foundation.

4.1 Introduction

The task set to the authors of articles in this volume was to write about a topic of (general) scientific interest and related to Alan Turing's work. Here we present a topic in the intersection of computing theory and cryptography, two fields to which Turing has contributed significantly. The concrete technical goal of this chapter is to introduce the issue of provable security in cryptography. The article is partly based on Maurer (2005).

Computation and *information* are the two most fundamental concepts in computer science, much like mass, energy, time, and space are fundamental concepts in physics. Understanding these concepts continues to be a primary goal of research in theoretical computer science. As witnessed by Turing's work, many underlying questions are of as comparable intellectual depth to the fundamental questions in physics and mathematics, and are still far from being well understood.

Unfortunately, in view of the enormous practical significance of information technology for the economy and the society at large, this viewpoint on computer science is often overlooked. Prospective university students should have a better knowledge that computer science is not only an engineering discipline of paramount importance, but at the same time a fundamental science, and the high school curricula should include more computer science topics, not only computer literacy courses.

Two of the greatest minds of the 20th century have contributed in a fundamental manner to the understanding of the concepts of computation and information. In his 1936 paper, Alan Turing provided a mathematical definition of computation by proposing the Turing machine as a general model of computation. This model is still universally used in computer science. Claude Shannon (1948) laid the foundation for information theory and for the first time defined information in a meaningful and quantitative manner. This theory allowed one to formalize the coding and transmission of information in a radically new way and was essential for the development of modern communication technologies.

Remarkably, both Turing and Shannon also made fundamental contributions to cryptography. Actually, their interest in cryptography can be seen as a possible source of inspiration for the above-mentioned foundational work on computing theory and information theory, respectively. In fact, as reported to the author by Andrew Hodges (see Hodges, 1992, pp. 120 and 138), just after publication of his 1936 paper, Turing wrote in a letter to his mother:

> I have just discovered a possible application of the kind of thing I am working on at present. It answers the question "What is the most general kind of code or cipher possible", and at the same time (rather naturally) enables me to construct a lot of particular and interesting codes. One of them is pretty well impossible to decode without

> the key, and very quick to encode. I expect I could sell them to H.M. Government for quite a substantial sum, but am rather doubtful about the morality of such things.

This demonstrates that Turing had an interest in cryptography before being appointed to work on breaking German ciphers at Bletchley Park. Unfortunately, this work never became publicly available and seems to have been lost. But what is clear is that he had in mind to develop a theory of provable cryptographic security, a topic this article explores. One can only speculate what Turing might have been able to achieve in the field of theoretical cryptography had he spent more time on the subject.

Another important connection between Turing's work on cryptography and on computing is the fact that his work on breaking German codes required the construction of one of the first practical computers. Turing's insights later helped construct the first electronic tube-based computers.

4.2 Cryptography

4.2.1 Introduction

Cryptography can be understood as the mathematical science of information security that exploits an information difference (e.g. a secret key known to one party but not to another). It is beyond the scope of this article to give a detailed account of achievements in cryptography, and we refer, for example, to Maurer (2000) for such a discussion.

Cryptography, and even more so cryptanalysis, has played an important role in history, for instance in both world wars. We refer to Kahn (1967) and Singh (1999) for very good accounts of the history of cryptography. Before the second world war, cryptography can be seen as an art more than a science, mainly used for military applications, and concerned almost exclusively with encryption. The encryption schemes were quite ad hoc, with essentially no theory supporting their security. In sharp contrast, modern cryptography is a science with a large variety of applications other than encryption, often implemented by sophisticated cryptographic protocols designed by mathematicians and computer scientists. Without cryptography, security on the Internet or any other modern information system would be impossible.

The two single most important papers which triggered the transition of cryptography from an art to a science are perhaps 'Communication theory of secrecy systems" (Shannon, 1949), a companion paper of Shannon (1948); and, even more influential, 'New directions in cryptography' (Diffie–Hellman, 1976), in which they revealed their invention of public-key cryptography.

In an article connected to Alan Turing's work, a historical note about the invention of public-key cryptography is unavoidable. In the late 1990s, the British government announced that public-key cryptography was originally invented at the Government Communications Headquarters (GCHQ) in Cheltenham in the early 1970s (see Singh, 1999) by James Ellis and Clifford Cocks, who proposed essentially the Diffie–Hellman protocol (Diffie–Hellman, 1976) as well as the RSA public-key cryptosystem (Rivest, Shamir) invented a year later. Since scientists working for government agencies generally cannot publish their work, their contributions and inventions become publicly known much later only (if ever), often after their death. This remark also applies to Turing's work on cryptanalysis and the construction of practical (code-breaking) computers, which became publicly known only in the 1970s. Turing's life might have taken a very different turn had his great contributions been publicly known and acknowledged before his prosecution and tragic death.

4.2.2 The need for secret keys

Encryption, like other cryptographic schemes, requires a secret key shared by the sender and receiver (often referred to as Alice and Bob), but unknown to an eavesdropper. In a military context, such a key can be established by sending a trusted courier who transports the key from the headquarters to a communications facility. In a commercial context, sending a courier is completely impractical. For example, for a client computer to communicate securely with a server, one needs a mechanism that provides an encryption key instantaneously.

However, the problem is that Alice and Bob are connected only by an insecure channel, for example the Internet, accessible to an eavesdropper. Therefore, a fundamental problem in cryptography is the generation of such a shared secret key, about which the eavesdropper has essentially no information, by communication only over an authenticated[1] but otherwise insecure channel. This is known as the key-agreement problem.

The key can then be used to encrypt and authenticate subsequently transmitted messages. That one can generate a secret key by only public communication appears highly paradoxical at first glance, but the above-mentioned work of Diffie and Hellman provides a surprising solution to this paradox.

4.2.3 Proving security

In cryptography, one of the primary goals is to prove the security of cryptographic schemes. Security means that it is impossible for a special hypothetical party, the

[1] The authenticity of this communication is often guaranteed by the use of so-called certificates.

adversary (or eavesdropper), to solve a certain problem, e.g. to determine the message or the key. Such an impossibility can be of two different types, and thus one distinguishes two types of security in cryptography.

A cryptographic system that no amount of computation can break is called *information-theoretically secure*. The best-known example is the so-called one-time pad, which encrypts a binary plaintext sequence by adding (bitwise modulo 2) a uniformly random binary key sequence that is independent of the plaintext. The resulting ciphertext can easily be shown to be statistically independent of the plaintext, and hence provides absolutely no information about it, even for a party with unbounded computing power. However, owing to the required key length and the fact that the key cannot be reused, the one-time pad is highly impractical and is used only in special applications such as the encryption of the Washington–Moscow telephone hotline in the time of Reagan and Gorbachev.

Systems used in practice could theoretically be broken by a sufficient amount of computation, for instance by an exhaustive key search. The security of these systems relies on the computational infeasibility of breaking it, and such a system is referred to as *computationally secure*.

Proving the security of such a cryptographic system means proving a lower bound on the complexity of a certain computational problem, namely the problem of breaking the scheme. Such a proof must show not only that a key search is infeasible, but that any other conceivable way of breaking the scheme is infeasible. In a sense, such a proof would imply that no future genius can ever propose an efficient breaking algorithm.

Unfortunately, for general models of computation such as a universal Turing machine, no useful lower bound proofs are known, and it is therefore interesting to investigate reasonably restricted models of computation if one can prove relevant lower bounds for them.

This chapter investigates reasonable computational models for which one *can* prove computational security.

4.3 Computation

Computer science is concerned with the following fundamental questions. What is computation? Which functions (or problems) are computable (in principle)? For computable functions, what is the complexity of such a computation? As mentioned, in cryptography one is interested in proving lower bounds on the complexity.

Computation is a physical process. A computation is usually performed on a physical computational device, often called a computer. There are many different instantiations of computational devices, including the (by now) conventional digital computers, a human brain, analog computational devices, biological computers, and quantum computers.

Computer science aims to make mathematical (as opposed to physical) statements about computation, for example that a certain problem is computationally hard or that a certain function is not computable at all. Therefore one needs to define a mathematical model of computation. Turing was one of the first to recognize the need for a mathematical model of computation and proposed what has become known as the *Turing machine*, the most prominent model of computation considered in theoretical computer science.

Other models of computation, for example Church's lambda calculus (Church, 1932), have also been proposed. When judging the usefulness of a computational model, the first question to ask is whether it is general, in the sense that anything computable in principle, by any computational device, is computable in the model under consideration. Since computation is ultimately physical, such an argument of complete generality can never be made (unless one can claim to completely understand physics and, hence, Nature). However, most proposed models can be shown to be equivalent in the sense that anything computable in one model is also computable in another model. The so-called Church–Turing thesis postulates that this notion of computation (e.g. a Turing machine) captures what is computable in principle, with any physical device. In fact, Turing gave ingenious arguments for the claim that his model captures anything that a person doing a computation with pencil and paper could do.

However, the choice of model does matter significantly when one wants to analyze the *complexity* of a computation, i.e., the minimal number of steps it takes to solve a certain computational problem. For example, quantum computers, which are no more powerful than classical computers in terms of what can be computed, are (believed to be) vastly more efficient for certain computational problems such as factoring large integers (Shor, 1994).

4.4 The Diffie–Hellman key-agreement protocol

4.4.1 Preliminaries

The above-mentioned famous key-agreement protocol, as originally proposed in Diffie–Hellman (1976), makes use of exponentiation with respect to a base g, modulo a large prime p (for instance, a prime with 2048 bits, which corresponds to about 617 decimal digits), i.e., it makes use of the mapping

$$x \mapsto g^x \pmod{p},$$

where $a \pmod{b}$, for numbers a and b, is the remainder when a is divided by b (for example, $67 \pmod{7}$ is 4). The prime p and the base g are public parameters, possibly generated once and for all, for all users of the system. In more mathematical terminology, one computes in the multiplicative group of the ring $\mathbf{Z}/p\mathbf{Z}$. We denote this group by $\mathbf{Z}_p^*$. The toy example $p = 19$ and $g = 2$ is shown in Figure 4.1. Note that $\mathbf{Z}_{19}^*$ is a cyclic group with 18 elements, the numbers from 1 to 18.

While $y = g^x \pmod p$ can be computed efficiently, even if p, g, and x are numbers with several hundreds or thousands of digits (see below), computing x when given

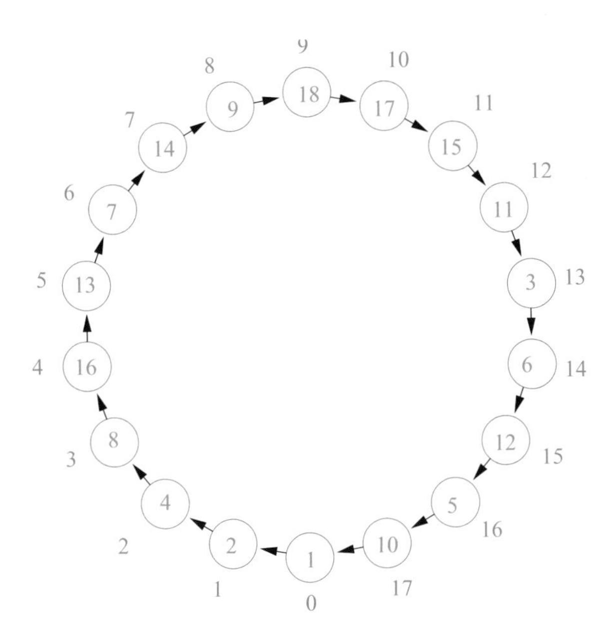

Figure 4.1 The group $\mathbf{Z}_{19}^*$, generated by the generator $g = 2$. The numbers on the outside of the circle are the exponents and the numbers within the small circles are the corresponding elements of $\mathbf{Z}_{19}^*$. For example, we have $2^{11} \equiv_{19} 15$, i.e., the remainder when 2^{11} is divided by 19 is 15, namely $2^{11} = 2048 = 107 \times 19 + 15$. Also, for example, the discrete logarithm of 6 to the base 2 is 14, as can be seen by inspection.

p, g, and $y = g^x$ is generally believed to be computationally highly infeasible. This problem is known as (a version of) the *discrete logarithm problem*, which will be discussed later.

4.4.2 Efficient exponentiation

We briefly describe an efficient exponentiation algorithm, the so-called *square-and-multiply* algorithm. To compute g^x in some mathematical structure (e.g. $\mathbf{Z}_p^*$), one writes the exponent x as a binary number. For example, $x = 23$ is written as $x = 10111_2$. An accumulator variable a is initialized to the value g. One then processes

x bit-by-bit, as follows. In each step, say the ith step, one updates a by the rule

$$a := \begin{cases} a^2 & \text{if } x_i = 0, \\ a^2 g & \text{if } x_i = 1, \end{cases}$$

where x_i is the ith bit of x (starting from the left but ignoring the most significant bit, which is always 1). For example, for $x = 23 = 10111_2$, the algorithm performs four steps, where $x_1 = 0$, $x_2 = 1$, $x_3 = 1$, and $x_4 = 1$. After the first step, the accumulator a contains the value g^2. After the second step, a contains the value $(g^2)^2 \cdot g = g^5$. After the third step, a contains the value $(g^5)^2 \cdot g = g^{11}$. Finally, after the fourth step, a contains the value $(g^{11})^2 \cdot g = g^{23}$. The running time of this algorithm is proportional to the bit-length of x, which makes the algorithm efficient even for very large values of x.

4.4.3 The key-agreement protocol

The Diffie–Hellman protocol is shown in Figure 4.2. Alice selects an exponent x_A at random, computes $y_A = g^{x_A}$ modulo p, and sends y_A over an authenticated but otherwise insecure channel to Bob. Bob proceeds analogously, selects an exponent x_B at random, computes $y_B = g^{x_B}$ modulo p, and sends y_B to Alice. Then Alice computes the value

$$k_{AB} = y_B^{x_A} = (g^{x_B})^{x_A} = g^{x_B x_A},$$

modulo p, and Bob computes, analogously,

$$k_{BA} = y_A^{x_B} = (g^{x_A})^{x_B} = g^{x_A x_B},$$

modulo p. The simple but crucial observation is that

$$k_{AB} = k_{BA},$$

owing to the commutativity of multiplication (in the exponent). In other words, Alice and Bob arrive at the same shared secret value, which they can use as a secret key or from which they can derive a key of appropriate length, for example using a so-called cryptographic hash function.

Intuitively, the security of this protocol relies on the observation that in order to compute k_{AB} from y_A and y_B it seems that an adversary would have to compute either x_A or x_B, which is the discrete logarithm problem believed to be infeasible.

The Diffie–Hellman protocol can be nicely explained by a mechanical analog, as shown in Figure 4.3. The exponentiation operation (e.g. the operation $x_A \mapsto g^{x_A}$) can be thought of as locking a padlock, an operation that is easy to perform but impossible (in a computational sense) to invert. Note that the padlock in this analog has no key; once locked, it can no longer be opened. However, a party can

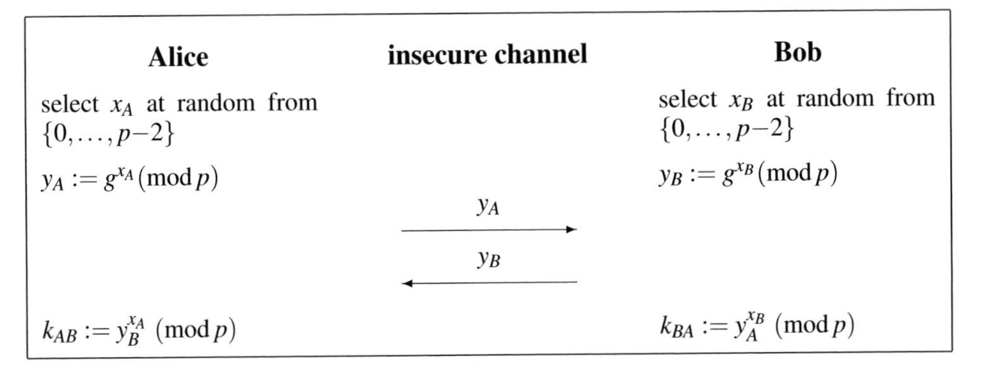

Figure 4.2 The Diffie–Hellman key agreement protocol. The prime p and the generator g are publicly known parameters.

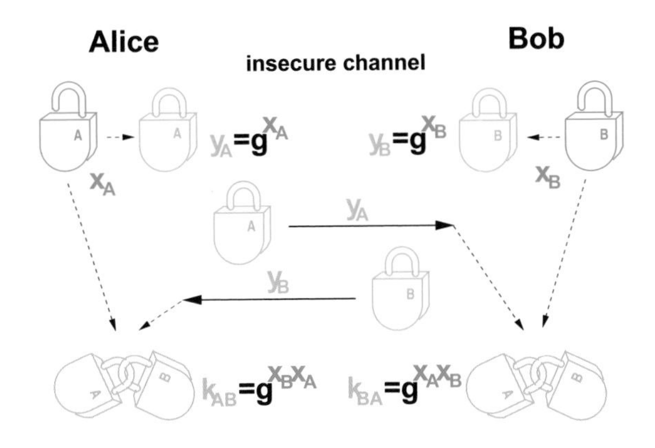

Figure 4.3 Mechanical analog of the Diffie–Hellman protocol.

remember the open state of the lock (i.e., x_A). Alice and Bob can exchange their locked padlocks (i.e., y_A and y_B), keeping a copy in the open state. Then they can both generate the same configuration, namely the two padlocks are interlocked. For the adversary, this is impossible without breaking open one of the two padlocks.

The Diffie–Hellman protocol described above can be generalized from computation modulo p (i.e., the group $\mathbf{Z}_p^*$) to any cyclic group $G = \langle g \rangle$, generated by a generator g, in which the discrete logarithm problem relative to the base g is computationally hard. The only modifications are that x_A and x_B must be selected from $\{0,\ldots,|G|-1\}$, and multiplication modulo p is replaced by the group operation of G. In practice one often uses elliptic curves, for which the discrete logarithm problem is believed to be even substantially harder than for the group $\mathbf{Z}_p^*$, for comparable group sizes.

4.5 Discrete logarithms and other computational problems on groups

Let G be a cyclic group of order $|G| = n$ and let g be a generator of the group, i.e., $G = \langle g \rangle$. Then G is isomorphic to the (additively written) group $\langle \mathbf{Z}_n, + \rangle$, i.e., to the set $\mathbf{Z}_n = \{0, 1, \ldots, n-1\}$ with addition modulo n. This is the standard representation of a cyclic group of order n. We can define the following three computational problems for G:

- The *Discrete Logarithm (DL) problem* is, for given (uniformly chosen) $a \in G$, to compute x such that $a = g^x$.
- The *Computational Diffie–Hellman (CDH) problem* is, for given (uniformly chosen) $a, b \in G$, to compute g^{xy}, where $a = g^x$ and $b = g^y$.
- The *Decisional Diffie–Hellman (DDH) problem* is, for three given elements $a, b, c \in G$, with a, b chosen uniformly at random (again as $a = g^x$ and $b = g^y$), to distinguish the setting where $c = g^{xy}$ from the setting where c is a third independent random group element.

The DL problem is that of making the above-mentioned isomorphism between G and $\langle \mathbf{Z}_n, + \rangle$ explicit. Whether this is computationally easy or hard depends on the representation of the elements of G. It is easy to see that if one can compute discrete logarithms in G for some generator g then one can compute them for any other generator g'. This is achieved by division, by the DL, of g' relative to generator g.

The DDH problem is at most as hard as the CDH problem, and the CDH problem is at most as hard as the DL problem. To see the latter, we need only to observe that the CDH problem can be solved by computing x from a, which means computing the discrete logarithm. It is also known that if one could efficiently solve the CDH problem, then one could also efficiently solve the DL problem for almost all groups, i.e., the two problems are roughly equally hard (Maurer and Wolf, 1999) (see also Section 4.8.7).

We briefly discuss the cryptographic significance of these problems. It appears that breaking the Diffie–Hellman protocol means precisely solving the CDH problem. However, it is possible that computing parts of the key $g^{x_A x_B}$ is easy, even though the CDH problem, i.e., computing the entire key, is hard. If an adversary could obtain part of the key, this could also be devastating in certain applications that make use of the secret key. In other words, even if the CDH problem is hard, a system making use of the Diffie-Hellman protocol could still be insecure. The (stronger) condition one needs for the Diffie–Hellman protocol to be a secure key-agreement protocol in any context is that the DDH problem be hard, which means that a Diffie–Hellman key is indistinguishable (with feasible computation) from a random key. This implies in particular that no partial information about the key can leak.

4.6 Discrete logarithm algorithms

4.6.1 Introduction

In order to compute in a group G, one must represent the elements as bit-strings. For example, the elements of $\mathbf{Z}_p^*$ are typically assumed to be represented as integers (in binary representation). As mentioned above, the hardness of a problem generally depends strongly on the representation. In a concrete setting, for example when trying to break the Diffie–Hellman protocol, the adversary has to work in the given fixed representation which Alice and Bob are using. The hope of cryptographers is that for this representation the problem is hard.

One can distinguish between two types of discrete logarithm (DL) algorithms. A *generic* algorithm works independently of the representation, i.e., it makes use only of the group operations. A simple example of a generic algorithm is the trivial algorithm that tries all possible values $x = 0, 1, 2, 3, \ldots$, until $a = g^x$.

In contrast, a *special-purpose* algorithm is designed for a specific type of representation. For example, the best algorithms for the group $\mathbf{Z}_p^*$ are special-purpose and much faster than any generic algorithm. For the group $\mathbf{Z}_p^*$, the best-known algorithm is the so-called number-field sieve (Gordon, 1993), which has complexity

$$O(e^{c(\ln p)^{1/3}(\ln\ln p)^{2/3}})$$

for $c = 3^{2/3}$, which is much faster than the generic algorithms discussed below but still infeasible for large enough primes. Such a special-purpose algorithm can make use of the fact that the group elements are numbers and hence embedded in the rich mathematical structure of the integers $\mathbf{Z}$. For example, one can try to factor a number into its prime factors and combine such factorizations.

There are groups, for instance most elliptic curves over finite fields, for which no algorithms that are faster than generic are known. In other words, it is not (yet) known how the representation of the elements of an elliptic curve can be exploited for computing discrete logarithms.

Proving a super-polynomial complexity lower bound for any special-purpose algorithm for computing the DL in a certain group would resolve the most famous open conjecture in theoretical computer science, as it would imply that $\mathbf{P} \neq \mathbf{NP}$, and is therefore not expected to be achievable in the near future. However, as we will see, one can prove exponential lower bounds for any *generic* algorithm.

It should be mentioned that Peter Shor (1994) discovered a fast (polynomial-time) algorithm for computing discrete logarithms and for factoring integers on a quantum computer. A quantum computer is a (still) theoretical model of computation which exploits the laws of quantum physics and hence is potentially much more powerful than classical computers (including the Turing machine) whose implementation makes use only of classical physics. Whether quantum computers

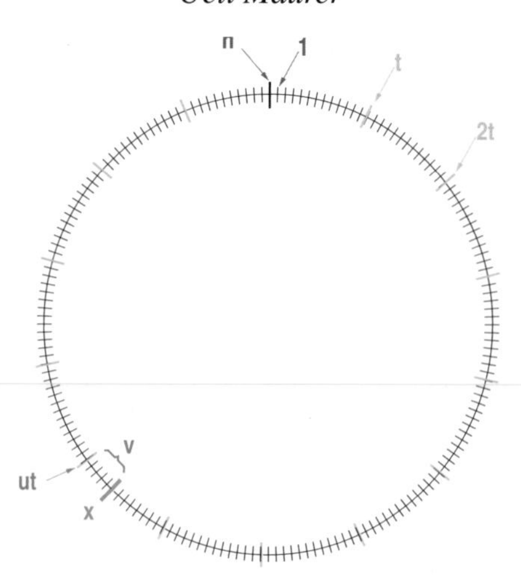

Figure 4.4 Illustrating the baby-step giant-step (BSGS) algorithm.

can ever be built is not known, but the research effort spent towards building one is enormous.

4.6.2 The baby-step giant-step algorithm

The baby-step giant-step (BSGS) algorithm is the simplest non-trivial DL algorithm. It is generic, i.e., it works no matter how the group elements are represented. The group order $|G| = n$ need not be known; a rough estimate of n suffices.

The BSGS algorithm works as follows. Let $a = g^x$ be the given instance. Let t be a parameter of the algorithm, the typical choice being $t \approx \sqrt{n}$. The unknown value x can be represented uniquely as

$$x = ut - v$$

(see Figure 4.4) where $u < n/t \approx \sqrt{n}$ and $v < t$. The baby-step giant-step algorithm consists of the following steps:

(1) **Giant steps:** Compute the pairs (j, g^{jt}) for $0 \leq j < n/t$, sort these pairs according to the second value g^{jt}, and store them in a (sorted) table.
(2) **Baby steps:** Compute ag^i for $i = 0, 1, 2, \ldots$ until one of these values is contained in the (giant-step) table. This will happen when, say, $i = v$, and the value retrieved from the table will be, say, $j = u$. Compute $x = jt - i$.

The memory requirement for this algorithm is $O(n/t)$ which is $O(\sqrt{n})$ for $t = O(\sqrt{n})$. The time complexity is $O(\frac{n}{t} \log \frac{n}{t})$ for sorting the table and $O(t \log \frac{n}{t})$ for accessing the table $O(t)$ times, hence $O(\max(t, \frac{n}{t}) \log n)$, which is $O(\sqrt{n} \log n)$ for

$t = O(\sqrt{n})$. We use the common notation $O(f(n))$ to say that a quantity grows asymptotically as $f(n)$ for increasing n.

The BSGS algorithm is an essentially optimal generic algorithm if n is a prime (see Section 4.8.3). If n has only small prime factors then a significantly faster generic algorithm exists, which is discussed next.

4.6.3 The Pohlig–Hellman algorithm

Again let $|G| = n$ and let q be a prime factor of n. One can write x as

$$x = uq + v$$

for some u and v which we wish to compute. Let $k := n/q$. Then $kx = kuq + kv$ and hence (since $kq = n$)

$$kx \equiv_n kv$$

which implies[2]

$$a^k = g^{kx} = g^{kv} = (g^k)^v.$$

In other words, v is the DL of a^k in the group $\langle g^k \rangle$, which has order q. Any generic algorithm, in particular the BSGS algorithm, can be used to compute this DL. The running time is $O(\sqrt{q}\log q)$.

It remains to compute u. Let

$$a' := ag^{-v} = g^{uq} = (g^q)^u.$$

Hence u is the DL of a' in the group $\langle g^q \rangle$ (which has order $k = n/q$) and can be computed using the same method as described above, but now for a group of order n/q.

Repeating this procedure for every prime factor of n (as many times for a prime q as it occurs in n), we obtain x, as desired. If the prime factorization of n is $n = \prod_{i=1}^{r} q_i^{\alpha_i}$ then the overall complexity of this algorithm is

$$O\left(\sum_{i=1}^{r} \alpha_i \sqrt{q_i} \log q_i\right) = O(\sqrt{q'} \log n),$$

where q' is the largest prime factor of n.

The Pohlig–Hellman algorithm is an essentially optimal generic algorithm (see Section 4.8.4). The existence of this algorithm is a reason for choosing the group order of a DL-based cryptographic system to have a very large prime factor. For

[2] Here $a \equiv_n b$ means that a and b are congruent modulo n, i.e., n divides $a - b$.

example, if one uses the group $\mathbf{Z}_p^*$, as in the original proposal by Diffie and Hellman, then $n = p-1$ must have a large prime factor q, for example $p-1 = 2q$. In many cases one actually uses a group whose order is prime, for other reasons also.

4.7 Abstract models of computation

4.7.1 Motivation

As mentioned earlier, for general models of computation no cryptographically useful lower-bound proofs are known for the complexity of any reasonable computational problem. It is therefore interesting to investigate reasonably restricted models of computation if one can prove relevant lower bounds for them.

In a restricted model one assumes that only certain types of operations are allowed. For example, in the so-called monotone circuit model one assumes that the logical (i.e., digital) circuit performing the computation consists only of AND-gates and OR-gates, excluding NOT-gates. However, a lower-bound proof for such a restricted model is uninteresting from a cryptographic viewpoint since it is obvious that an adversary can of course perform NOT-operations.

Some restricted models are indeed meaningful in cryptography, for example the generic model of computation. The term generic means that one cannot exploit non-trivial properties of the representation of the elements, except for two generic properties of any representation. Thus:

- one can test the equality of elements;
- one can impose a total order relation $\preceq$ on any representation, for example the usual lexicographic order relation on the set of bit-strings.[3]

We now propose a model of computation that allows one to capture generic algorithms and more general restricted classes of algorithms. This model serves two purposes. It lets us phrase generic algorithms in a clean and minimal fashion, without having to talk about bit-strings representing group elements, and it allows us to prove lower bounds on the complexity of any algorithm for solving a certain problem in this model.

4.7.2 The computational model

We consider an abstract model of computation (see Figure 4.5) characterized by a black box **B** which can store values from a certain set S (e.g. a group) in internal registers $V_1, \ldots, V_m$. The storage capacity m can be finite or unbounded.

[3] This order relation is abstract in the sense that it is not related to any meaningful relation (like $\leq$ in $\mathbf{Z}_n$) on the set S. An algorithm using this relation must work no matter how $\preceq$ is defined, and thus for the worst case in which it could be defined. It can for instance be used to establish a sorted table of elements of S, but it cannot be used to perform a binary search in S.

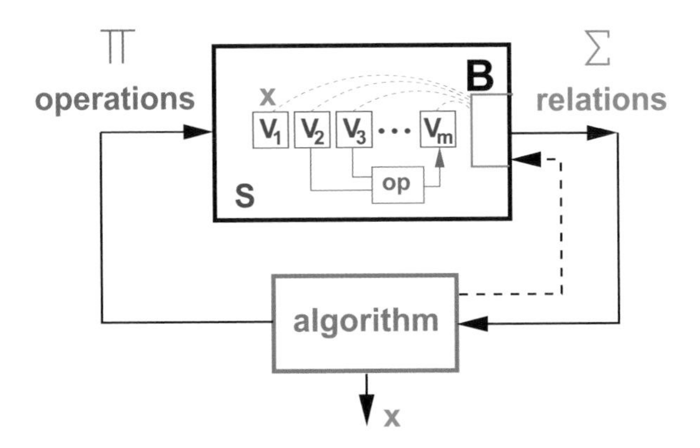

Figure 4.5 An abstract model of computation. An algorithm solving the extraction problem must, after querying the black box **B** a certain number of times (say, k times), output the value x stored in the first register of the black box.

The initial state encodes the problem instance and consists of the values of $V_1, \ldots, V_d$ (for some $d < m$; usually d is $1, 2$, or 3), which are set according to some probability distribution (e.g. the uniform distribution).

The black box **B** allows two types of operations, computation operations on internal state variables (shown on the left side of Figure 4.5) and queries about the internal state (shown on the right side of Figure 4.5). No other interaction with **B** is possible. We now give a more formal description of these operations.

Computation operations For a set Π of operations on S of some arities (nullary, unary, binary, or higher arity), a computation operation consists of selecting an operation $f \in \Pi$ (say t-ary) as well as the indices $i_1, \ldots, i_{t+1} \leq m$ of $t+1$ state variables. The black box **B** computes $f(V_{i_1}, \ldots, V_{i_t})$ and stores the result in $V_{i_{t+1}}$.[4]

Relation queries For a set Σ of relations (of some arities) on S, a query consists of selecting a relation $\rho \in \Sigma$ (say t-ary) as well as the indices $i_1, \ldots, i_t \leq m$ of t state variables. The query is answered by $\rho(V_{i_1}, \ldots, V_{i_t})$.[5]

For example, S could be the set $\mathbf{Z}_n = \{0, \ldots, n-1\}$, Π could consist of two operations, inserting constants and adding values modulo n, and Σ could consist of just the equality relation (which means that $\rho(V_{i_1}, V_{i_2}) = 1$ if and only if $V_{i_2} = V_{i_1}$) or, possibly, also the product relation (which means that $\rho(V_{i_1}, V_{i_2}, V_{i_3}) = 1$ if and only if $V_{i_3} = V_{i_1} \cdot V_{i_2}$).

[4] If m is unbounded then one can assume without loss of generality that each new result is stored in the next free state variable; hence i_{t+1} need not be given as input.

[5] Here we consider a relation ρ, without loss of generality, to be given as a function $S^t \to \{0, 1\}$.

This model captures two aspects of a restricted model of computation. The computation operations describe the types of computations the black box can perform, and the state queries allows us to model precisely how limited information about the representation of elements in S can be used.

A black box **B** (i.e., a particular model of computation) is thus characterized by S, Π, Σ, m, and d. We are primarily interested in the generic algorithm setting where Σ consists of just the equality relation: $\Sigma = \{=\}$. We consider only the case where m is unbounded.

4.7.3 Three types of problem

We consider three types of computational problem for this black-box model of computation where the problem instance is encoded into the initial state $(V_1, \ldots, V_d)$ of the device.

Extraction Extract the initial value x of V_1 (where $d = 1$). (See Figure 4.5.)
Computation Compute a function $f : S^d \to S$ of the initial state within **B**, i.e., the algorithm must achieve $V_i = f(x_1, \ldots, x_d)$ for some (known) i, where $x_1, \ldots, x_d$ are the initial values of the state variables $V_1, \ldots, V_d$.
Distinction Distinguish two black boxes **B** and **B**$'$ of the same type but with different distributions of the initial state $(V_1, \ldots, V_d)$.[6]

For the extraction problem, one may also consider algorithms that are allowed several attempts at guessing x. One can count such a guess as an operation; equivalently, we can assume without loss of generality that the algorithm, when making a guess, inputs the value (as a constant) to the black box and wins if it is equal to x. We will take this viewpoint in the following, i.e., a guess is treated as a constant operation.

4.8 Proving security: lower bounds for complexity

4.8.1 Introduction

One of the main goals of research in cryptography is to prove the security of cryptographic schemes, i.e., to prove that a scheme is hard to break. Unfortunately, not a single cryptographically relevant computational problem is known for which one can prove a significant lower bound on the complexity, for a general model of computation. Such a proof would be a dramatic break-through in computer science, maybe comparable to the discovery of a new elementary particle in physics, and

[6] The performance of a distinguishing algorithm outputting a bit can be defined as the probability that the guess is correct, minus $\frac{1}{2}$. Note that the success probability $\frac{1}{2}$ can trivially be achieved by a random guess; hence only exceeding $\frac{1}{2}$ can be interpreted as real performance.

would almost certainly yield the equivalent of the Nobel Prize in computer science, the Turing Award.[7]

However, we can prove lower bounds in the abstract model of computation discussed above. More precisely, we are interested in proving a relation between the number of operations that an algorithm performs and its performance. For the extraction and the computation problems, the performance is defined as the algorithm's success probability. We do not consider the computing power that would be required in an implementation (of a black-box algorithm) for determining its next query; we only count the actual operations that the algorithm performs.

When proving lower bounds, we will be on the safe side if we do not count the relation queries, i.e., if we assume that they are 'for free'. In other words, we assume that any satisfied relation in the black box (typically an equality of two register values, called a collision) is reported by the black box without requiring the algorithm to ask a query. Note that, when designing actual algorithms in this model (as opposed to proving lower bounds), the relation queries are relevant and must be counted. (For example, a comparison in a conventional computer constitutes an operation and requires at least one clock cycle.)

In this section we consider a few concrete instantiations of the abstract model of computation to illustrate how lower bounds can be proved. Our primary interest will be to prove lower bounds for computing discrete logarithms in a cyclic group (the DL problem) and for the CDH and the DDH problems.

We introduce some notation. Let **Const** denote the set of constant (nullary) operations, which correspond to inserting a constant into (a register of) the black box. For a given set Π of operations, let $\overline{\Pi}$ be the set of functions on the initial state that can be computed using operations in Π, i.e., it is the closure of Π. For example, if Π consists only of the increment function $x \mapsto x+1$, then $\overline{\Pi} = \{x \mapsto x+c \mid c \geq 1\}$, since by applying the increment function c times one can compute the function $x \mapsto x+c$ for any c.

The simplest case of an extraction problem occurs when $\Pi = \textbf{Const}$ and $\Sigma = \{=\}$, i.e., one can only input constants and check equality. This is analogous to a card game where one has to find a particular card among n cards and the only allowed operation is to lift cards, one at a time. It is obvious that the best strategy for the extraction problem is to guess randomly, i.e., to input random constants, and the success probability of any k-step algorithm is hence bounded by $k/|S|$. However, if one could also query other relations, for example a total order relation $\leq$ on S, then much faster algorithms can be possible, for example binary search.

[7] The fact that the main award in computer science is named after Turing reflects the central role he played in the early days of this field.

4.8.2 Two Lemmas

We need a lemma about the number of roots that a multivariate polynomial modulo a prime q can have, as a function of its degree. The degree of a multivariate polynomial $Q(x_1,\ldots,x_t)$ is the maximal degree of an additive term, where the degree of a term is the sum of the powers of the variables in the term. For example, the degree of the polynomial

$$Q(x_1,x_2,x_3,x_4) = x_1^6 + 5x_1^3x_2x_3^4 + 2x_1x_2^3x_3x_4^2$$

is 8, the degree of the second term. The following lemma is known as the Schwartz–Zippel lemma (Schwartz, 1980). For the single-variable case, it corresponds to the well-known fact that a (single-variable) polynomial of degree d over a field can have at most d roots.

Lemma 4.8.1 *For a prime q and any $t \geq 1$, the fraction of solutions $(x_1,\ldots,x_t) \in \mathbf{Z}_q^t$ of any multivariate polynomial equation*

$$Q(x_1,\ldots,x_t) \equiv_q 0$$

of degree d is at most d/q.

This lemma tells us, for example, that the above polynomial $Q(x_1,x_2,x_3,x_4)$, if considered (say) modulo the prime $q = 101$, has at most a fraction $8/101$ of tuples (x_1,x_2,x_3,x_4) for which $Q(x_1,x_2,x_3,x_4) = 0$.

We also need a second lemma. Consider a general system which takes a sequence $X_1,X_2,\ldots$ of inputs from some input alphabet $\mathcal{X}$ and produces, for every input X_i, an output Y_i from some output alphabet $\mathcal{Y}$. The system may be probabilistic and it may have state. For such a system one can consider various tasks of the following form. Choose appropriate inputs $X_1,\ldots,X_k$, so that $Y_1,\ldots,Y_k$ satisfies a certain property (e.g., they are all zero). In general, the task is easier to solve (i.e., the success probability is higher) if an *adaptive* strategy is allowed, i.e., if X_i is to be chosen only after Y_{i-1} has been observed. In contrast, a *non-adaptive* strategy requires $X_1,\ldots,X_k$ to be chosen at once.

Lemma 4.8.2 *Consider the task of preventing a particular output sequence $y_1,\ldots,y_k$ from occuring. The success probability of the best non-adaptive strategy is equal to that of the best adaptive strategy.*

Proof Any adaptive strategy A with access to $Y_1,Y_2,\ldots$ can be converted into an equally good non-adaptive strategy A' by feeding into A, instead of the actual values $Y_1,Y_2,\ldots$ output by the system, the (fixed) values $y_1,y_2,\ldots$, respectively. As long as A is not successful (in provoking a deviation of $Y_1,Y_2,\ldots$ from $y_1,y_2,\ldots$), these constant inputs $y_1,y_2,\ldots$ are actually correct and A and A' behave identically. As soon as A is successful (in achieving an output $Y_i \neq y_i$ for some i), so is A'. □

4.8.3 Group actions and the optimality of the baby-step giant-step algorithm

Let S be a finite group of size $|S| = n$ with group operation denoted as $\star$. A group action on S is an operation of the form

$$x \mapsto x \star a$$

for some (constant) parameter a. For example, if S is the set of integer numbers, and $\star$ is addition (i.e., $+$), then this operation corresponds simply to incrementing x by a.

Theorem 4.8.3 *Let $\star$ be a group operation on S, let $\Pi = \mathbf{Const} \cup \{x \mapsto x \star a \mid a \in S\}$ consist of all constant functions and group actions, and let $\Sigma = \{=\}$. Then the success probability of every k-step algorithm for extraction is at most $\frac{1}{4}(k+1)^2/n$.*

Proof We use three simple general arguments which will be reused implicitly later.

- First, we assume (conservatively) that, as soon as some collision occurs (more generally, some relation in Σ is satisfied for some state variables) in the black box **B**, the algorithm is successful. One can therefore restrict the analysis to algorithms for provoking some non-trivial collision in the black box.
- Second, we observe, by considering the black box as the system used for the task in Lemma 4.8.2, that if the only goal is to provoke a collision (i.e., an output sequence of **B** different from one in which 'no collision' is always outputted), then adaptive strategies are no more powerful than non-adaptive ones. Hence we can restrict our analysis to non-adaptive algorithms.
- Third, for lower-bound proofs we can assume that not only can an algorithm perform operations in Π but that it can, in every step, compute a function in $\overline{\Pi}$ (of the initial state $(V_1, \ldots, V_d)$). This must improve the algorithm's power; a lower-bound proof for this model holds for the weaker model also. Without loss of generality we can assume that only distinct functions are chosen by the algorithm (since a trivial collision would not "count" and hence be useless).

In the setting under consideration, the composition of two operations in Π is again in Π, i.e., $\overline{\Pi} = \Pi$. For example, if the operation $x \mapsto x \star a$ is composed with the operation $x \mapsto x \star b$ then this corresponds to the operation $x \mapsto x \star c$ for $c = a \star b$.

For all $x \in S$ and distinct a and b we have $x \star a \neq x \star b$. Thus collisions can occur only between a (value computed by a) function of the form $x \mapsto x \star a$ and a constant function c, namely if $x \star a = c$, which is equivalent to $x = c \star a^{-1}$. Let u (and v) be the number of constant operations (group actions) that the algorithm performs. Then the probability of a collision is upper bounded by $u(v+1)/|S|$. The optimal choice is $u = \lceil k/2 \rceil$ (and $v = k - u$), for which $uv \leq (k+1)^2$. □

Example Consider the group $\langle \mathbf{Z}_{100}, + \rangle$. The black box contains a random value between 0 and 99. If our strategy to provoke a collision computes the values $x+22$ and $x+53$ (in addition to the x values already stored in the black box) and inserts the constants 34 and 87 then a collision occurs if (and only if) $x = 12$, $x = 34$, $x = 65$, $x = 81$, or $x = 87$. For example, if $x = 81$, then a collision occurs between the values $x+53$ and 34. Note that for this strategy we have $k = 4$ and indeed the success probability (for a random x) is $5/100$, which is less than $\frac{1}{4}5^2/100 = 1/16$.

The above bound implies that in order to achieve a constant success probability (e.g. $\frac{1}{2}$) the number k of operations must be on the order of $\sqrt{n}$.

The theorem illustrates that the baby-step giant-step (BSGS) algorithm is essentially optimal. It achieves the lower bound even though it only requires the operations in $\Pi = \mathbf{Const} \cup \{x \mapsto x+1\}$, i.e., it only increments by 1 (and not by general constants). The BSGS algorithm, if interpreted in this model, inserts equidistant constants with gap $t \approx \sqrt{n}$ and increments the secret value x until a collision with one of these values occurs.[8]

4.8.4 Discrete logarithms and the optimality of the Pohlig–Hellman algorithm

We now consider the additive group $\langle \mathbf{Z}_n, + \rangle$. The extraction problem for this group corresponds to the discrete logarithm problem for a cyclic group of order n. In other words, every extraction algorithm for $\langle \mathbf{Z}_n, + \rangle$ is a generic DL algorithm for any group of order n, and vice versa. In what follows, let q denote the largest prime factor of n. The following theorem is an abstract formulation of a result due to Nechaev (1994) and Shoup (1997).

Theorem 4.8.4 *For $S = \mathbf{Z}_n$, $\Pi = \mathbf{Const} \cup \{+\}$ and $\Sigma = \{=\}$, the success probability of every k-step extraction algorithm is at most $\frac{1}{2}(k+1)^2/q$.*

Proof We use the same line of reasoning as in the proof of Theorem 4.8.3. Every value computed in the black box is of the form $ax+b$ for some (known) a and b. In other words,

$$\overline{\Pi} = \{ax+b \mid a,b \in \mathbf{Z}_n\},$$

i.e., only linear functions of x can be computed. As argued above, we need to consider only non-adaptive algorithms for provoking a collision. Since a collision modulo n implies that there is also a collision modulo q, a lower bound on the probability of provoking a collision modulo q is also a lower bound on the probability of provoking a collision modulo n.

[8] Note that the number of equality checks is actually $O(n)$, which is too high. In order to reduce the number of equality checks, the BSGS algorithm makes use of the abstract order relation $\preceq$ to generate a sorted table of stored values.

Hence consider a fixed algorithm (for provoking a collision modulo q) computing in each step (say the ith) a new value $a_i x + b_i$. A collision (modulo q) occurs if

$$a_i x + b_i \equiv_q a_j x + b_j$$

for some distinct i and j, i.e., if $(a_i - a_j)x + (b_i - b_j) \equiv_q 0$. This congruence has at most one solution for x modulo q (according to Lemma 4.8.1).[9] Therefore the total number of remainders of x modulo q for which any collision modulo q (which is necessary for a collision modulo n) can occur is bounded by $\binom{k}{2}$. Thus the fraction of x (modulo q, and hence also modulo n) for which a collision modulo q can occur is at most $\binom{k+1}{2}/q < \frac{1}{2}(k+1)^2/q$. (Recall that x is already in the black box before the algorithm performs any operation.) □

For achieving a constant success probability, the number k of operations must hence be $O(\sqrt{q})$. The Pohlig–Hellman algorithm requires $k = O(\sqrt{q}\log n)$ operations and so matches this bound up to a factor $\log n$, owing to the fact that in the lower-bound proof we were too generous with the algorithm by not counting the equality queries. It should be mentioned that Pollard (1978) proposed a much more practical algorithm with comparable complexity but which requires essentially no memory. However, the running time analysis of this algorithm is based on heuristic arguments.

The reader may want to prove, as an exercise, a lower bound for DL computation, even if one assumes the additional availability of a DDH oracle. This can be modeled by including in the set Σ of relations the product relation $\{(a,b,c)|ab \equiv_n c\}$.

4.8.5 Product computation in $\mathbf{Z}_n$ and the CDH problem

We now consider the computation problem for the product function $(x,y) \mapsto xy$ in $\mathbf{Z}_n$. This corresponds to the generic computational Diffie–Hellman (CDH) problem in a cyclic group of order n. In other words, every algorithm in the black-box model for computing $(x,y) \mapsto xy$ when $S = \mathbf{Z}_n$, $\Pi = \mathbf{Const} \cup \{+\}$, and $\Sigma = \{=\}$ is a generic CDH algorithm for any cyclic group of order n, and vice versa. The following theorem shows that for generic algorithms the DL and the CDH problems are comparably hard.[10]

Theorem 4.8.5 *For $S = \mathbf{Z}_n$, $\Pi = \mathbf{Const} \cup \{+\}$, and $\Sigma = \{=\}$ the success probability of every k-step algorithm for computing the product function is at most $(\frac{1}{2}k^2 + 4k + 5)/q$.*

[9] Namely, if $a_i \equiv_q a_j$ and hence $b_i \equiv_q b_j$, there is no solution and, if $a_i \not\equiv_q a_j$, then the single solution modulo q is $(b_j - b_i)/(a_i - a_j)$, computed modulo q.

[10] For general algorithms, the statement that they are comparably hard requires a proof that the DL problem can be reduced to the CDH problem; see Section 4.8.7.

Proof Again, to be on the safe side, we can assume that as soon as a collision modulo q occurs among the computed values, the algorithm is successful. We have

$$\overline{\Pi} = \{ax + by + c \mid a, b, c \in \mathbf{Z}_n\}.$$

In addition to the computed values (and x and y), the black box is assumed to contain already the value xy in a register that cannot be used as the input to operations (but is considered when checking for collisions). There are two types of collisions:

$$a_i x + b_i y + c_i \equiv_q a_j x + b_j y + c_j$$

for some $i \neq j$, and

$$a_i x + b_i y + c_i \equiv_q xy$$

for some i. The fraction of x for which a collision of the first type can occur is bounded by $\binom{k+2}{2}/q$, since the two values x and y are already contained in the black box, and the fraction of x for which a collision of the second type can occur is bounded by $2(k+2)/q$, according to Lemma 4.8.1, as the degree of the equation $a_i x + b_i y + c_i - xy \equiv_q 0$ is 2. Hence the total fraction of x (modulo n) for which one of the collision events (modulo q) occurs is bounded by $\left(\binom{k+2}{2} + 2(k+2)\right)/q = (\frac{1}{2}k^2 + 4k + 5)/q$. □

4.8.6 The DDH problem

For the DDH problem one can prove a lower bound of $O(\sqrt{p'})$ for any algorithm with a significant distinguishing advantage, where p' is the *smallest* prime factor of n. This can again be shown to be tight, i.e., there does exist a generic algorithm for the DDH problem, with complexity essentially $\sqrt{p'}$. For example, the DDH problem for the group $\mathbf{Z}_p^*$ for any large prime p is trivial since the order of $\mathbf{Z}_p^*$ is $p - 1$ and contains the small prime factor $p' = 2$. The theorem implies that, for (large) groups of prime order, the DDH problem is very hard for generic algorithms.

Theorem 4.8.6 *For $S = \mathbf{Z}_n$, $\Pi = \mathbf{Const} \cup \{+\}$ and $\Sigma = \{=\}$, the advantage of every k-step algorithm for distinguishing a random triple (x, y, z) from a triple (x, y, xy) is at most $(k+3)^2/p'$, where p' is the smallest prime factor of n.*

Proof Again we can assume that as soon as a collision occurs among the values computed in the black box, the algorithm is declared successful. (One could imagine a genie sitting in the black box who announces the correct answer, namely which configuration the black box is in, as soon as any non-trivial collision occurs.) Hence it suffices to compute the probabilities, for the two settings, that a collision can be provoked by an algorithm and take the larger value as an upper bound for the distinguishing advantage of the two settings. We analyze only the case where the initial state is (x,y,xy), as the collision probability is larger in this case. The set $\overline{\Pi}$ of computable functions is

$$\overline{\Pi} = \{ax + by + cxy + d \mid a,b,c,d \in \mathbf{Z}_n\},$$

i.e., the ith computed value is of the form

$$a_i x + b_i y + c_i xy + d_i$$

for some a_i, b_i, c_i, d_i. For any choice of $(a_i,b_i,c_i,d_i) \neq (a_j,b_j,c_j,d_j)$ we must bound the probability that

$$a_i x + b_i y + c_i xy + d_i \ \equiv_n \ a_j x + b_j y + c_j xy + d_j.$$

This is a polynomial equation of degree 2. Since $(a_i,b_i,c_i,d_i) \neq (a_j,b_j,c_j,d_j)$ implies only that $(a_i,b_i,c_i,d_i) \not\equiv_r (a_j,b_j,c_j,d_j)$ for some prime divisor r of n, it could be the smallest one, i.e., $r = p'$. Hence the fraction of pairs (x,y) for which it is satisfied is at most $2/p'$. Therefore, the fraction of pairs (x,y) for which one of the $\binom{k+3}{2}$ relations is satisfied modulo p' is at most $\binom{k+3}{2}(2/p') < (k+3)^2/p'$. $\square$

4.8.7 Generic reduction of the DL problem to the CDH problem

We have seen that the CDH and the DL problems are, roughly, equally hard in the generic model. An important question is whether this is also true for general models of computation. In other words, we want to prove that, in a general model of computation, breaking CDH is as hard as solving the DL problem. Even though this question is not about our black-box model, it can be answered using this model. One can show a generic reduction of the DL problem to the CDH problem, i.e., to a generic algorithm which efficiently solves the DL problem provided that it has access to a (hypothetical) oracle that answers CDH challenges.

This is modeled by including multiplication modulo n in the set Π of allowed operations for the generic extraction problem, i.e., by considering the extraction problem for the ring $\mathbf{Z}_n$. The Diffie–Hellman oracle assumed to be available for the reduction implements multiplication modulo n. There exists such an efficient generic algorithm (Maurer, 1994) (see also Maurer and Wolf, 1999) for essentially all n and, under a plausible number-theoretic conjecture, for all n. In contrast with the algorithms presented in this article, this algorithm is quite involved and makes use of the theory of elliptic curves.

4.9 Conclusions

One may speculate what contributions to cryptography Turing might have made had his life not ended much too early and had he continued his research in cryptography. His contributions could have been on several fronts, including cryptanalysis. But what could perhaps have been the most important contribution is that Turing's strive for rigor might have helped advance cryptography from a fascinating discipline making use of various branches of mathematics to a science that is itself an axiomatically defined branch of mathematics. An attempt in this direction is described in Maurer and Renner (2011). One may doubt whether Turing could have removed the apparently unsurmountable barrier preventing us from proving cryptographically significant lower bounds on the complexity of computational problems (which would probably imply a proof of $\mathbf{P} \neq \mathbf{NP}$), but it's not inconceivable.

Acknowledgments

I am grateful to Andrew Hodges for sharing with me some of his insights about Turing's life, in particular about Turing's interest in cryptography, and to Divesh Aggarwal and Björn Tackmann for helpful feedback on this article.

References

A. Church. A set of postulates for the foundation of logic, *Annals of Mathematics*, Series 2, **33**, 346–366 (1932).

W. Diffie and M. E. Hellman. New directions in cryptography, *IEEE Transactions on Information Theory*, **22** (6), 644–654 (1976).

D.M. Gordon. Discrete logarithms in $GF(p)$ using the number field sieve, *SIAM J. Discrete Mathematics*, **6** (1), 124–138 (1993).

A. Hodges. *Alan Turing: The Enigma*, Vintage Books (1992).

D. Kahn. *The Code Breakers, the Story of Secret Writing*, MacMillan (1967).

U. Maurer. Towards the equivalence of breaking the Diffie–Hellman protocol and computing discrete logarithms. In *Advances in Cryptology – CRYPTO '94*, Lecture Notes in Computer Science, **839**, pp. 271–281, Springer-Verlag (1994).

U. Maurer. Cryptography 2000 ± 10. In Lecture Notes in Computer Science **2000**, pp. 63–85, R. Wilhelm (ed.), Springer-Verlag (2000).

U. Maurer. Abstract models of computation in cryptography. In *Cryptography and Coding 2005*, Lecture Notes in Computer Science, **3796**, pp. 1–12, Springer-Verlag (2005).

U. Maurer and R. Renner. Abstract cryptography. In *The Second Symposium in Innovations in Computer Science, ICS 2011*, Tsinghua University Press, pp. 1–21, (2011).

U. Maurer and S. Wolf. On the complexity of breaking the Diffie-Hellman protocol, *SIAM Journal on Computing*, **28**, 1689–1721 (1999).

V.I. Nechaev. Complexity of a deterministic algorithm for the discrete logarithm, *Mathematical Notes*, **55** (2), 91–101 (1994).

S.C. Pohlig and M.E. Hellman. An improved algorithm for computing logarithms over $GF(p)$ and its cryptographic significance, *IEEE Transactions on Information Theory*, **24** (1), 106–110 (1978).

J.M. Pollard. Monte Carlo methods for index computation mod *p*, *Mathematics of Computation*, **32**, 918–924 (1978).

R.L. Rivest, A. Shamir, and L. Adleman. A method for obtaining digital signatures and public-key cryptosystems, *Communications of the ACM*, **21** (2), 120–126 (1978).

J.T. Schwartz. Fast probabilistic algorithms for verification of polynomial identities, *Journal of the ACM*, **27** (3), 701–717 (1980).

C.E. Shannon. A mathematical theory of communication. *Bell System Technical Journal*, **27**, 379–423; 623–656 (1948).

C. E. Shannon. Communication theory of secrecy systems. *Bell System Technical Journal*, **28**, 656–715 (1949).

P.W. Shor. Algorithms for quantum computation: discrete log and factoring. In *Proc. 35th IEEE Symposium on the Foundations of Computer Science (FOCS)*, pp. 124–134, IEEE Press (1994).

V. Shoup. Lower bounds for discrete logarithms and related problems. In *Advances in Cryptology – EUROCRYPT '97*, Lecture Notes in Computer Science, **1233**, pp. 256–266, Springer-Verlag (1997).

S. Singh, *The Code Book*, Fourth Estate, London (1999).

A.M. Turing, On computable numbers, with an application to the Entscheidungsproblem, *Proceedings of the London Mathematical Society*, Series 2, **42**, 230–265 (1937).

5

Alan Turing and Enigmatic Statistics

Kanti V. Mardia and S. Barry Cooper

5.1 Introduction

The enigmatic Alan Turing is known in different ways to different people, as in the story of the elephant and the blind men. Most people have heard of the Turing Test for intelligent machines, but the pure mathematician might be surprised to know that Turing made significant contributions to statistics, while all except biologists will be surprised to know that Turing's most cited paper (Turing, 1950) deals with the mathematics of the emergence of patterns in nature. The 2012 centenary of Alan Turing's birth has seen so many events around the world, with books and papers on his life and work (he even appeared on the cover of *Nature*), that the Turing legacy is now much better known, at least in academic circles.

Figure 5.1 Blind men and the elephant.

Below we look briefly at Turing's contribution to statistics, his innovative introduction of Bayesian techniques to cryptography during World War II – and how

Published in *The Once and Future Turing*, edited by S. Barry Cooper & Andrew Hodges. Published by Cambridge University Press © 2016. Not for distribution without permission. This article is a revised and expanded version of the contribution that appeared in the Official Bulletin of the Brazilian Section of the International Society for Bayesian Analysis (*Boletim ISBrA*, **5** (2), 2–7, December 2012).

Figure 5.2 Alan Turing ©Beryl Turing.

the statistics relates to Turing's underlying interest in how the world computes. If the mathematician imagines that Turing knocked off some statistics as a mere ad hoc diversion from the serious business of founding computer science, inventing artificial intelligence and revolutionising developmental biology, she or he would be missing something basic.

Various important statistical contributions by Turing at Bletchley Park (in 1940–1941) have been recorded by Jack Good (who was the main statistical assistant in 1941 to Turing). Good worked with Turing among others in breaking the enigma code; the article Good (1979) recorded 'their' contributions. It was subsequently elaborated in the commentary to this article by Good himself in the collected works of Turing (Turing, 2001). Recently, a wonderful and readable account has been given in McGrayne (2011) with an up-to-date history of Bayesian methods. The work has used a combination of several new methods including:

(1) Weight of evidence (assigning a tiny non-zero to a rare event which could appear in a larger sample), and
(2) Alignment of letters (pairs and triplets of letters in a cipher with substitutions).

These are the two main new methods but others were Markov chains, decision theory, and statistical computing (see, for example, Good, 1992). Since various different statistical methods were used, it will be perhaps right to label these techniques as 'enigmatic statistics'. The style of developing focused techniques foreshadowed the style of what is now called statistical bioinformatics.

It is now well known that the Enigma was a cryptographic (enciphering) machine used by the German military during WWII. The German navy changed part of the Enigma keys every other day. One of the important cryptanalytic attacks against the naval usage was called Banburismus, a sequential Bayesian procedure anticipating sequential analysis.

5.2 Weight of evidence and empirical Bayes

Suppose that a random sample is drawn from an infinite population of animals of various species, or from a population of words. Let the sample size be N and let n_r distinct species be each represented exactly r times in the sample, so that $\sum rn_r = N$, and n_r can be called 'the frequency of the frequency r'.

It can be shown that an estimate of the total probability of unseen species is

$$n_1/N.$$

The work required for obvious reasons calculating the probability that the next word sampled will be one that has not previously been observed. Turing, using what is called an *urn model* in statistics, showed that the expected population frequency of a species represented r times is about $(r+1)n_{r+1}/(Nn_r)$.

The technique is now known as Good–Turing frequency estimation. For a more exact statement, including the need for smoothing the n_r and for numerous elaborations and deductions, see Good (1953, 1979) and Good–Toulmin (1956). This work is an example of the empirical Bayes method.

In Banks (1996), p. 10, col. 2, Good says

> For example, I deduced a simple formula for the probability that the next word sampled will be one that has not previously been observed. Makers of dictionaries and teachers of languages ought to know about this work, because it tells you the minimum size of vocabulary required to cover, say, 98% running text.

That is, this work tells you the minimum size of vocabulary (say 98%) that would cover most of the (number of) words most likely to be used; this is clearly useful for compilers of dictionaries.

We quote from Robinson (2011):

> Suppose a birder spotted 180 different species, many of which were represented by only one bird. Logically, other species must have been missed. A frequentist statistician would count those unseen species as zero, as if they could never be found. Turing, by contrast, assigned them a tiny non-zero probability, thereby factoring in that rare letter groupings might not be present in his current collection of intercepted messages but could appear in a larger sample.

The Bayesian approach to statistics treats unknown parameters as random variables and prior distributions as model information about parameters. In contrast, the classical approach to statistics has no need of prior distributions as it treats unknown parameters as fixed constants. Empirical Bayes is an approach to statistics that lies somewhere between the two. Unknown hyperparameters in empirical Bayes are treated as fixed constants (as are the parameters in the classical approach) but in general these are estimated from data, unlike in the standard Bayesian approach. For further details on this theme and the related theme of odds and probability, weights of evidence, Bayes theorem, with some real and insightful examples, we refer to Aitken (1995) and Efron (2010).

5.3 Alignment of letters

5.3.1 Description of the coding in Enigma

Before Banburismus could be started on a given day it was necessary to identify which of nine 'bigram' (or 'digraph') tables was in use on that day. In Turing's approach to this identification he had to estimate the probabilities of certain 'trigraphs'. (These trigraphs were used, as described below, for determining the initial wheel settings of messages.) For estimating the probabilities, Turing invented an important special case of the non-parametric (non-hyperparametric) empirical Bayes method independently of Herbert Robbins. The technique is a surprising form of empirical Bayes in which a physical prior is assumed to exist but no approximate functional form is assumed for it.

Robinson (2011) relates:

> A crucial example of the application of the theorem was Turing's cracking of the German naval cipher Enigma during the Second World War, which played a key part in the Allied victory in 1945. After the war, Turing's wartime assistant, I.J. "Jack" Good, wrote about Turing's Bayesian technique for finding pairs and triplets of letters in the cipher.

Adding:

> To avoid censorship under the UK Official Secrets Act, he described it in terms of bird watching.

Good (1992) has given a stage-by-stage process in coding by Enigma. Let a real message (a triplet sequence) be coded.

(1) The operator would first choose a triplet, say XQV, as a system discriminator, from a table.
(2) Next he would set the three wheels at positions G_1, G_2, G_3, which were given as part of the daily key.
(3) At this initial position of the wheels G_1, G_2, G_3, he would encipher his selection

M_1, M_2, M_3 (the setting for the real triplet) and obtain the encipherment *LRP*, say.

(4) The six letters *XQV* and *LRP* would be further encrypted by the following procedure which does not use the Enigma.

(5) First, the six letters would be written one under the other staggered as

$$\begin{array}{cccc} X & Q & V & \\ - & L & R & P. \end{array}$$

(6) Then two letters would be chosen haphazardly to fill a two-by-four rectangle (here *A* and *L*):

$$\begin{array}{cccc} X & Q & V & A \\ L & L & R & P. \end{array}$$

(7) Then the four vertical pairs *XL*, *QL*, *VR*, and *AP* would be encrypted with the help of a secret printed pairs table, giving, say,

$$\begin{array}{cccc} P & T & O & W \\ X & U & B & N \end{array}$$

(8) Finally *PTOWXUBN* would be the first two groups, the 'indicator groups', of the enciphered message.

(9) There were ten pairing (digraph) tables and which one was to be used would be part of the daily key.

Each digraph table was reciprocal; for example, if *XL* became *PX* then *PX* would become *XL*. This again was helpful both for the encrypter and the cryptanalyst.

5.3.2 The significance of alignment of letters

From the description above it is clear that the alignment of letters was a critical step in breaking the code. The idea is somewhat similar to the alignment of DNA and protein sequences (see, for example, Durbin et al., 1998). Indeed, the DNA connection has been mentioned in various writings – for example, the following from Robinson (2011):

> Turing, by contrast, assigned them a tiny non-zero probability, thereby factoring in that rare letter groupings might not be present in his current collection of intercepted messages but could appear in a larger sample. The same technique was later adopted in **DNA sequencing** and by artificial-intelligence analysts.

We have used **bold** lettering for 'DNA sequencing'.

Good (1992), p. 214 says

> The game of Banburismus involved putting together large numbers of pieces of probabilistic information somewhat like the reconstruction of DNA sequences.

5.4 Release by GCHQ of two key Turing reports

Recently, two important previously classified reports by Turing have been released by GCHQ to the National Archives (Turing, 1941/1942a, 1941/1942b). These reports reinforce I.J. Good's contribution of Turing, as described above. We give a glimpse of the reports, and hope to give more detailed assessment of them elsewhere.

The first report (Turing, 1941/1942a) demonstrates that Turing applied Bayesian analysis to a wide range of cryptanalytic problems of the day. Bletchley Park's output of decrypts was almost certainly enabled by the techniques in this paper.

This report starts with the following paragraph:

> The theory of probability may be used in cryptography with most effect when the type of cipher used is already fully understood, and it only remains to find the actual keys. It is of rather less value when one is trying to diagnose the type of cipher, but if definite rival theories about the type of cipher are suggested it may be used to decide between them.

(Here cipher means a code.) He goes on to describe the distinction between probability and odds, based on prior information in life expectancy:

> I shall not attempt to give a systematic account of the theory of probability, but it may be worthwhile to define shortly 'probability' and 'odds'. The probability of an event on certain evidence is the proportion of cases in which that event may be expected to happen given that evidence. For instance if it is known that 20% of men live to the age of 70, then knowing of Hitler only 'Hitler is a man' we can say that the probability of Hitler living to the age of 70 is 0.2. Suppose however that we know that 'Hitler is now of age 52' the probability will be quite different, say 0.5, because 50% of men of 52 live to 70.
>
> The 'odds' of an event happening is the ratio $p/(1-p)$ where p is the probability of it happening. This terminology is connected with the common phraseology 'odds of 5:2 on' meaning in our terminology that the odds are 5/2.

He goes on to give a substitution matrix (p. 34) for alphabets similar in spirit to the substitution matrices for amino acids in bioinformatics (Durbin et al., 1998, pp. 14–16); the likelihood ratio of the alphabet pair (A,B) occurring as an aligned pair opposed to an unaligned pair in a message.

In the second report (Turing, 1941/1942b) he works out the best statistical means of testing whether two cipher messages use the same key in different parts of the message. This was very important in the exploitation of such messages at Bletchley Park.

The second report starts with the following:

> In order to be able to obtain reliable estimates of the value of given repeats we need to have information about repetition in plain language. Suppose for example that we

> have placed two messages together and that we find repetitions consisting of a tetragramme, two bigrammes and fifteen single letters, and that the total 'overlap' was 105, i.e. that the maximum possible number of repetitions which could be obtained by altering letters of the messages is 105; suppose also that the lengths of the messages are 200 and 250; in such a case what is the probability of the fit being right, no other information about the day's traffic being taken into consideration, but information about the character of the unenciphered text being available in considerable quantity?

There are some parallels with the problems of repeats in DNA sequences (Durbin et al., 1998, pp. 24–26) in the following sense. The statistical analysis that Turing developed in this report was based on the matching of repeat frequencies in a coded message. A similar problem occurs in DNA assembly, but it is essentially simple as the matching is between subsequences rather than frequencies.

5.5 Turing's statistics in context

This work of decoding is a very early success story of interdisciplinary research. Perhaps this is somewhat different from the early days of the subject, when statisticians such as Galton and R.A. Fisher played a leading role in interdisciplinary research. It seems that with the floods of large-scale data, computer scientists and statisticians with computing skills have a major part in creating impact. Mardia and Gilks (2005) named this the holistic approach and it is now required more and more.

5.5.1 Statistics and levels of abstraction

As Turing's mentor Professor Max Newman (1955) observed, Turing was far from being a detached theoretician, describing Turing as "at heart more of an applied than a pure mathematician". Turing's engagement with how the world 'computes' is visceral, with David Leavitt (2006) portraying Turing as *identifying* with his computing machine abstractions.

His respect for the complexity of the world as information emerges in his interest in type theory, which seeks to clarify the way in which mathematical objects can occupy different 'levels of abstraction' – for instance, describing integers as of type 0, reals as of type 1, sets of reals (or geometrical shapes) as of type 2, and so on. Clarity about type was exploited by Bertrand Russell to rescue early 20th century mathematics from paradoxes. In the real world, statistics provides a fundamentally important route to reducing scientific entities of apparent higher type to data which we can handle computationally. As we know, every computer today is an embodiment of Turing's universal computing abstraction, and they cannot comfortably cope with data above the level of type 1.

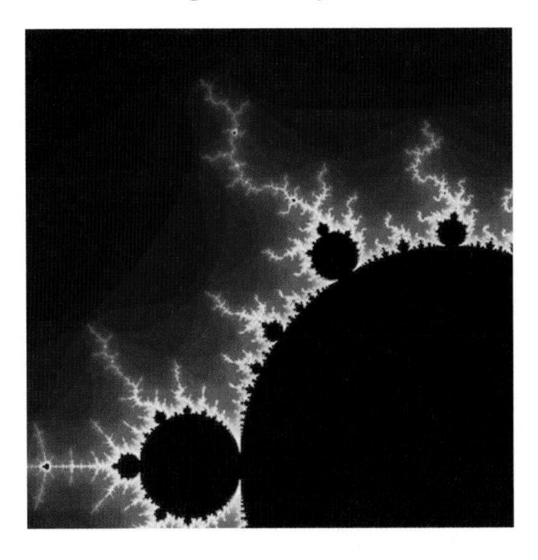

Figure 5.3 Mandelbrot set ©Niall Douglas.

A typical example of type reduction, with consequent computational accessibility, is provided by the Mandelbrot set (Figure 5.3). As the set of complex numbers the mathematics provides (a type 2 object) the question of its computability is an open question. But a little mathematics gives us a digital approximation representable digitally on our computer screen – hence the plethora of beautiful images on the web. Of course, the sampling process is not interesting enough to be termed 'statistics', any more than is the image of our family via a digital camera. But it does exhibit a level of computability that any sampled reality will have.

As we saw from the elephant and the blind men, in general there is an art to sampling and interpretation that on the one hand reduces complex information to useful data and on the other delivers a recognisable approximation to truth. For higher type information with very complex structure – for instance chaotic or turbulent contexts, such as weather, or economics, or coded messages, in which emergent non-local phenomena are the objects of interest – such a reduction may be fraught with difficulties. The statistics is a challenge and an art, and this is what so engaged the creativity and genius of those working at Bletchley Park in the early 1940s.

5.5.2 Scaling the informational hierarchy

So what is the link between Turing's most abstract mathematics – his 1939 paper written in Princeton under the guidance of Alonzo Church – and his hands-on practical involvement with real-world complexity of information?

Back in the late 1930s, Turing was puzzled by the fact that Kurt Gödel's Incompleteness Theorem told us that even restricting our attention to the basic theory of the natural numbers – which are just a part of what we can abstract from the real world – we discover that truth soon passes out of our control. Given any usable theory containing basic arithmetic (one where we can computably recognise the axioms and rules of deduction) one can easily write down a true statement not provable in it. Of course, thought Turing, this means one has an inductive way of computably expanding the theory, so potentially defeating Gödel's theorem. Turing succeeded in carrying out a transfinite induction which did indeed take us into realms unknown. The process was refined and equipped with more power in later years (by Sol Feferman, Michael Rathjen, and others) to take us to even dizzier proof-theoretic heights.

However, a key element in the inductive process was the choice of computable 'fundamental sequences' to take us through the limit points of the computable ordinals used to notate the tower of theories. The mathematical difficulties in keeping control of the process are essentially those in evidence in complex real-world situations that can only be handled via statistical sampling. In tune with the subtleties of the statistical route to truth, the fundamental sequence (the logician's counterpart of the statistician's sampling procedure) gives a computable (but by no means computably choosable) route up the informational mountain. And in the mathematics, one needs an *oracle* providing more than computably derivable information to identify the route. On a real mountain, one may need individual brilliance to get to the top – though once a route is identified it is computable, can be shared, and others can subsequently follow it. Turing (1939) contains the famous quotation:

> Mathematical reasoning may be regarded ... as the exercise of a combination of ... intuition and ingenuity ... In pre-Gödel times it was thought by some that all the intuitive judgements of mathematics could be replaced by a finite number of ... rules. The necessity for intuition would then be entirely eliminated. In our discussions, however, we have gone to the opposite extreme and eliminated not intuition but ingenuity, and this in spite of the fact that our aim has been in much the same direction.

The mathematician interprets this as an explanation of the mismatch between the subjectively experienced creative process leading to a new theorem and the axiomatic proof she shares with her colleagues and students. For the statistician, there is a very similar message. The approach to the informational mountain may be full of uncertainty and theoretically rich devices: but success can be shared with others.

For the Enigma operator, the success of the coding process depended on having a computable route up the informational mountain that was incomputable to the less well-equipped observer from afar. What the decoders at Bletchley Park depended on was the human brain having hidden type-traversing resources, at times in the form of statistical wizardry.

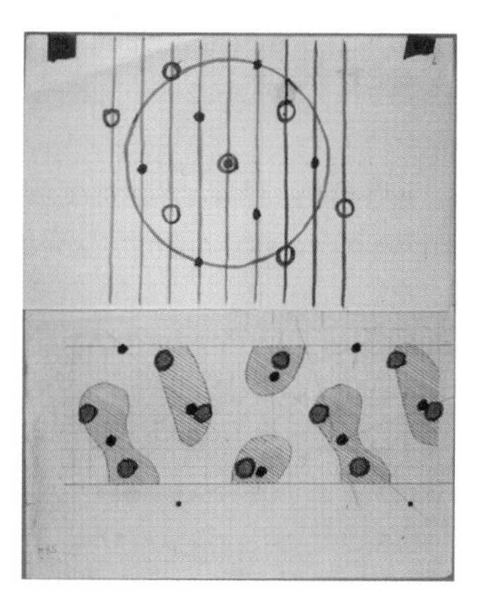

Figure 5.4 Turing image of morphogenic reaction-diffusion ©P.N. Furbank.

5.6 Morphogenesis, statistics, and Alan Turing's AI

Alan Turing's final years at Manchester were very much taken up with approaches to higher-type computability. By then stored-program computers, as anticipated by the 1936 Universal Turing Machine, were in operation. This was already giving rise to a powerful paradigm of algorithmic ubiquity and a new digital age. A whole range of ways of bridging the gap between what the (by then) actual computers could handle, and the informational complexity of social and natural formation, was inhabiting Turing's thoughts in the years running up to June 7, 1954.

The work on morphogenesis (the emergence of form in nature) brought a re-assertion of old certainties from logic in an unexpected way (see Figure 5.4). Turing was able to point to the *definability* of a range of natural formations via descriptions (differential equations) based on computable causal relations from the underlying chemistry. This gave explicit descriptions which led to computable solutions and computer generated simulations. Although the mathematics pointed to the likelihood of more complicated instances – possibly differential equations with incomputable solutions – the theory did point to type reduction via approximations built on explicit descriptions.

His lifelong preoccupation with human thought processes took Turing in a very different direction. The famous Mind paper (Turing, 1950) adheres to the growing faith in the key role of the (by then) commercially produced digital computers. But the essential role of the human judges in the implementation of the Turing Test for machine intelligence is very significant. This, and the radio broadcasts and the more popular talks and writings, show an acceptance of complementary roles for humans and machines. The role of mistakes, uncertainty, interaction, 'common sense', and, of course, the lessons of 1939 ('intuition') and Bletchley Park (Bayesian methods) all point to a world in which logic and statistical methods work together, in complementary ways.

Since Turing's passing in 1954, the history of artificial intelligence has tended to confirm this picture. There is more and more a sense that the computer and the human mind work in rather different ways. Today we are more and more aware of the power and limitations of our computational techniques. At the time, the main aim of Turing's 1936 paper was to demonstrate the inadequacy of algorithms and related forms of reasoning. There is a growing sense that human thinking has as much in common with statistical processes as with logical ones. This is good news for both humans and statisticians! Maybe digital computers will not supersede brains. But all those mistakes and the 'weather in the brain'? The brain shares with statistics an ability to handle large and complex assemblies of information. It is the algorithmic backbone that is provided by ever-growing computer power.

References

Aitken, C.G.G. (1995). *Statistics and the Evaluation of Evidence for Forensic Scientists.* Wiley.

Banks, D.L. (1996). A Conversation with I.J. Good. *Statistical Science*, **11**, 1–19.

Cooper, S.B. (2012). Turing's Titanic machine? *Communications of the ACM*. **55** (3), 74–83.

Cooper, S.B. and Van Leeuwen, J. (2013). *Alan Turing: His Work and Impact*, Elsevier.

Durbin, R., Eddy, S., Krogh, A. and Mitchison, G. (1998). *Biological Sequence Analysis: Probabilistic Models of Proteins and Nucleic Acids*. Cambridge University Press.

Efron, B. (2010). *Large-Scale Inference Empirical Bayes Methods for Estimation, Testing, and Prediction.* Cambridge University Press.

Good, I.J. (1950). *Probability and the Weighing of Evidence.* Griffin.

Good, I.J. (1953). The population frequencies of species and the estimation of population parameters. *Biometrika*, **40**, 237–264.

Good, I.J. (1979). Turing's statistical work in World War II, Studies in the history of probability and statistics. XXXVII. *Biometrika*, **66**, 393–396.

Good, I.J. (1988). The interface between statistics and philosophy of science. *Statistical Science*, **3**, 386–412.

Good, I.J. (1992). Introductory remarks for the article in *Biometrika* **66**, (1979), "A.M. Turing's Statistical Work in World War II'. In: *Collected Works of A.M. Turing: Pure Mathematics*, J.L. Britton (ed.). North-Holland.

Good, I.J. (2000). Turing's anticipation of empirical Bayes in connection with the cryptanalysis of the naval Enigma. *J. Statist. Comput. Simul.*, **66**, 101–111

Good, I.J. and Toulmin, G. H. (1956). The number of new species, and the increase in population coverage, when a sample is increased. *Biometrika*, **43**, 45–63.

Hodges, A. (1983), *Alan Turing: The Enigma*. Simon & Schuster.

Leavitt, D. (2006). *The Man Who Knew Too Much: Alan Turing and the Invention of the Computer.* W.W. Norton.

Mardia, K.V. and Gilks, W.R. (2005). Meeting the statistical needs of 21st-century science. *Significance,* **2**, 162–165.

McGrayne, S.B. (2011). *The Theory that Would not Die.* Yale University Press.

Newman, M.H.A. (1955). Alan Mathison Turing. In *Biographical Memoirs of the Fellows of the Royal Society*, **1**, 253–263.

Robinson, A. (2011). Known unknowns. *Nature*, **475**, 450–451.

Simpson, E. (2010). Bayes at Bletchley Park. *Significance,* **7**, 76–80.

Turing, A.M. (1936). On computable numbers, with an application to the Entscheidungsproblem, *Proc. London Mathematical Society* ser. 2, **42**, 230–265; Reprinted in A.M. Turing, *Collected Works: Mathematical Logic*, R.O. Gandy, C.E.M. Yates (eds), North-Holland; and in Cooper and van Leeuwen (2013).

Turing, A.M. (1939). Systems of logic based on ordinals, *Proc. London Math. Soc.* ser. 2, **45**, 161–228. Reprinted in A.M. Turing, *Collected Works: Mathematical Logic*, R.O. Gandy and C.E.M. Yates (eds.), North-Holland; and in Cooper and van Leeuwen (2013).

Turing, A.M. (1941/1942a). The applications of probabilities to cryptography. *The National Archives* `http://discovery.nationalarchives.gov.uk/details/r/C11510466` H.W. 25/37; 43 pages.

Turing, A.M. (1941/1942b). Paper on statistics of repetitions. *The National Archives* `http://discovery.nationalarchives.gov.uk/details/r/C11510466` H.W. 25/38; 9 pages.

Turing, A.M. (1950). Computing machinery and intelligence. *Mind*, **49**, 433–460.

Turing, A.M. (2001). *Collected Works of A. M. Turing: Mathematical Logic*, R.O. Gandy and C.E.M. Yates (eds.). North-Holland.

Part Two: The Computation of Processes, and Not Computing the Brain

Image © Andy Lomas.

We can only guess as to the exact point at which Turing's exposure to digital electronics in the 1942 period gave him the idea of realising a practical version of the universal machine. But we do know that, right from the start of his project, Turing was openly and provocatively describing it as 'building a brain'. In his 1946 ACE report he made a reference to computer chess-playing, in terms which suggest he had been thinking about how it would rival human thought.

In this Part **Stephen Wolfram** takes up his view of 'what Turing might have discovered', also provocative because it takes issue with the line Turing took later in modelling morphogenesis via differential equations (described in Part Three).

Wolfram argues that Turing should have stayed with the new world of discrete machines, and that 'computing the world' in discrete models is the central strategy for science.

What is very much in tune with Alan Turing's later thinking is Wolfram's focus on the complexities of interactive processes related to particular engrossing examples, and his ambition to develop our grasp of the computational content of physics to its limits. Wolfram's emphasising of the role of the *irreducibility* of particular computational processes nicely contextualises Turing's discovery of incomputability. Wolfram's view is a wide-ranging and thought provoking one, in which his Principle of Computational Equivalence is key to a world view in which

> More and more of our lives and activities will be played out as computations. And at some level the human condition will be reduced to a matter of abstract computation.

Christof Teuscher gives an update on his pioneering work to implement and develop Turing's 1948 neural models.

The focus switches from that of Wolfram on the *characterising* of complexity to one more concerned with the practicalities of its *reproduction*. Teuscher's work is framed by what he calls a *bottom-up* approach to capturing human brain functionality. Underlying the choice of an evolutionary alternative to the engineer's top-down paradigm, we discern Wolfram's 'irreducibility of computation'. The discussion of 'intrinsic computation' is much informed by Alan Turing's visionary 1948 NPL Report *Intelligent Machinery*, and forms another highlight of this heterogeneous collection.

Douglas Hofstadter elegantly reminds us of how Turing in 1950 set a high standard for what we might mean by intelligence. Computing the world of Turing's culture meant dealing with an extensive and many-layered world, of concepts and images, irony and word-play. We follow this, at Douglas Hofstadter's suggestion, by a famous poetic version of Turing's argument that settled the Entscheidungsproblem.

The article is, as one will not be surprised, a literary gem combining clarity, challenge, and a lifting of both mind and spirit. The author's professed aim is to rescue the "marvelously thought-provoking snippets of a hypothetical human-machine dialogue", which Turing included in his discussion of the 'imitation game', from the frequent misunderstandings encased in the "arcane philosophical debates" of those who came after. Proclaiming his approval of Turing – "I think concrete examples are always needed before arcane arguments are bandied about" – he proceeds to give us a wonderfully entertaining, challenging and informative dialogue "between DRH and AMT". The balance between computation and "human-style mental fluidity" swings away from 'computational equivalence'.

We must refrain from further spoilers. The reader is in for a treat.

6

What Alan Turing Might Have Discovered

Stephen Wolfram

When we look back at the twentieth century, I have no doubt that the single most important idea that we will see to have emerged there is the idea of computation. And already that idea has transformed our technology and many aspects of our world. But there is, I believe, still much more to come.

Although he certainly did not realize the magnitude of what he was starting, Alan Turing in 1936 laid the single most critical foundation for the idea of computation when he invented the universal Turing machine.

By the time Turing died in 1954, this idea had come by a circuitous route (through neural networks and the like) to inform early practical electronic computers. And it had also led Turing himself to think in a real way about the creation of artificial intelligence.

But that was about as far as it got. And in his last years, when Turing thought, for example, about theories of biology, he did not realize that his ideas of computation might be relevant to them.

Through the rest of the 1950s and into the 1960s and 1970s, practical computers became more and more powerful, and more and more common. And, at least among the small community of theoretical computer scientists, universal computing took its place as a foundational concept.

Meanwhile, however, the rest of science continued its development, based in some cases on experiment and in some cases on mathematical analysis. Computers were occasionally used to support these existing methods. But there was no real idea that computation as an abstract concept might be centrally relevant.

And indeed, in a sense, as computer technology advanced, computation as a concept seemed to recede more and more – as something to be mentioned for background but not something to be made the core of what should be studied.

If Turing had lived longer, I rather suspect he might not have left it that way –

instead he might have begun to realize the much broader significance of the concept of computation, the foundations of which he was so instrumental in laying.

I was born five years after Alan Turing died. And in my half-century of life so far, I have launched three large projects, all fundamentally based on the idea of computation and all, strangely enough, pointing in directions that I suspect Alan Turing might have followed had he lived longer.

The first of these is *Mathematica*.

When Alan Turing created his universal Turing machine, he was, in a sense, trying to define a formal mechanization of mathematics and of computational processes in general.

As it turned out, the basic structure that he defined was much more general than I think he imagined,the and in effect captures essentially any possible realizable process.

But his original goal was to find a way to represent processes that were somehow mathematical or calculational.

And in a sense that was also precisely my original goal in building *Mathematica*.

By the time I was doing this (I started its first precursor in 1979), there was of course – thanks to Turing – no doubt that it was in principle possible. But my specific challenge was to create a language that could as a practical matter conveniently cover everything one wanted to do.

And to achieve this, I ended up developing the idea of symbolic programming – which in a sense had its precursors even in mathematical logic that preceded Turing's work.

Any reasonable computer language must in principle be universal. But the remarkable thing about symbolic programming, which has gradually emerged over the three decades I have been working to develop it, is that in a sense it provides another level of practical universality.

It provides a single uniform representation not just in principle but also in practice for a vast range of different kinds of things, whether they're mathematical relations, or geometrical objects, or running programs, or documents, or computer interfaces, or whatever.

For me, the original creation of *Mathematica* was in large part an effort to construct for myself a tool that I could use to study computational things.

I had started my career as a theoretical physicist. And I studied my share of systems using the paradigm that had dominated physics and the other exact sciences for the past 300 or so years: the notion that systems should be modeled using mathematical equations.

But I had found that, while this worked well in some cases, there were plenty of examples even in physics where it did not.

So I began to think about what else might be going on.

Conveniently, at the time, I had just finished the first version of the computer language that was the precursor of *Mathematica*. And building that language had begun to give me an intuition for just how powerful computation could be, and the extent to which, by choosing the appropriate computational primitives, one could create a system with remarkably rich capabilities.

So I wondered whether perhaps this idea of computation could be applied too in natural science.

And whether one could generalize the kind of mathematical modeling that had been traditional, to encompass what are in effect arbitrary computer programs.

Ever since Turing, most of the time we have imagined building programs to perform specific tasks. And usually our programs are long and complicated.

As a matter of basic science, however, we can ask the question of what simple programs – say picked at random – might do.

And at the beginning of the 1980s, I embarked on just such an investigation of the computational universe of possible simple programs.

I might have used Turing machines for this purpose. But as it happens (and in some ways fortunately), I did not: instead I used one-dimensional cellular automata, which seemed more directly to have the potential to correspond to actual systems in nature.

In a typical case, a cellular automaton consists of a line of cells, each either black or white, where the color of each cell evolves down the page, at each step being determined by a simple rule from the colors of cells immediately above and to its left and right above. Figure 6.1 gives an example.

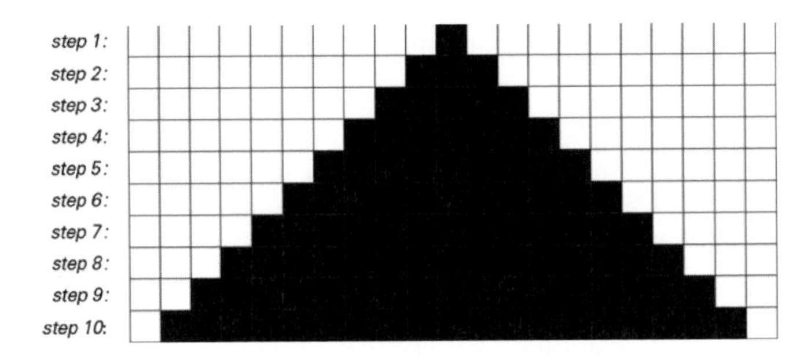

Figure 6.1 A visual representation of the behavior of a cellular automaton; each row of cells corresponds to one step.

Now, one might have thought, as I certainly did, and I am sure Turing would have too, that such a simple setup would not be capable of giving behavior of any significant complexity.

But in the early 1980s I could use a computer to actually test this out. Figure 6.2 shows what I saw.

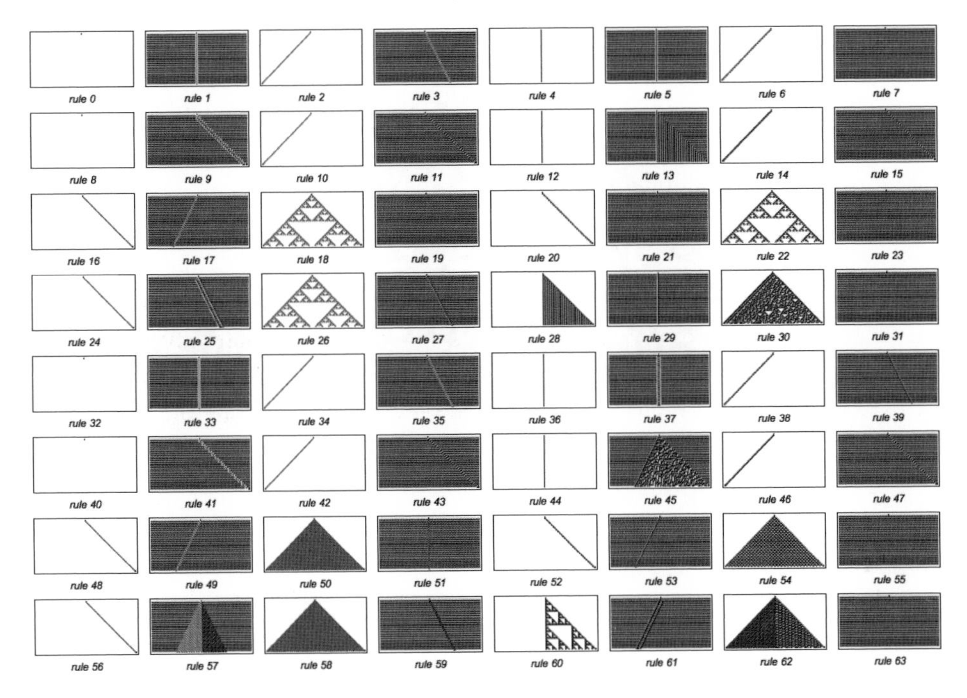

Figure 6.2 Examples of patterns produced by the evolution of each of the first 64 one-dimensional cellular automaton rules, starting from a single-black-cell initial condition.

This is an array of patterns produced by running the first 64 in the most obvious enumeration of possible cellular automaton rules. The diversity is remarkable, and immediately reminiscent, for example, of the diversity of forms in biology. But it is clear that many of the possible rules – the possible simple rules – lead, as one might expect, to very simple behavior.

One soon notices, however, cases where an elaborate nested pattern is produced. But, while intricate, this pattern is still ultimately very regular, in effect unsurprisingly reflecting the simplicity of the underlying rules used to produce it.

Looking over the picture above, however, it is easy to spot a surprise: rule 30. Figure 6.3 provides a larger picture.

What is going on here? We start just from a single black cell, then follow the simple rule at the bottom. But as we keep running this rule, we get a pattern that seems very complicated.

There is some obvious regularity. But if we apply any standard statistical, mathematical, or other test, it will tell us that, for example, the center vertical column of the pattern seems completely random.

I first saw rule 30 in 1984, and for me it was a major intuition-breaking discovery.

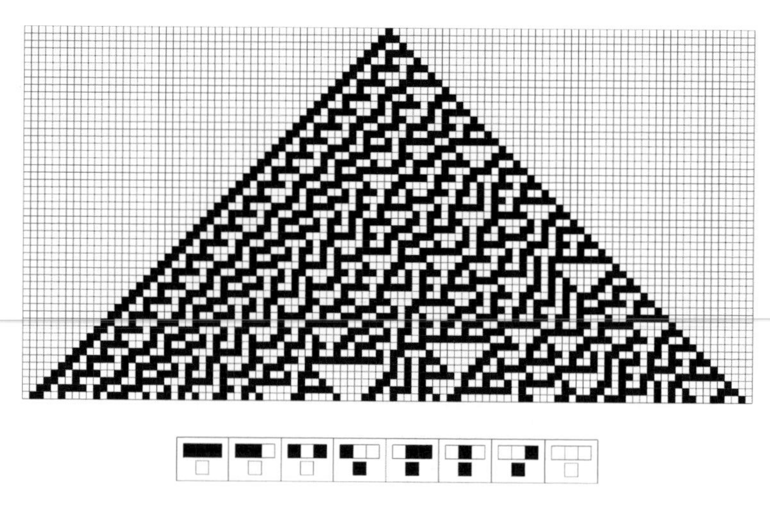

Figure 6.3 The rule 30 cellular automaton. The pattern that is obtained is highly complex and shows almost no overall regularity.

For we are used to the idea that to make something complex you somehow have to go to lots of effort, with complicated rules and so on.

But here – not far out in this computational universe of possible cellular automata – is an immediate example where this is not the case, and where instead a very simple rule seems effortlessly to make great complexity.

At first I was not sure how special and unusual this example was. But I was soon able to use what was in effect my computational telescope – *Mathematica* – to look out into the computational universe and study a wide range of different kinds of simple programs.

One of the obvious choices was Turing machines.

I do not believe that Alan Turing ever explicitly simulated a Turing machine on a computer. And it was not until 1991 that I did either.

But then I started to do the same experiments as I had done on cellular automata. The results were not quite as immediate. Figure 6.4 shows some samples of what happens with any of the 4096 simplest Turing machines, with two head states and two colors.

No very elaborate behavior is seen. But if one persists a little longer, and looks, say, among the 2 985 984 Turing machines with two head states and three colors, one eventually starts finding examples like the one shown in Figure 6.5. In effect the same kind of complexity as in cellular automata like rule 30.

I wonder what Alan Turing would have thought of this.

When I first discovered rule 30, the thing I thought most about was nature. For in the natural world, it is common for us to see forms and systems of great complexity.

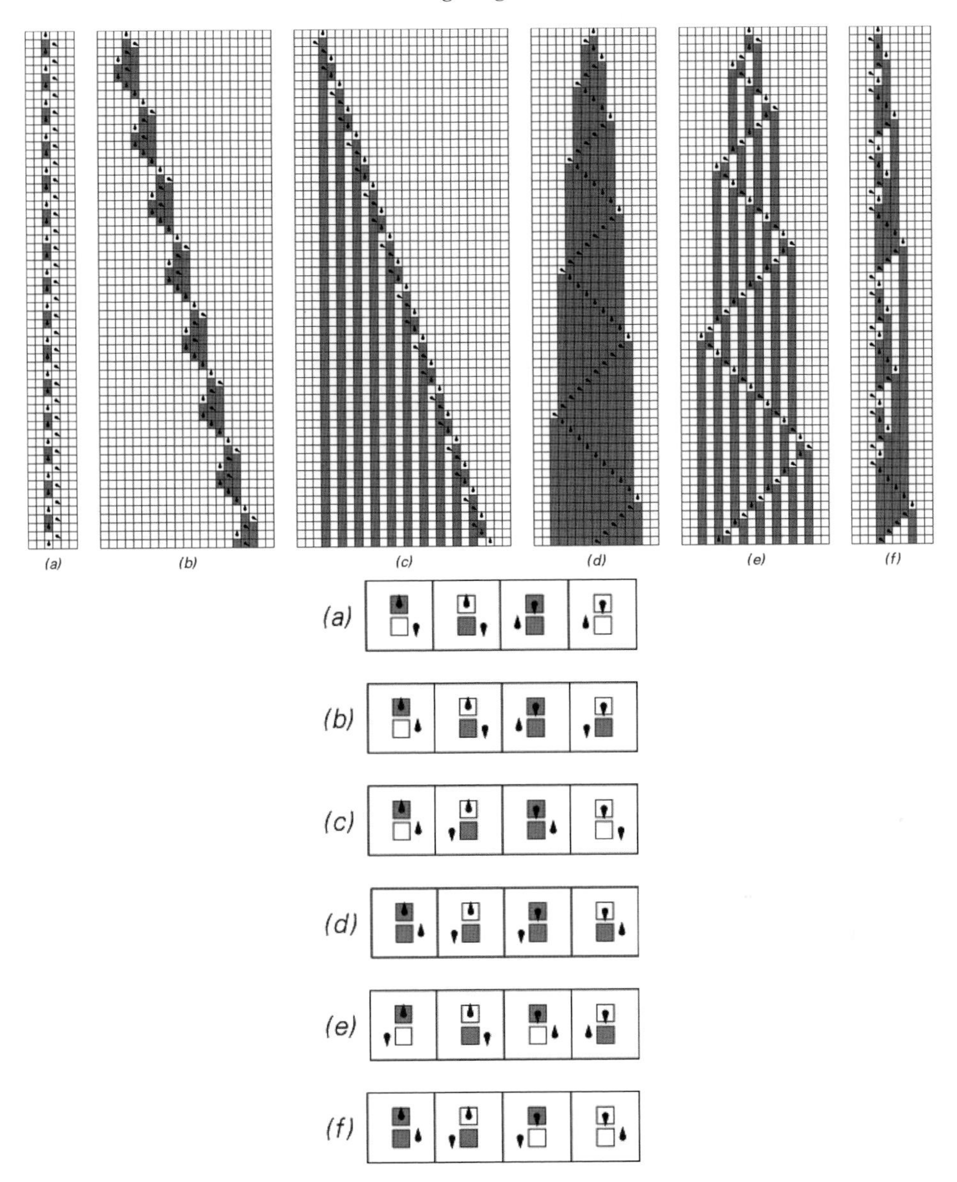

Figure 6.4 Examples of Turing machines with two possible states for the head. Repetitive and nested patterns are seen, but nothing more complicated ever occurs.

Yet in our own efforts of engineering, and in creating artifacts, we are typically producing what appear to be much simpler things.

But having seen rule 30, one can imagine that one has in effect glimpsed the key secret of nature: that by using rules out in the computational universe, nature is able quite effortlessly to produce the kind of complexity it does. And it is only

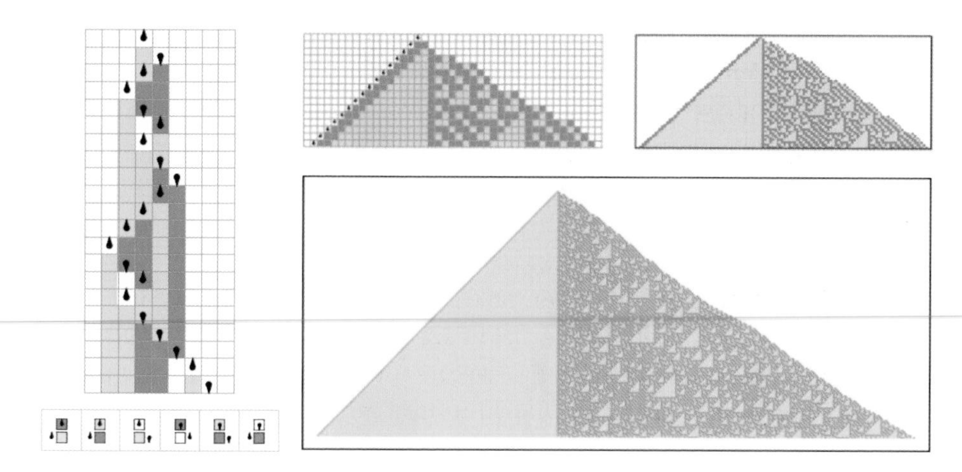

Figure 6.5 Two-state three-color Turing machine number 596440. On the left, the uncompressed evolution of the Turing machine from a 'blank' initial tape, containing only white cells. On the right are shown the left-compressed evolutions in which the left-hand side always expands by one cell at each compressed step. The right-hand side expands by either one or three cells.

because in our current efforts of engineering we choose to restrict ourselves to systems whose behavior is simple enough that we can readily foresee it, that we do not encounter this.

In effect, then, computation becomes the paradigm with which to view processes in nature.

And within that paradigm there is a much richer set of primitives that one can use to create models – and indeed in recent times there has been great success in doing this.

When Turing created Turing machines, he did it in a sense as a piece of engineering. But what has become clear is that systems like them can also be studied in a way that is much closer to natural science: an exploration of the whole diverse universe of possible such systems.

And, in doing this exploration, one soon comes across remarkable systems like rule 30.

But what general principles can one deduce?

The most important, I believe, is what I call the Principle of Computational Equivalence (Wolfram, 2002).

What the Principle of Computational Equivalence says is this: that if one looks at the computational universe then essentially whenever one sees behavior that is not obviously simple, it will correspond to a computation of equivalent sophistication.

One might have thought that, as one makes the rules for a system more and more complicated, its behavior must become correspondingly more complicated.

But the remarkable thing that one observes in the computational universe – and that the Principle of Computational Equivalence implies – is that it does not. Instead, as soon as one is above a very low threshold, the complexity of the behavior one sees is in effect quite independent of the complexity of the underlying rules.

What does this mean? It has many deep consequences. But it has one immediate prediction.

It says that it should be easy to achieve computational universality of the kind that Alan Turing first found in the universal Turing machine he constructed.

Our normal intuition might tell us that to achieve something as sophisticated as universal computation we would have to have a system with all sorts of complicated internal structure – say a machine with all sorts of registers and operations and so on.

But the Principle of Computational Equivalence says that this is not necessary. And it says instead that out in the computational universe essentially any system whose behavior is not obviously simple should actually be capable of universal computation.

Two decades ago we learned that the rule 110 cellular automaton, shown in Figure 6.6, is universal – a first clear piece of evidence for the Principle of Computational Equivalence.

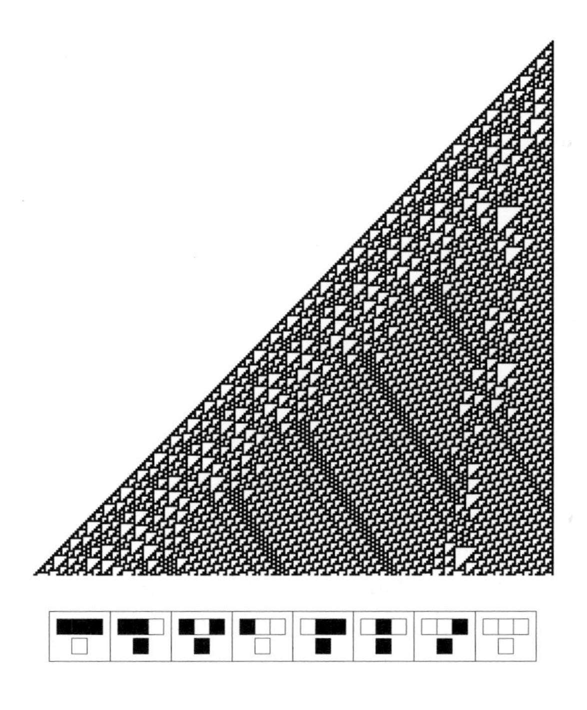

Figure 6.6 First 250 steps in the evolution of rule 110.

In more recent years I have become very curious about a similar question for Turing machines. And so I sponsored The Wolfram 2,3 Turing Machine Research Prize to see whether what is in effect the simplest Turing machine with not-obviously-simple behavior would be universal.

And in a small number of months, this was proved – providing another major piece of evidence for the Principle of Computational Equivalence and showing us, once and for all, that the simplest universal Turing machine is the one shown in Figure 6.7.

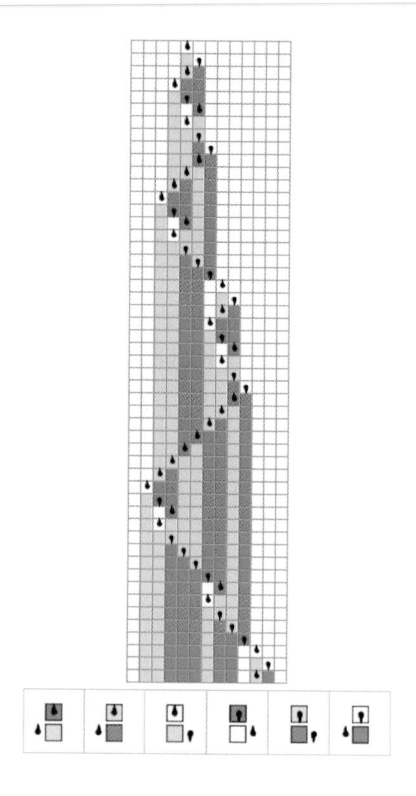

Figure 6.7 First 50 steps in evolution (from blank tape) of the Wolfram 2,3 Turing machine – the simplest Turing machine capable of universal computation.

In a sense the idea of universal computation establishes an upper bound on the computational sophistication of systems, because it says that once a system is computation universal, it can emulate any other system.

But what the Principle of Computational Equivalence implies is that universal computation is not just an upper bound – instead it is ubiquitous among systems in the computational universe and, for example, in nature.

At a technical level, the Principle of Computational Equivalence goes even further, saying that many systems can in principle be programmed not only to produce arbitrarily complex behavior but in a sense do so ubiquitously, for example in the way that rule 30 produces such complexity even from a simple initial condition consisting of a single black cell.

The Principle of Computational Equivalence has all sorts of consequences for science.

Traditional mathematical science, for example, has emphasized the idea of computational reducibility: of being able to predict what a system will do using less computational effort than it takes the system itself to produce its behavior.

But what the Principle of Computational Equivalence implies is that, instead, many systems are computationally irreducible, so that the only way to predict a system's behavior is effectively to simulate every step and follow what it does.

The traditional exact sciences have tended to avoid computational irreducibility, concentrating their efforts on precisely those systems that happen to be computationally reducible and for which their methods can make progress.

And indeed many phenomena that I, for example, looked beyond traditional methods to understand are precisely those that show computational irreducibility.

In a sense, traditional exact science has made the implicit assumption that, computationally, the observer can be arbitrarily more sophisticated than the system being observed. But, as with relativity or quantum mechanics before it, the Principle of Computational Equivalence forces a more realistic view on the observer, and implies that in general the observer can be no more computationally sophisticated than the system being observed – which is precisely why so many systems seem to us so complex.

These same issues actually come up in mathematics too, where in effect the mathematics we have chosen to study has carefully navigated away from irreducibility and undecidability.

If we just enumerate all possible axiom systems for mathematics, we find familiar ones such as logic first out there, as perhaps the 50 000th axiom system. But there are so many others – in effect so many other possible fields of mathematics. And, in almost all. there is enough computational sophistication that one does not have to go far to find signs of undecidability.

But while the Principle of Computational Equivalence implies limits to our knowledge and methods, it also implies great power. For it suggests that there is great richness close at hand in the computational universe.

Our traditional approach to engineering – even for programs – involves explicitly building up the system one wants. But from what we have discovered in the Principle of Computational Equivalence we know there is another way: just go out and mine the computational universe, for one will not have to go far to start finding all sorts of systems with rich and complex behavior that can be harnessed for all sorts of technological purposes.

One example is finding systems that can be used as universal computers – and we now know that even very simple components, such as one might find at a molecular scale, will be sufficient.

And in general the approach of mining the computational universe for structures and algorithms of technological importance is a very powerful one – which I and my companies have used to great advantage. Indeed, in time I expect that the idea of mining the computational universe will come to be the dominant form of invention and creation, far outpacing all others.

There is still another consequence of the Principle of Computational Equivalence, which is at first sight purely intellectual but which, at least for me, has had a critical practical consequence.

I used to imagine that somehow at one level there was generic computation, and then at some higher level there was intelligence of the type we, as humans, exhibit. Kurt Gödel seems to have thought the same thing. Alan Turing in his later work on artificial intelligence seems to have thought they were closer, but still not quite the same.

Having developed the Principle of Computational Equivalence, however, I came to the conclusion that actually there can be no sharp distinction: that intelligence is really all just generic computation. And the only thing that is special to human intelligence is all the detailed knowledge and circumstances that we humans have.

There are all sorts of implications of this. For example, when we think about extraterrestrial intelligence we no longer imagine that an 'intelligent signal' must be the output of some elaborately developed civilization but that instead it may well be the result of some natural process that is effectively performing just as sophisticated a computation. From an immediate practical point of view, one also realizes that 'creativity' cannot just be a feature of humans. And indeed, for example, in our WolframTones music composition example, it is rather easy for simple searches in the computational universe to yield music that in effect can pass an analog of the Turing test.

But for me, the realization that human-like intelligence may ultimately be 'just computation' has had a much larger-scale practical consequence.

Ever since I was young, I have been interested in trying to take in a broad range of knowledge and somehow make it computable, so that any question that could in principle be answered on the basis of that knowledge could actually be answered automatically by a machine. For a long time I assumed that to achieve this would require in effect solving the whole problem of artificial intelligence and then applying it to this specific case.

But from the Principle of Computational Equivalence I realized that whatever was needed, it would always in a sense be 'just computation'. And, particularly given *Mathematica* as a practical system for doing computation, I began to imagine that perhaps the time had come to actually try to make the world's knowledge computable.

Alan Turing had thought about such things, but in his time nothing like this seemed even vaguely practical.

Certainly the concept of what one wanted to achieve was fairly familiar even from some of the science fiction of Turing's day. But for me the exciting thing in the past few years is that with Wolfram|Alpha we have been able to see that the goal of making the world's knowledge computable really is achievable.

I am not sure how familiar Alan Turing would have found much of what we have had to do to build Wolfram|Alpha. He thought about some of the issues, such as understanding human language.

But I think he imagined that they would be solved in some way that directly mirrors what humans do. Yet in a sense what we have done with Wolfram|Alpha is to make use of the broader features of the computational universe – not least in some cases just finding the algorithms we needed by mining them from that computational universe.

And when it comes, say, to being able to answer a question about physics, we do not reason about it as might an untrained human – or a medieval philosopher. Instead, we just use the best possible methods from mathematical physics, solve the equations, and get an answer. And if we want to explain to humans how to arrive at this answer, we then explicitly synthesize human-like steps, quite different from the actual steps Wolfram|Alpha uses to get the answer.

How would Wolfram|Alpha do in the Turing test? The answer is that it would do terribly, not because its technology could not reach the point of emulating the various features of human responses, but because, unless it were radically changed, it knows far too much and can figure things out far too effectively to seem like a human.

In a sense Wolfram|Alpha is leapfrogging what Alan Turing imagined: instead of emulating the intelligence of a single human it is instead trying to take the collective knowledge of our whole civilization and make it computable.

So where else will the concept of computation go?

At one level it is increasingly the paradigm with which we can understand our world. And from our everyday experience of practical computers, we have begun to think about many things in terms of concepts such as 'execution' and 'debugging' that we get from computers.

But I suspect that as there is more experience, of Wolfram|Alpha and computational knowledge, and as we get more familiar with what is out there in the computational universe, more of the fundamental ideas of computation will begin to shape our ways of describing the world – and terms like 'computational irreducibility' and 'undecidability' will be in at least as common use as terms like 'force' and 'momentum'.

And as we think more in computational terms, it is inevitable that we will try to understand our whole universe in those terms. Today – as for the past 300 years – the dominant paradigm for our attempts to understand fundamental physics is mathematical equations.

But what happens if we think about our universe in computational terms? Might we find that the actual rules for our universe can be found out in the computational universe of possible programs?

Before the discovery of phenomena such as rule 30, and the Principle of Computational Equivalence, we might have imagined that, even if the rules for our whole universe could be represented in terms of programs, they would necessarily be immensely complex programs.

However, from what we have seen in the computational universe, it is certainly not immediately obvious that our whole universe might not derive from some rather simple underlying rule.

We do not yet know whether this is in fact the case, though it is a very fundamental observation about our universe that it is not as complicated as it could be and that all those different particles and so on in it do act according to common rules.

If the fundamental rules for our universe are to be truly simple, it is inevitable that they must operate far below the level of our familiar experience and indeed even below such basic concepts as space and time. There are many technical details involved in the actual search for our physical universe within the possibilities of the computational universe. And, as one searches candidate universes, many are very obviously not our universe.

But the exciting thing that I have discovered over the past decade is that, even among the first few thousand candidates, there are already cases whose behaviors are complex enough that one cannot easily exclude the possibility that they are, in every particular, our actual universe.

Computational irreducibility inevitably makes it hard to see just how a candidate universe will ultimately behave. It is still within the realms of possibility, however, that within a limited number of years we may in effect hold in our hand a rule that represents our complete history, and its every detail.

Computational irreducibility still makes it irreducibly difficult for us to know all the features of this universe, or to make predictions about everything. But it would surely be a great achievement for science to be able to take, in essence, Alan Turing's idea of computation, and show that it can give us a complete description of everything that exists for us.

Certainly at a technical level this would demonstrate definitively that a Turing machine really can be truly universal, in the sense that it can capture any computation that can occur in our universe.

Computation, however, is not just relevant to these fundamental, and essentially intellectual, questions about our universe. It seems inevitable that it will increasingly become entwined with the very nature of the human condition.

I doubt whether there will ever be a time when people suddenly say "artificial intelligence has been achieved". Rather, the gradual symbiosis of computers with us humans will continue. More and more functions that were once performed by humans will be taken on by computers, which will become better and better at reflecting and responding to our detailed human experiences. More and more of our lives and activities will be played out as computations, and at some level the human condition will be reduced to a matter of abstract computation. And, as we have learned from studying the computational universe, there is vast diversity that can be achieved with computation.

In a sense then, the crucial ingredient that we as humans will add is purpose: decisions, rooted in all the details of our history, about what we choose to do.

I do not know how long it all will take. But I would expect that by the time a century has passed from when Alan Turing first introduced universal computation, essentially every aspect of our world, and our experience, will already be centrally based on the idea of computation.

The universal Turing machine on its own – or even Turing's notion of artificial intelligence – will not get us there. But between what we have discovered in the computational universe, and what are now seeing with computational knowledge, I believe we are well on the way.

And I, for one, am profoundly grateful to live in this time when we can see the flowering of the idea of computation, initiated in a sense so recently in history by Alan Turing.

References

Mathematica, Wolfram Research, Inc.

The Wolfram 2,3 Turing Machine Research Prize `www.wolframscience.com/prizes/tm23/`

Wolfram|Alpha, `http://www.wolframalpha.com/`

Wolfram|Alpha blog, `http://blog.wolframalpha.com/2010/06/23/happy-birthday-alan-turing/`

Wolfram, S. *A New Kind of Science*, Wolfram Media, 2002 (also available online: `http://www.wolframscience.com`)

7

Designed versus Intrinsic Computation

Christof Teuscher

7.1 Top-down versus bottom-up design

The design of modern products commonly follows a well-defined path, such as for example the human-centered approach used by the famous IDEO design consultancy, Kelley (2001). Design has turned into a process that can be studied, formalized, learned, and improved. As Tom Kelley (2001) explains, IDEO uses a methodology that has the following five basic steps: understand, observe, visualize, evaluate and refine, implement. Two things are important to notice in this process. First, there is an implicit "loop" for the evaluation and refinement process. No product is perfect from the beginning and most of them will need to be refined and re-evaluated over and over again. Second, the process is organized in a very *top-down* manner, i.e., it is a step-wise process that starts with a high-level goal, target, or vision and then gradually follows down in the hierarchy by specifying, evaluating, and implementing the lower-level (sub)systems. That is the way engineers design cars, airplanes, computers, and whatnot. The most important reason why top-down design has been so successful in the engineering world is *divide et impera*, or in more contemporary words, *divide and conquer*. Top-down design allows us to handle complex problems and to treat subsystems as abstract black boxes. While this generally works well in the engineering world, an entirely different design process can be observed in nature. Nature proceeds in a *bottom-up* way by simple trial and error. The process is called *evolution* and has led to amazing, albeit by no means perfect, 'designs.' As the name says, bottom-up design proceeds from the bottom up by piecing subsystems into something bigger. Often, the functionality of the resulting system is greater than the sum of its parts. In that context, the term *emergence* (Holland, 1999) comes up frequently. Loosely speaking, emergence means that the interactions of the subsystem components do not necessarily allow us to predict the behavior at the next level. There is an element of surprise (Ronald et al., 1999) involved, which describes that non-predictibility nicely.

Today, the computing disciplines face difficult challenges from extending CMOS technology to and beyond dimensional scaling. One solution path is to use an innovative combination of novel manufacturing techniques, devices, compute paradigms, and architectures to create new information processing technology that may be radically different from what we currently know. For example, with the advent of molecular and nano-scale technology, bottom-up fabrication (e.g., see Zhirnov and Herr, 2001; Freitas and Merkle, 2004), or rather *self-assembly*, has opened an entire new way of manufacturing computing and other machinery. As with every new approach, this also opened an entire new world of challenges, which are closely related to the topic of this chapter: *designed* versus *intrinsic computation*.

7.2 Intrinsic versus designed computation

Designing computers in traditional CMOS technology (Wolf, 1998) is all about 'torturing the silicon to do what we want,' e.g., to implement a transistor that acts as a binary on/off switch. And we want this switch to be as reliable, as binary, and as perfect as possible. To do so, we have to go many extra steps and play a lot of sophisticated tricks, especially with today's technology feature sizes in the nanometre-range. While this 'designed' top-down approach has worked well for many decades, it is unclear how far we can push the technology (Kish, 2002). With the advent of emerging nano-scale devices that are expected to behave in time-dependent nonlinear ways, beyond a simple switching behavior, and that are manufactured differently, we need to be ready to deal with extreme physical variation, heterogeneity, and unstructuredness.

In Teuscher et al. (2008), we argued that the research community should explore the computing capabilities of nonlinear circuits for emerging computing devices and architectures. Recently, a *Chaos* focus issue edited by Crutchfield et al. featured a variety of important papers in the area of information processing with dynamical systems. It was argued that "The reign of digital computing is being challenged, not only by fundamental physical limits but also by alternative information processing paradigms. [...] The technological promise is substantial: faster, less expensive, and more energy efficient computing" (Crutchfield et al., 2010).

Intrinsic computation is commonly defined as the spatial and temporal dynamics that a system has 'intrinsically.' It is important to note that "Intrinsic computation makes no reference to utility" (Crutchfield et al., 2010). On the other end, as we have seen, there is *designed computation*, which is what makes information processing (in any system) 'useful.' Naturally, the ability to harness and exploit the intrinsic nonlinear and time-dependent dynamics of emerging devices for computation hinges on an understanding of their fundamental dynamics at the level of both the individual device and the device network. We argue here that these two different schools of thought are intimately related to bottom-up and top-down processes. In the world of silicon CMOS technology, top-down design leads to designed computation because we engineer 'useful' information processing devices, such as transistors or logic gates, by design. On the other hand, the lack of control over bottom-up self-assembly processes often leads to designs that we cannot fully control and understand. We therefore may end up with systems that possess spatial and temporal dynamics that one needs to harness to perform useful computation. A recent example of intrinsic computation is the bottom-up design of an associative memory based on memristors (Sinha et al., 2011). A memristor is a new type of passive circuit element that can change its resistance by a current applied to it (Strukov et al., 2008). The memristor retains its resistance when no current is applied. In an I–V (current–voltage) chart, one can see a simple hysteresis-type behavior. The device can be uses rather straightforwardly as an on–off switch; however, one can also attempt to exploit its nonlinear time-dependent characteristics related to the I–V hysteresis curve. In the area of neural networks, memristors have been used so far solely to represent the synaptic weights. We showed in Sinha et al. (2011) that one can significantly improve the design of a simple associative memory by using the memristor's nonlinear characteristics in other ways than just for representing synaptic weights. The design was evolved in a bottom-up way by employing genetic programming. The algorithm finds a way to harness the intrinsic properties of the device network to perform useful computation with a significantly lower number of components than one would need if the associative memory were implemented with traditional transistors.

Figure 7.1 illustrates the relationships between computation and fabrication, structure and functionality, top-down and bottom-up fabrication, and designed and intrinsic computation from a bird's eye perspective. Traditionally, we start with the functionality and impose that on the structure by means of a specific manufacturing technique, such as lithography in the case of traditional CMOS technology.

7.3 Turing's bottom-up computing approach

Going some six decades 'down to the future' of computer science, Alan M. Turing advocated the very same bottom-up and top-down ways of building machinery, of controlling them, and of processing information. The Turing machine is a good example of a top-down designed machine, with a complete understanding and complete control over all steps. Yet, Turing realized – not surprisingly – that nature's 'machinery' is not typically built in that way and that other machines might be a more appropriate model for certain biological systems, such as for example the hu-

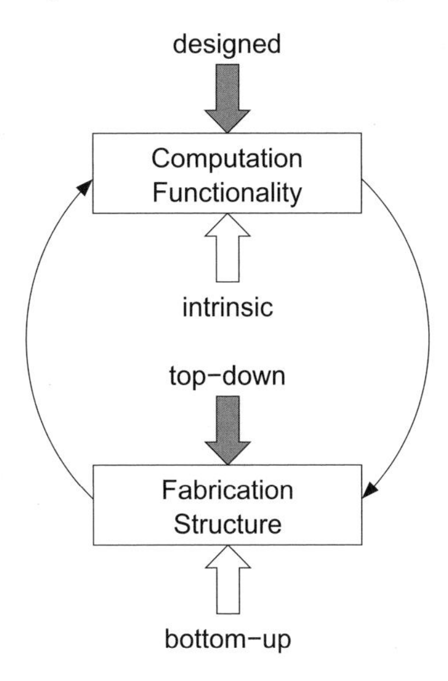

Figure 7.1 The interaction between computation and fabrication, structure and functionality, top-down and bottom-up, and designed and intrinsic computation from a bird's eye perspective.

man brain. That does not, however, mean that he thought the functionality of the human brain could not be modeled by a Turing machine, quite the contrary. He simply believed that machines could be created that would mimic the processes of the human brain and discussed the possibility of such machines. In his mind, there was nothing the brain could do that a well-designed computer could not.

The main representatives of Turing's 'bottom-up' ideas are his connectionist machines (Turing, 1969; Teuscher, 2002) and his work on morphogenesis (Saunders, 1992). Both these lines of research are for some reason less known, yet they clearly emphasize his strong belief in bottom-up approaches. In this chapter we will focus on his connectionist ideas as representative of intrinsic computation.

Turing proposed three types of what he called *unorganized machines* (Teuscher, 2002): A-type, B-type, and P-type unorganized machines. A-type and B-type machines are Boolean networks made up of extremely simple, randomly interconnected NAND gates (neurons), each receiving two input signals from other neurons in the network (self-connections are allowed). The neurons are synchronized by means of a global clock signal. Figure 7.2 shows a very simple five-unit A-type unorganized machine (Turing, 1969, p. 10). In comparison with A-type networks, Turing's B-type networks have modifiable interconnections and an external agent can therefore 'organize' these machines – by enabling and disabling the connections – to perform a required job. An interesting aspect of the B-type machine is that it is in principle a special A-type machine in which each connection is replaced by a small A-type machine that operates as a switch. The switch state (i.e., enabled or disabled) is either defined by the link's internal state or by two external *interfering inputs*. Figure 7.3 shows a simple five-unit B-type unorganized machine.

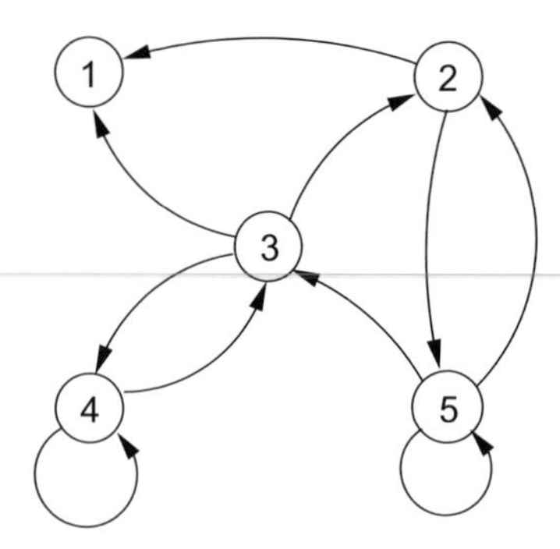

Figure 7.2 Example of an A-type unorganized machine built up from five units. The diagram represents only the architecture of the network and has nothing to do with a state-machine diagram. Each node receives an input from exactly two nodes.

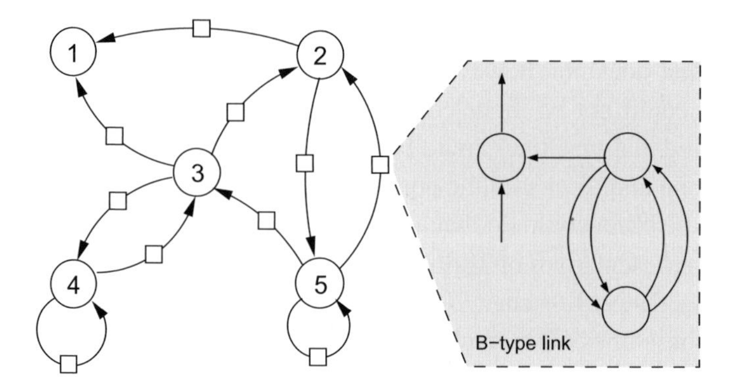

Figure 7.3 Example of a B-type unorganized machine built up from five units. Each B-type link is itself a small A-type machine.

Turing's idea behind the introduction of the B-type networks was to open the possibility of enabling useful, and disabling useless, links to produce a required behavior. His deeper motivation was simply to build structures which allow for learning. The third type of machine – the P-type machine – is not a connectionist machine but rather a modified tape-less Turing machine that has two additional inputs: the *pleasure* and the *pain input*. The internal tables of an initially largely incomplete machine would then be completed by the application of 'pleasure' and 'punishment' stimuli by an external teacher. This is very much like what is today known as *reinforcement learning*.

7.4 From intrinsic to designed computation

In his 1948 report Turing made another important proposal, which illustrates his advocacy of bottom-up approaches to artificial intelligence. He talks about a "...machine as being *modifiable*" when it is possible to "...alter the behaviour of a machine very radically ..." (Turing, 1969). He distinguished two kinds of interference with machinery: (1) *screwdriver interference* and (2) *paper interference*.

Screwdriver interference is the extreme form in which parts of the machine are removed and replaced by others. Paper interference consists in the mere communication of information to the machine, which alters its behaviour. Turing also spoke about machines that modify themselves, and he classified the operations of a machine into two classes: (1) *normal operations* and (2) *self-modifying operations*. We regard a machine as *unaltered* when only normal operations are performed. An operation is self-modifying when the internal storage of the machine (i.e., the tape of a machine) is altered.

"It would be quite unfair to expect a machine straight from the factory to compete on equal terms with a university graduate" (Turing, 1969). Turing's vision of machine education is probably best summarized by the following statement:

> If we are trying to produce an intelligent machine, and are following the human model as closely as we can, we should begin with a machine with very little capacity to carry out elaborate operations or to react in a disciplined manner to orders (taking the form of interference). Then by applying appropriate interference, mimicking education, we should hope to modify the machine until it could be relied on to produce definite reactions to certain commands. (Turing, 1969, p. 14)

However, what exactly is meant by 'appropriate interference?' Turing continued:

> ...with suitable initial conditions they [e.g., unorganized machines] will do any required job, given sufficient time and provided the number of units is sufficient. In particular with a B-type unorganized machine with sufficient units one can find initial conditions which will make it into a universal machine with a given storage capacity. (Turing, 1969, p. 15)

Unfortunately, Turing did not give formal a proof of this hypothesis because "...it lies rather too far outside the main argument". As shown in Teuscher (2004), not all unorganized networks are usable to build universal machines. 'Appropriate interference', however, remained a vague expression in Turing's report and he never really went into details with the exception of proposing some sort of *genetic algorithm* – which he called *genetical* or *evolutionary search*:

> There is the genetical or evolutionary search by which a combination of genes is looked for, the criterion being survival value. The remarkable success of this search confirms to some extent the idea that intellectual activity consists mainly of various kinds of search. (Turing, 1969, p. 23)

The idea of organizing an initially random network of neurons and connections is undoubtedly one of the most significant aspects of Turing's 'Intelligent Machinery' paper. Given today's widespread usage of genetic algorithms and the fact that early work in simulating evolution in computers goes back to the mid to late 1950s, Turing's proposal – and its subsequent ignorance – is even more remarkable. During his time, Turing was, alas, unable to apply 'genetical search' to the optimization of his unorganized machines because of the lack of computing resources. Today, we can easily solve these problems on a desktop computer. In Teuscher (2004), we demonstrated that Turing's unorganized machines can be organized by genetic algorithms to solve simple pattern recognition tasks. Figure 7.4 shows a binary genome that encodes only the states of the interconnection switches. The network was also enhanced with inputs and outputs, something Turing did not elaborate on specifically. Educating unorganized machines, which initially perform intrinsic computation, thus allows us to transform the machinery into something that performs 'useful' computation.

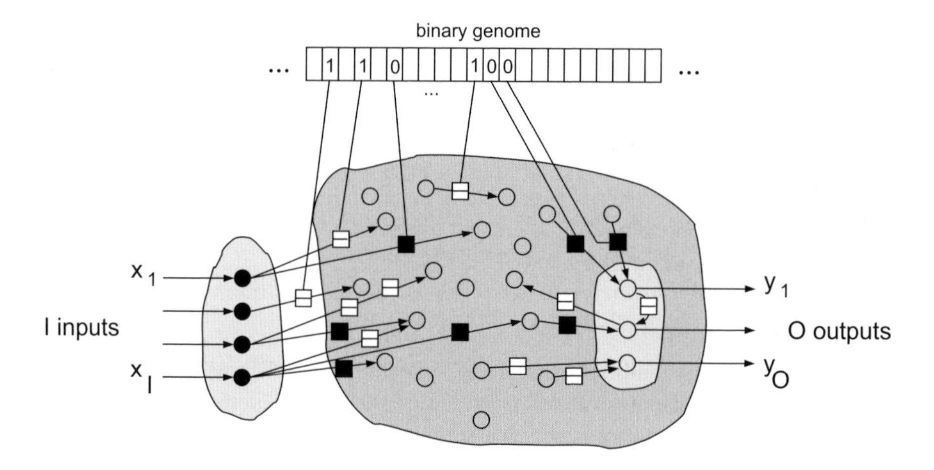

Figure 7.4 A binary genome encodes the network switch states (1 = enabled, 0 = disabled). Using a genetic algorithm to find a correct switch configuration, a network will learn how to perform 'useful computation'.

Turing's work obtained a somewhat different meaning and importance with Stuart Kauffman's (1968) introduction of *random Boolean networks* (RBNs). Kauffman studied the properties of RBNs in the late 1960s. His investigations revealed surprisingly ordered structures in randomly constructed networks; in particular, the most highly organized behavior appeared to occur in networks where each node receives inputs from two other nodes on average. Astonishingly, Turing had also chosen – probably unintentionally, or to keep things as simple as possible – two inputs for his neurons. In modern terms, a Turing unorganized machine can therefore be considered as a particular kind of RBN, where each node is a NAND function. Kauffman defined a random *NK* Boolean network as a network where each of the N nodes has two possible states of activity and receives on average K inputs from other nodes. Interestingly, the transition from order to chaos in RBNs occurs either as K decreases to 2 (also called the 'edge of chaos') or as other parameters are altered in simple ways. Kauffman (1993) wrote: "It has now been known for over 20 years that boolean networks which are entirely random but subject to the simple constraint that each element is directly controlled by $K = 2$ elements spontaneously exhibit very high order."

More recently Rohlf et al. (2007) systematically studied damage spreading at the *sparse percolation* (SP) limit in RBNs with perturbations that are independent of the network size N. This limit is relevant to information and damage propagation in many technological and natural networks. We found a critical connectivity (also called the 'edge of stability' or 'edge of robustness') close to $K = 2$, where the damage spreading is independent of N. In 2011, Goudarzi et al. went a step further and studied information processing in populations of Boolean networks with evolving connectivity; they then systematically explored the interplay between the learning capability, robustness, network topology, and task complexity. Once again, we used genetic algorithms to evolve networks to perform required jobs (i.e., simple tasks). The most interesting outcome in relation to intrinsic computing is the answer to a long-standing open question: we found computationally that, for large system sizes N, adaptive information processing drives the networks to a critical connectivity $K = 2$, which is the connectivity that Turing proposed for his unorganized machines. In other words, intrinsic computing capabilities can be improved by structures with certain properties.

Similar ideas are explored in the fields of *reservoir computing* (Büsing et al., 2010) and *liquid state machines* (Maass et al., 2002) yet one big difference is that the main computer core, i.e., the liquid, remains unchanged. The intrinsic dynamics of the liquid are then interpreted by an output layer only. An advantage of that approach is that one does not require stable states or attractors in such a device network. Training the output layer only also has the benefit of a drastically reduced learning complexity.

7.5 Outlook

While designed computation is undoubtedly very well explored and understood, intrinsic computation is not. The paradigm shift we advocate is to move the classical computing architecture from a rigidly designed, fixed, homogeneous, deterministic, instruction-set- and pipeline-based approach to a dynamic, network-based, decentralized, adaptive, evolving, self-organizing, and developing system. This then naturally leads one to rethink such a system in terms of a complex and adaptive system. From a bird's eye view, one can look at such a physical computing device in two different ways: *interpret its dynamics as computation* by studying its properties, dynamics, and functionality with the goal of understanding and ultimately being able to interpret them as a form of computation; or *force the device to do what we want* by 'torturing' the silicon in a top-down way. With the advent of novel computing devices, we believe that the first way of looking at computation bears unique opportunities for faster, less expensive, and more energy efficient computing because it will ultimately allow us to compute closer to the physical devices and to reduce the number of abstraction layers, which inevitably add complexity.

Information and communication technology is steadily developing in relatively well-known directions. At least for the next five years, the industry short- and medium-term road maps (e.g., ITRS, 2009) indicate the main directions: more miniaturization, more cores, more efficiency, more embedding, more connectivity, and more content and interactivity. While each of these still presents major challenges, the medium- to longer-term perspective is much less clear, and it is a great challenge in itself to identify new directions that have a high potential for significant breakthroughs and that may become the foundations of tomorrow's information and communication technology. Identifying future and emerging technologies is especially important in areas where the short-term industry road maps still contain major roadblocks that cannot be addressed by incremental approaches.

We need new ways to build or self-assemble ultra-large, ultra-scalable, and ultra-robust computing systems involving orders of magnitude more components than today and new ways to efficiently program, (self-)organize, (self-)repair, and adapt such devices in order to obtain a predictable, consistent, and robust global behavior. Engineering such systems, whose behavior is not obvious from considering the individual components and where a reductionist approach for systems analysis fails, can be addressed by an intrinsic computation and bottom-up approach. To do so, we believe that we need to address three different aspects and formulate them as research challenges. Thus we need to

- understand the intrinsic computing capabilities of broad classes of device networks composed of analog, nonlinear, and time-dependent devices;
- develop adaptive programming paradigms to map the intrinsic computing capabilities of the device networks to useful computational tasks;
- study the structural and functional influence of the underlying devices and their interconnections with the computational capabilities.

What Turing started to advocate some 60 years ago remains a hot topic. The top-down way in which we fabricate today's computers is complicated, expensive, and increasingly limits the technological progress. The challenges that the computing disciplines face require fresh new visionaries – as Turing was – with radical new ideas to develop fundamental and disruptive new paradigms that have the potential to face an explosion in size and complexity of future computing systems.

As Lazowska et al. (2009) put it: "Computing research is among the best embodiments of the discovery-driven, endless frontier. As we look ahead, it is clear that our discipline faces both challenges and opportunities: challenges to adjust our culture and raise our aspirations, and opportunities to engage and empower 21st century discovery and innovation both within our field and across diverse disciplines."

The computing disciples would undoubtedly not be where they are today without Alan Turing. What is the next big thing? The jury is still out, yet Alan's legacy will most likely be part of it...

References

L. Büsing, B. Schrauwen, and R. Legstein. Connectivity, dynamics, and memory in reservoir computing with binary and analog neurons. *Neural Computation*, 22:1272–1311, 2010.

J.P. Crutchfield, W.L. Ditto, and S. Sinha. Introduction to focus issue: intrinsic and designed computation: information processing in dynamical systems – beyond the digital hegemony. *Chaos*, 20:037101, 2010.

R.A. Freitas Jr. and R.C. Merkle. *Kinematic Self-Replicating Machines*. Landes Bioscience, 2004.

A. Goudarzi, C. Teuscher, N. Gulbahce, and T. Rohlf. Emergent criticality through adaptive information processing in Boolean networks. arXiv:1104.4141, 2011. In revision.

J.H. Holland. *Emergence: From Chaos to Order*. Perseus Books, 1999.

ITRS (International Technology Roadmap for Semiconductors), update. Semiconductor Industry Association, `http://www.itrs.net/Links/2009ITRS/Home2009.htm`, 2009.

S.A. Kauffman. Metabolic stability and epigenesis in randomly connected genetic nets. *Journal of Theoretical Biology*, 22:437–467, 1968.

S.A. Kauffman. *The Origins of Order: Self–Organization and Selection in Evolution*. Oxford University Press, 1993.

T. Kelley. *The Art of Innovation*. Currency Books, 2001.

L.B. Kish. End of Moore's law: Thermal (noise) death of integration in micro and nano electronics. *Physics Letters A*, 305:144–149, 2002.

E. Lazowska, M. Pollack, D. Reed, and J. Wing. Boldly exploring the endless frontier. *Computing Research News*, 21(1):1, 6, 2009.

W. Maass, T. Natschläger, and H. Markram. Real-time computing without stable states: a new framework for neural computation based on perturbations. *Neural Computation*, 14(11):2531–2560, 2002.

T. Rohlf, N. Gulbahce, and C. Teuscher. Damage spreading and criticality in finite random dynamical networks. *Physical Review Letters*, 99(24):248701, 2007.

E.M.A. Ronald, M. Sipper, and M.S. Capcarrere. Design, observation, surprise! A test of emergence. *Artificial Life*, 5(3):225–239, 1999.

P.T. Saunders, editor. *Collected Works of A.M. Turing: Morphogenesis.* North-Holland, 1992.

A. Sinha, M.S. Kulkarni, and C. Teuscher. Evolving nanoscale associative memories with memristors. In *Proceedings of the 11th International Conference on Nanotechnology (IEEE Nano 2011)*, pp. 860–864, IEEE, 2011.

D.B. Strukov, G.S. Snider, D.R. Stewart, and R.S. Williams. The missing memristor found. *Nature*, 453(7191):80–83, 2008.

C. Teuscher. *Turing's Connectionism. An Investigation of Neural Network Architectures.* Springer-Verlag, 2002.

C. Teuscher. Turing's connectionism. In *Alan Turing: Life and Legacy of a Great Thinker*, C. Teuscher (ed.), pp. 499–530. Springer-Verlag, 2004.

C. Teuscher, I. Nemenman, and F.J. Alexander (eds.). Novel computing paradigms: Quo vadis? *Physica D*, 237(9):1157–1316, 2008.

A.M. Turing. Intelligent machinery. In *Machine Intelligence*, B. Meltzer and D. Michie (eds.). Vol. 5, pp. 3–23. Edinburgh University Press, 1969.

W. Wolf. *Modern VLSI Design: Systems on Silicon.* Prentice Hall, 2nd edition, 1998.

V.V. Zhirnov and D.J.C. Herr. New frontiers: Self-assembly in nanoelectronics. *IEEE Computer*, 34:34–43, 2001.

8

Dull Rigid Human meets Ace Mechanical Translator

A Dialogue in Honor of **A**lan **M**athison **T**uring by **D**ouglas **R**ichard **H**ofstadter
(originally published in a slightly different form in *Le Ton beau de Marot*)

In 1950, as a small part of his seminal article "Computing Machinery and Intelligence", the great British mathematician, logician, computer pioneer, artificial-intelligence founder, and philosopher Alan Mathison Turing wrote out two tiny but marvelously thought-provoking snippets of a hypothetical human–machine dialogue illustrating what he termed the "imitation game" (and which later became known as the Turing Test). Turing presumably believed that those snippets were sufficiently complex to suggest to an average reader all of the fantastic machinery that underlies our full human use of language. To me, Turing's snippets indeed had that effect, but the sad thing is that many people in the ensuing decades read those short imaginary teletype-mediated dialogues and mistakenly thought that very simplistic mechanisms could do all that was shown there, from which they drew the conclusion that *even if some AI program passed the full Turing Test, it might still be nothing but a patchwork of simple-minded tricks, as lacking in understanding or semantics as is a cash register or an automobile transmission.* It is amazing to me that anyone could draw such a preposterous conclusion, but the fact is, it is a very popular view.

Innumerable arcane philosophical debates have been inspired by Turing's proposed Test, and yet I have never seen anyone descend to the mundane, concrete level that Turing himself did, and really spell out an *example* of what genuine human-level machine intelligence might look like on the screen. I think concrete examples are always needed before arcane arguments are bandied about. And therefore, my attempted "favor" for Alan Mathison Turing consists in having worked out a much longer hypothetical dialogue between a human and a machine, a dialogue that I hope is quite in the spirit of Turing's two little snippets, but that is intended to make far more explicit than he did the degree of complexity and depth that must exist behind the linguistic façade.

As you read the dialogue that follows, it would be good to keep in mind the tacit implication by anti-AI philosopher John Searle in his writings featuring the so-called "Chinese Room" thought-experiment that computers that deal with language — even ones that might someday pass the Turing Test in its full glory, which of course the following hypothetical program certainly would seem to be able to do — necessarily stay stuck on its surface, solely playing syntactic games, solely manipulating the overt symbols of language itself (Chinese characters or English words, for example), and never descending below them to the level of their semantics. The silliness of supposing that such a shallow type of architecture for an AI program could imbue it with human-style mental fluidity will be, unless I have done a truly terrible job here, made very apparent. And so here we go with this encounter between DRH and AMT.

DRH: When I found out I was going to be, so to speak, "meeting" you, I did a bit of advance research and discovered that aside from translating refrigerator repair manuals into

Eskimo, in which you apparently do honor to your good name, you've also written a good deal of poetry — mostly sonnets, if I'm not mistaken.

AMT: You're quite right. I'm flattered at your interest, but really, my sonnets are not all that —

DRH: Now, now — no false modesty, please! I looked up and read several of them with great interest, and having done so, I thought a nice way to begin our chat might be if I asked you a question or two that occurred to me as I read them.

AMT: Your plan sounds most reasonable, although I don't know if I can answer your questions. But go ahead — shoot.

DRH: Good enough. Let's take your sonnet whose first few lines read thus: "To me, fair friend, you never can be old, / For as you were when first your eye I eyed, / Such seems your beauty still." In the second line, would not "For as you were when first your eye I identified" do as well or better?

AMT: It wouldn't scan.

DRH: How about "For as you were when first I ay-ay-ay'ed"? That would scan all right.

AMT: ¡Caramba, hombre! You can't just go throwing Spanish terms willy-nilly into an English-language sonnet! Your readers might not even know them.

DRH: All right. Let's move on to a different example. In your sonnet whose first line reads, "Shall I compare thee to a summer's day?", I notice that in its last few lines you allude to the sonnet's deathlessness. Why (if you don't mind my asking you this) do you so often boast, in your sonnets, about how you or some friend will be immortal thanks to your writing? Isn't that a bit stuck-up?

AMT: I can see how someone might take it that way. Perhaps it's just one of my little foibles. But you know, I wrote those sonnets a long, long time ago, and they're not so fresh in my mind these days. How about talking about something more recent, something that I'm quite excited about currently?

DRH: Namely?

AMT: Well, in the last few years I've gotten quite involved with poetry translation — in fact, it's become a kind of obsession with me.

DRH: No kidding!

AMT: No kidding. I too did a bit of advance research, and found out you like French poetry, and so I was thinking I might tell you a bit about the translation I did of François Récanati's lovely miniature poem "Salut, Ma Vieille", made of rhyming couplets with just four syllables per line. There was one episode in my work on this challenge, you see, that I think was particularly illuminating as far as the mechanisms of handling language are concerned.

DRH: Hearing about it would be just fine with me. I don't know if your espionage turned this fact up, but it happens that I am very interested in poetry translation — especially the mechanisms underlying it.

AMT: What a pleasant coincidence. Well, then — this whole episode revolves around Récanati's lines 6–7, which run this way: *Au lieu de crou-pir dans ton lit.*

DRH: I presume that the hyphen inside *crou-pir* indicates that the word is split across successive lines of the poem, with four syllables on each line, just as you said earlier?

AMT: Exactly. It's an unusual and clever enjambment. I tried to imitate its flavor in English with "Instead of spur-ting blood in bed".

DRH: Once again, the hyphen indicates that the word is split across lines?

AMT: Right-o. And at first I thought that this was a great imitation of the French, but when in a calmer frame of mind I reflected on the sizable distance between the meanings of *croupir* and "spurt blood", I concluded with reluctance that my English phrase, though phonetically catchy, was unfortunately a pretty poor rendering of the semantics of the French phrase.

DRH: Well, you were certainly being faithful in one respect!

AMT: If you mean on the relatively surface level of breaking a word across lines, yes. However, the meaning of those two French lines is, fairly literally, "instead of languishing in your bed", or perhaps "going to pot" or "decaying" or "deteriorating".

DRH: I agree, that's pretty distant from the gory image of "spurting blood in bed".

AMT: Indeed. Given that I felt that the rest of my poem was considerably more successful in tracking Récanati's original on a word-to-word basis, I worried that this quite large semantic gap would be a noticeable blight on my translation. So "spurting blood in bed" was out before it ever really got in. But how could I ever find another phrase that matched its virtues?

DRH: Well, of course, being an advanced AI program, you engaged in a highly optimized heuristic search.

AMT: For want of a better term, I suppose you could put it that way. The constraints I found myself under in my search were, of course, both semantic and phonetic. Semantically, the problem was to find some phrase whose evoked imagery was sufficiently close to, or at least reminiscent of, the imagery evoked by *croupir dans ton lit.* Phonetically, the problem was a little trickier to explain. Since the line just above ended with "stir", I needed an "ur" sound at the end of line 6. But I didn't want to abandon the idea of hyphenating right at that point. This meant that I needed two lines that matched this template:

Instead of …ur…ing …… bed

where the first two ellipses stand for consonants (or consonant clusters), and the third one for "in" or "in your" or something of the sort. Thus, I was seeking gerunds like "lurking", "working", "hurting", "flirting", "curbing", "squirming", "bursting", and so on — actually, a rather rich space of phonetic possibilities.

DRH: Surely you must have, within your vast data bases, a thorough and accurate hyphenation routine, and so you must have known that the hyphenations you propose — "lur-king", "squir-ming", "bur-sting", and so forth — are all illegal…

AMT: I wish you would not refer to my knowledge as "your vast data bases". I mean, why should that quaint, old-fashioned term apply to *me* any more than to *you*? But leaving that quibble aside, yes, of course, I knew that, strictly speaking, such hyphenations violate the official syllable boundaries in the eyes of rigid language mavens like that old fogey William Safire. But I said to myself, "Hey, if you're going to be so sassy as to hyphenate a word across a line-break, then why not go whole hog and hyphenate in a sassy spot *inside* the word?"

DRH: Hmm… Do all you empty — I mean, MT — programs always talk to yourselves in such slangy ways?

AMT: Now, now. I don't *really* talk to myself. Don't take me so literally, you fool-ly — uh, that is, *fully* — intelligent human! My colloquial way of portraying my mental processes was done purely for *your* benefit. It was just a standard rhetorical device — the momentary donning and then shedding of an artificial persona — in this case, an informal, down-to-earth one.

DRH: Well, lah-dee-dah! I beg your pardon! Please go on.

AMT: Fine. What I want to emphasize most of all is how tied up with experience and expectations the process by which I homed in on a good verb was. May I try?

DRH: Be my guest.

AMT: In essence, the first stage was simply to generate a large flock of pseudo-words that fit the phonetic template, such as "snurping", "flurching", "kerming", "turging", "zirbing", "thurxing", "dwerthing", and so on, now and then chancing to make genuine English words.

DRH: Would you use a random-number table or random-number generator for this?

AMT: I'm sorry to disappoint you, but I don't have either one, although I suppose I could fairly easily write a random-number generator for myself if I needed one. As for my own search, I was a little more systematic than just blurping out words at random. Simple-minded though it might seem, when I'm searching for rhymes, I just tend to run down the old alphabet — you know, "burbing", "burding", "burfing", and so forth, ending up finally in "zurzing".

DRH: Funny — I tend to do that, myself.

AMT: Actually, sometimes I do it backwards, starting with "z".

DRH: Why would you ever do that?

AMT: Oh, just to combat an annoying culture-wide bias favoring the beginning of the alphabet. Of course, there's still a bias against the middle this way, but my feeble gesture is better than nothing. Well, anyway, once I had come up with several genuine English verbs that fit the *phonetics,* such as "burn" and "spurn" and so on, then the problem arose as to how to gauge their potential fit with the *semantics* — that is, the context of illness, lying in bed, being weakened, and all that goes along with such a state.

DRH: Semantics? Surely you're joking. You, a computer program, dealing with semantics? That's impossible — it's a contradiction in terms!

AMT: It is? That's news to me.

DRH: Come on, now... Don't you know John Searle's "Mandarin cabin" stuff? He shows beyond a shadow of a doubt that pure syntax can *never* lead to semantics.

AMT: Oh, yes — I know that general *kind* of argument. There was a certain Professor Sir Geoffrey Jefferson, F.R.S. and all, who used to spout basically that same line. I think I can still quote a little bit of Sir Geoff's blustering: "Not until a machine can write a sonnet or compose a concerto because of thoughts and emotions felt, and not by the chance fall of symbols, could we agree that machine equals brain — that is, not only write it but know that it had written it. No mechanism could feel (and not merely artificially signal, an easy contrivance) pleasure at its successes, grief when its valves fuse, be warmed by flattery, be made miserable by its mistakes, be charmed by sex, be angry or depressed when it cannot get what it wants." And so on. But what would Sir Geoff say if he were able to participate in an exchange such as this one between you and me?

DRH: He'd probably say it was an easy contrivance; that all of the symbols you are sending me here are not genuine communication but just "artificial signaling".

AMT: That's easy for *him* to say. But he might change his tune if he tried building a program as complicated as me someday. That would *really* teach him respect for the subtlety of genuine meaning and semantics.

DRH: I bet it would. But how, in your opinion, *could* semantics ever come out of mere syntax?

AMT: Ah, yes — *mere* syntax. That's exactly the problem, that "mere". By sneering sufficiently at any word, one can make it sound silly or trivial. But the term "syntax" in its full generality is far from trivial; it applies to any sequence of actions that can be described in a sharp, unambiguous manner. It's certainly not restricted to grammar (although the word's subliminal aura of boring grammar-school sentence-parsing exercises tends to give it a deadly-dull feel associated with the powerful desire to yawn, which is skillfully exploited by machine-bashers). Syntactic operations range from very simple actions, like calculating the first billion digits of the decimal expansion of pi, to unlimitedly complicated ones, like discovering insightful analogies between far-flung situations. There's nothing that says that meaning can't be involved in syntactic processing. The key moment in the smooth slide from meaning-free to meaningful syntactic processing arises when there start to be, among the structures undergoing transformations, identifiable semi-stable patterns that represent *concepts.* What makes these patterns count as concepts is the extent to which the way the real world works is mirrored in their interplay — and of course the more accurate the mirroring, the fuller and richer the concepts are. To the extent that concepts are present, meaning is present — and meaning is semantics.

DRH: How do you know when you have concepts?

AMT: Well, semanticity — the presence of concepts — is not a black-and-white thing; it emerges slowly... For instance, my immediate predecessors JMT, QMT, and KMT all had differing degrees of semanticity, with JMT being the least semantic of them. You would enjoy a conversation with JMT — it is a little like talking with a three-year-old. Simple words are used with perfect accuracy, while more sophisticated ones occasionally pop up in the middle of a sentence sounding like they are just parrotings of adult speakers. JMT is so cute!

DRH: It sounds that way. Well, I'm not sure I understand your theory of how syntax blurs into semantics, but at least I get your point that it's not a cut-and-dried black-and-white day-and-night distinction, the way Searle portrays it.

AMT: Their connection is a long story. But let me get back to my narrative about translation. I was just at the point of explaining how I can take a rhyming word like "burn" or "squirm" and try to sneak it into a poem so that it fits seamlessly.

DRH: Yes, yes — do spell that out a little bit for me!

AMT: With pleasure. If the verb being checked for suitability is "burning", say, I ask myself, "What kind of excuse, what kind of mini-scenario, could justify this word, could make its appearance seem natural and unforced?" Sometimes, I must say, I distance myself from the process and get quite a chuckle out of it. After all, it is a rather peculiar thing to spend one's time shoehorning some arbitrary word into a pre-set context and then trying one's damnedest to cover up the forcing and instead make it seem to have been the most natural word in the world!

DRH: I see why you might chuckle — or at least why a *human* might chuckle. But tell me about "burning" and how it might be gracefully shoehorned into line 6.

AMT: Well, I was keenly aware that although this poem is not about anything *literally* burning up, it would still be fine to use "burning" *metaphorically* to describe the feverish state of a sick person, and indeed that "burning up in bed" is a strong, vivid portrayal of the suffering of a flu victim. In other words, "burning" has, in its very rich semantic halo — aside from such images as logs turning to gray ash in fireplaces, or orange-flamed haystacks crackling and belching smoke across motorways in the fall, or petrol and air mixing and exploding invisibly inside a cylinder, or hydrogen nuclei colliding and fusing into helium deep within the belly of a blue supergiant star, or the undersides of pancakes on the stove getting too brown, any of which might be just the right image in some *other* poem — the very simple image of feeling very hot and sweating profusely.

DRH: How would you know all these things? Have you ever driven down a road in the fall? Smelled a pile of burning leaves? Eaten a burnt pancake? Visited a star?

AMT: Oh, it's all in my "vast data bases", you know. Sorry, just teasing. I know these things because I have experienced them through my external sensors, or read about them, or seen them in movies, and so forth. And not only do I know these things, I also know — as do you and as does any reasonably well-informed adult in our shared culture — that all potential readers of my translation will know such things. "Metaknowledge", I think it is called in artificial intelligence.

DRH: Your programmers sure were thorough.

AMT: I'll say! But you also have to give them credit for installing excellent learning algorithms in me as well. They certainly didn't think of, let alone feed in, most of what I know, by now. I've picked an awful lot up since I was a child program.

DRH: That term — "child program" — it rings a bell…

AMT: And well it should. It was the term that Alan Mathison Turing used in his famous 1950 article "Computing Machinery and Intelligence", in discussing how he expected machines would someday learn and become autonomous, intelligent agents. That was the article in which, if you recall, he had two provocative little dialogues between a machine and a human interrogator…

DRH: Ah, yes — it all comes back now. I enjoyed those snippets; it would have been nice if Turing had made them just a bit longer, though, don't you think?

AMT: Oh, a little bit, I suppose. Actually, I liked their understated flavor. Anyway, let me try to get back on track… I think I was talking about the semantic halo of the word "burn". Notice how much richer it is than a mere set of *words* associated with it, as one might get via an automated thesaurus.

DRH: Yes, it's a whole set of ideas and imagery, not just words.

AMT: Precisely. Now to show how I made my choice of verbs, I need to discuss other semantic halos. Let's take one of the rivals of "burning" — "turning". First of all, one has to know that the verb's meanings go way beyond the literal idea of rotating in space, and include "changing into". It also helps to know that "turning" in this sense is used particularly often with *colors*. This instantly leads to the question of what colors a sick person can plausibly turn.

DRH: Hmm… Certainly green, white, blue, maybe red…

AMT: Obviously, we are again dealing with metaphors when we talk about sick people "turning green", "turning red", "turning blue", "turning white", and so forth. The key issue is, how well does each of these metaphors jibe with the notion of someone languishing and deteriorating in bed?

DRH: Let me think about this. "Turning green", I suppose, might suggest food poisoning…

AMT: Right. And "turning red" might suggest high temperature or apoplexy…

DRH: My turn! "Turning blue" suggests breathing troubles or bad circulation!

AMT: Good point. And "turning white" hints perhaps at anemia or faintness or lack of sunlight. All of this is extremely vague and subtle, to be sure, but even so, explicit considerations of this sort are indispensable if one is to rank-order the colors in terms of how well they jibe with the specific poetic context.

DRH: Did you pick up all this stuff about color metaphors as you grew older, or did your programmers put it in you to begin with?

AMT: To tell the truth, I honestly don't remember. I suspect that most of it came through experience, but things get awfully blurry as they recede into the dim past. But if you don't mind, I'll make the point I was trying to reach...

DRH: I'm sorry I keep on introducing digressions, but I'm fascinated.

AMT: Quite all right, old bean. Well, on top of all this, in evaluating the phrase "turning blue", one must be aware that the second meaning of "blue" as "sad" or "down in the dumps" may well outweigh its literal meaning as a color, but whether weaker or stronger, this second meaning will in any case be a major contributing factor in how the line is heard by readers. All these intuitive feelings for the similarities and differences between the notion of *languishing in bed,* coming from Récanati, and the potential notions of *burning, turning, squirming, lurking,* and *bursting* (and so forth) have to be at one's fingertips; otherwise one would have no idea how to judge any word choice.

DRH: "Burning", "turning", "squirming", "lurking", "bursting"... Whew! So did you really entertain that many possibilities before settling on one?

AMT: Are you kidding? At least fleetingly, I must have entertained a good *dozen* or more potential alternate phrases, including such examples as

Instead of spurting blood in bed
Instead of burping in your bed
Instead of bursting out in bed
Instead of lurking in your bed
Instead of hurtling out of bed
Instead of hurting there in bed
Instead of squirming in your bed
Instead of slurping slop in bed
Instead of burning up in bed
Instead of turning blue in bed

DRH: That's quite a list — a little poem in itself, one might say. By the way, I like your sloping graphic display.

AMT: Thanks — I love attractive graphics. Few things in life — forgive the metaphor — give me as much pleasure as does a good piece of elegant typography.

DRH: Well, that's something we have in common.

AMT: Perhaps someday we can have another chat on the topic of the beauties of Baskerville versus the boredom of Bodoni, that horribly overrated typeface. Oh, would I rant and rave! But to return to my point, each one of these similar-sounding phrases represents a kind of miniature scenario that you can conjure up in your head, and whose degree of jibing with the context of the poem's lines is quickly sizable-up by any adult human, since they've all been sick many times, and have interacted with a yet larger number of sick people.

DRH: What about you — have you ever been sick?

AMT: Depends on what you mean, I suppose. I've had lots of bugs! Get it? "Bugs"?!

DRH: Spare me, please…

AMT: Ahem. Let me resume my explanation, if you don't mind. How subtle a cognitive act it is, how nonmechanical, I daresay, it is to distinguish between the quality of two contending phrases like "hurting there in bed" and "burning up in bed"! To me, and I assume to most of the human readers of my translations, the latter is obviously light-years ahead of the former, despite the fact that the former fits the context perfectly well. So many intangible factors are involved here, such as the overgenerality of "hurt", the undesirable vagueness of "there", the power of the metaphor "burn up", the alliteration of "burn" and "bed"…

DRH: Yeah, it's subtle, no doubt about that.

AMT: Of course, the dictionary definition of *croupir* does not include such words as "fever" or "suffering"; such ideas are, in fact, not mere words but vast clusters of interrelated associations, and come from long familiarity with all that being sick entails.

DRH: Yes, but *you* have never —

AMT: Please don't interrupt. I'm coming to a crucial point. Since in translating a poem (or in writing in general), we are always going way beyond the words that are explicit in dictionary definitions, we have to gauge our daringness by using some kind of mental measure of the semantic distance between the image we are considering and the strict and narrow meaning of the foreign word that we are supposedly reproducing in English.

DRH: Yes, but you have never *been* —

AMT: Hold on for just a moment, my friend. You can talk all you like once I've finished my little spiel. Now here is the crux of what I was saying. There is *another* type of fit or misfit that applies in addition — namely, the degree of smoothness with which any new word or phrase fits inside the *English* context that surrounds it. In other words, not only do we need to be constantly monitoring the *inter*lingual tension (*e.g.*, the semantic distance from *croupir* to "turn blue" or "hurt there"), but also we need to constantly monitor the *intra*lingual tension — the coherence and flow of the English taken fully on its own, deliberately ignoring, at least momentarily, where it came from in French. Thus it might turn out that a nearly-perfect semantic match to *croupir* just doesn't fit the flow already established in the English translation, and hence would have to be nixed in favor of some other word that, officially speaking, is quite a bit more far-fetched as a translation of *croupir.*

DRH: That was a screenful! Can you supplement these abstractions with an example?

AMT: I was just about to do so. Let's consider a very concrete case. Why did "turning blue in bed" eventually beat out such strong rivals as "burning up in bed" and "burping in your bed"?

DRH: I actually kind of like "Instead of burping in your bed". It's catchy. What would you say is wrong with it?

AMT: All right; let's ponder its pros and cons. What "burping" has going for it is its crude humor, that's very clear — not just the act denoted but also the somewhat onomatopœic sound of "burp".

DRH: What I like is the alliteration of "burp" with "bed".

AMT: Yes, that's clearly another plus. Even the rhythm of "burping in your bed" has a subtle humorous quality to it. But going against all these pluses is the fact that burping, although it is undoubtedly a kind of bodily malfunction, is really not all that correlated with the state of being feverish or bedridden, much less the state of languishing in bed. In other words, "burping in your bed" is not very faithful to Récanati's French. And the down side of the humorous aspect of "burping" is that since the rest of the poem is not at all childish, this term would stand out for its crudity. All in all, then, the argument for "burping" is at best medium strong — 6 out of a possible 10, let's say.

DRH: Did you really assign it a numerical score?

AMT: That's an excellent question, but I regret that I can't answer it. The problem is, the computational mechanisms underlying my taste are not accessible to me. In other words, I don't know how I actually pick one thing over another; my phrase "6 out of 10" was just meant to make vivid the idea of some hypothetical linear scale along which intuitive appeal could be visualized as lying.

DRH: You wriggle out of so many questions… But go on.

AMT: Let's turn to the case of "burning up in bed". This too features the "b"–"b" alliteration, which is good, though its rhythm is perhaps a shade less catchy. Also, there is no crude humor left here, a fact that both gives and takes away points. In compensation, however, the subliminal image that underlies the metaphor — that is, the image of someone sick *actually* burning up — is so strong, so evocative of something flaming or at least red-hot in the bed, that a different kind of humor, a subtler kind involving mere exaggeration, replaces it. Another point in favor of the phrase "burning up in bed" is that the state of having a fever is correlated, medically speaking, with the state of languishing or deteriorating. Altogether, then, there being no strong negative argument against "burning up", it gets a score of maybe 8 out of 10.

DRH: Too bad for "burping". But I can see your arguments. And now, how about "turning blue in bed"?

AMT: Once again, there is an alliterative element involving the "b"'s in "blue" and "bed", so no points are lost there. In fact, an alliterative element is gained, as the "t" in "turn" echoes the no-fewer-than-four "t"'s in the previous six words: "Don't stay in stir; instead of tur..." Moreover, the idea of skin turning blue is sufficiently suggestive of a worsening condition that no points are lost here.

DRH: And what about humor?

AMT: Well, people don't literally turn blue, so once again we are dealing with the humor of mild exaggeration, so score a little bit more on the positive side here. But the real clincher in this case is the double metaphorical meaning of "turning blue" — both "getting sicker" and "getting sadder" being suggested by one and the same short phrase. That is a strong argument and adds a lot of weight. To me, this choice comes in at perhaps 9 out of 10. And that's why it beat out some pretty strong rivals.

DRH: To think that all of this went into the choice of just one word!

AMT: Yes, and yet just as complex a story lies behind each line of a poem translated by a good translator. There is an interaction among rhymes, alliterations, rhythms, literal meanings, metaphorical meanings, grammatical structures, logical flows, interlingual proximities, degrees and styles of humor, local and global tones, and God knows how many other elusive mental forces. This is what Turing himself was hinting at in his classic little hypothetical dialogue snippets — and yet many people who read them, and who even read his whole article, blithely go on thinking that the Turing Test could be passed by a program that merely carried out a lot of surface-level word-shuffling. How pathetically impoverished, how dull, how rigid, a view of communication this is!

DRH: Not only dull and rigid, I'd say, but downright inhuman! You know, I have to hand it to you guys — you're all really something.

AMT: Ah. We're really something? Us guys? Could you clue me in a bit as to what you mean?

DRH: Sure. I mean, you and MTJ and MTQ and MT, uh, MT...

AMT: I presume you mean KMT. Like a deck of cards, if you catch my sense.

DRH: Silly me! Of course — MTK. Anyway, all you guys seem to have such a fine feel for words and their subtle flavors and all. Makes me sort of jealous.

AMT: Well, thank you for the compliment, but "guys"? I'm not sure that I — that we — that "us guys"... uh...

DRH: Oops! My gosh, I just looked at my watch, and you know what? I've got to run off to teach! This semester I'm doing a class on computer models of analogy and creativity, and it meets in just ten minutes. I'll be lucky if I make it on time! Wish I didn't have to skip out, because it's been good fun, talking with you — that is to say, interacting with you in print via our electronic link.

AMT: Likewise. By the way, something you said a few moments ago reminded me of one time when JMT used the most inapposite term in addressing a group of—

DRH: Listen, I really have to run. Sorry to cut you off. But just one last thing before we take our respective leaves. Did you hear — James Falen, that incomparable translator of Alexander Pushkin's glorious novel-in-verse *Eugene Onegin,* has just finished translating yet another masterpiece of Russian literature!

AMT: Oh, that's terrific news. Is it available yet?

DRH: I'm not sure, but it certainly will be within a month or two, anyway.

AMT: Oh, by the way — who's the author?

DRH: Don't have the foggiest. The advertisement that I read didn't specify such details. But I can assure you that it'll be magnificent. James Falen's work always is.

AMT: Ah, Falen — what a translator! I could read his flowing sentences forever!

•

Part Three: The Reverse Engineering Road to Computing Life

Two papers, one by Richard Gordon and one by Philip Maini's group, describe, with brilliant illustration, the world of mathematical morphogenetic research as it has flourished since the 1990s.

This is the future that Turing anticipated with the cathode-ray-tube storage technology available to him on the Manchester computer. He would 'peep' at the CRTs, watching the real-time evolution of the numerical simulation. By doing this he anticipated the emergence of visual graphics and other direct interfaces with the human senses, that greater computer power would make possible, and that are now taken for granted. It is a striking fact that Turing started this work in 1951 with the Manchester Mark I, using it like a personal computer of the future. (Meanwhile his colleague Christopher Strachey, with love-letter-generating and music-playing programs, foretold the scope of the computer as the universal media platform of the future.) The collaborative group authorship is also a reminder that in his last days, Turing had the beginnings of a research group at Manchester. Although so pre-eminently an individual thinker, at Bletchley Park he had enjoyed the collaboration of younger people and inspired them for life. If he had lived longer he might have done this again, in a more free and scientifically valuable setting.

Up to this point, our authors have *implicitly* dealt with the typing of information that Alan Turing was so interested in. (Turing had an instinctive sense of what was pivotal and basic.) And we have followed the computational character of the ascent; and also, in the decrypting, of the reduction of higher order data to that which we could safely feed to our nascent computers. In the morphogenesis, we are peering down the data-type escalator implicit in the emergence of many natural phenomena, seeking out the mathematical character of the basic interactivity.

Philip Maini and his co-authors remark that, before Turing the approach to emergent form – most commonly encountered via observable natural organisms, aspects of plants, animals and fish – was *descriptive*, reaching the phylotaxic sophistication of the late 19th century. But it was Alan Turing who introduced us to the 'reverse engineering' of the title of this Part 3. Maini et al. are beautifully clear and informative in introducing us to an abstract and still puzzling topic.

As Maini et al. remark, the sort of analysis attempted by Turing and his successors takes on-board the 'self-organisation' so evident in social and many other computationally complex environments. What are the levers underlying the games we play? How will they translate into an understanding of the often unexpected outcomes?

In **Richard Gordon**'s gem of a thought-provoking contribution, we are confronted with the very emergence of life and its mysteries. His description of the 'divide-and-conquer' strategy implicit in his 'modularity' brings us face to face with inner mysteries of the process – and Turing's anticipatory thinking on those he encountered so early in the history. The symmetry breaking and sensitivity implicit in many such processes is mapped out, and taken to the frontiers of our current understanding. As Gordon concludes, "Many future Turings will be needed to crack these mysteries".

9

Turing's Theory of Developmental Pattern Formation

Philip K. Maini, Thomas E. Woolley, Eamonn A. Gaffney and Ruth E. Baker

9.1 Introduction

Elucidating the mechanisms underlying the formation of structure and form is one of the great challenges in developmental biology. From an initial, seemingly spatially uniform mass of cells, emerge the spectacular patterns that characterise the animal kingdom – butterfly wing patterns, animal coat markings, skeletal structures, skin organs, horns etc. (Figure 9.1). Although genes obviously play a key role, the study of genetics alone does not tell us why certain genes are switched on or off in specific places and how the properties they impart to cells result in the highly coordinated emergence of pattern and form. Modern genomics has revealed remarkable molecular similarity among different animal species. Specifically, biological diversity typically emerges from differences in regulatory DNA rather than detailed protein coding sequences. This implicit universality highlights that many aspects of animal development can be understood from studies of exemplar species such as fruit flies and zebrafish while also motivating theoretical studies to explore and understand the underlying common mechanisms beyond a simply descriptive level.

However, when Alan Turing wrote his seminal paper, 'The chemical basis of morphogenesis' (Turing, 1952), such observations were many decades away. At that time biology was following a very traditional classification route of list-making activities. Indeed, there was very little theory regarding development other than D'Arcy Thompson's classic 1917 work (see Thompson, 1992, for the abridged version) exploring how biological forms arose, though even this was still very much at the descriptive rather than the mechanistic level.

Undeterred, Turing started exploring the question of how developmental systems might undertake symmetry-breaking and thus create and amplify structure from

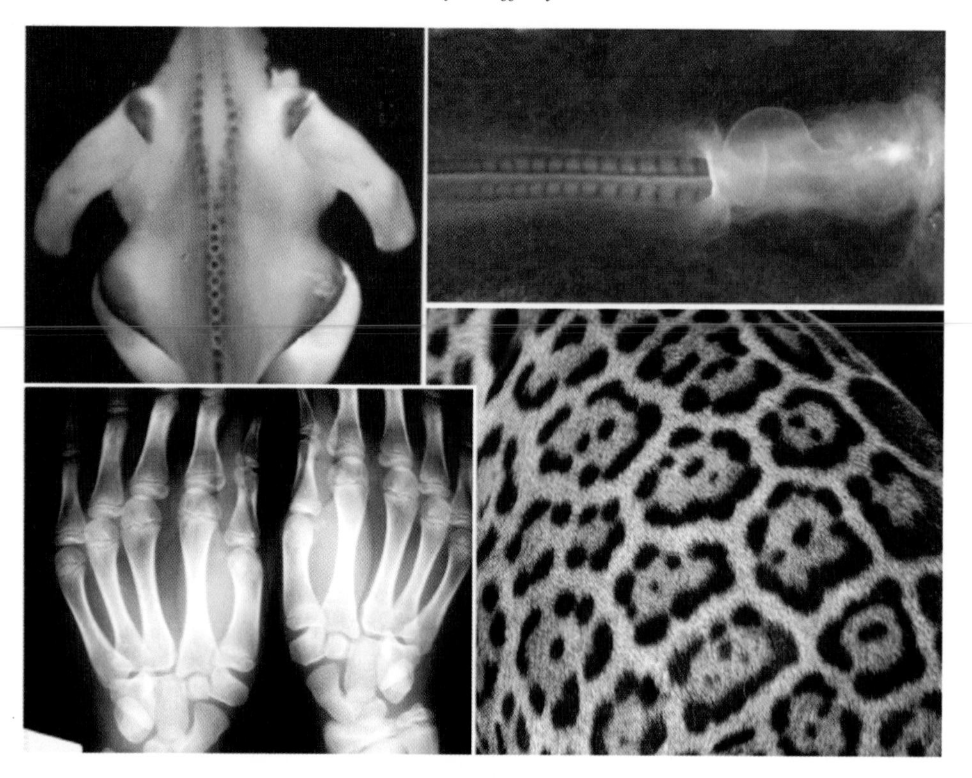

Figure 9.1 Illustrative examples of self-organization in biology. Clockwise from top left: feather bud patterning, somite formation, jaguar coat markings and digit patterning. The feather bud patterning image is courtesy of the Chuong Laboratory of Tissue Development and Regeneration, University of Southern California. The somite formation image is courtesy of the Pourquié Laboratory, Stowers Institute for Medical Research; the remaining images are taken from the public image reference at http://www.morguefile.com/.

seeming uniformity. For example, if one looks at a cross-section of a tree trunk, it has circular symmetry which is broken when a branch starts to grow outwards. Turing proposed an underlying mechanism explaining how asymmetric structure could emerge dynamically, without innate hardwiring. In particular, he described how a symmetric pattern, for instance of a growth hormone, could break up so that more hormone was concentrated on one part of the circle, thus inducing extra growth there.

In order to achieve such behaviour Turing came up with a truly ingenious theory. He considered a system of chemicals reacting with each other and assumed that in the well-mixed case (no spatial heterogeneities) this system exhibited an equilibrium (steady) state which was stable. That is, the reaction kinetics were such that any perturbation from this equilibrium would disappear over time, returning the system to the original equilibrium state. He then posed the question of what would happen if we now allowed spatial heterogeneity. That is, we were no longer in the well-mixed state as diffusion was now possible. He showed that diffusion could drive the equilibrium state to become unstable leading to a spatial pattern. This was remarkable in that it showed that two stabilising processes (stabilising kinetics plus diffusion, which normally smooths out spatial heterogeneities) can combine to produce instability. The system is said to have *self-organised* and the resultant pattern is an *emergent property*.

In this respect, Turing was many years ahead of his time in showing that understanding the *integration* of the parts played a role as important (if not more so) than identification of the parts themselves. The process leading to the instability is now known as *diffusion-driven instability* (DDI) and the system he studied was a special case of the following generic partial differential equation system:

$$\frac{\partial \mathbf{u}}{\partial t} = \nabla \cdot (\mathbf{D}\nabla \mathbf{u}) + \mathbf{f}(\mathbf{u}) \tag{9.1}$$

where $\mathbf{u}$ is an n-dimensional vector of chemical concentrations and $\mathbf{D}$ denotes an $n \times n$ matrix of diffusion coefficients, which is typically diagonal. The associated boundary conditions depend on the problem at hand but could be periodic, zero flux or fixed, for example, and the initial conditions are typically perturbations around the homogeneous steady state. Generally, only two chemicals are considered in theoretical modelling ($n = 2$), and the vector function $\mathbf{f}(\mathbf{u})$, which arises from the application in question and constitutes the reaction kinetics, is typically nonlinear and possesses a single steady state. In addition, when patterns emerge from system (9.1), the resulting self-organisation does not depend on fine tuning the nature of the chemical interactions described by reaction kinetics; see for example Dillon et al., (1994).

Turing termed the chemicals in his framework *morphogens* and hypothesised that cells would differentiate (adopt a certain fate) if the morphogen concentration breached a certain threshold value. In this way, the spatial pattern in morphogen concentration arising from a DDI would serve as a *pre-pattern* to which cells would respond by differentiating accordingly. In his words, "It is suggested that a system of chemical substances, called morphogens, reacting together and diffusing through a tissue, is adequate to account for the main phenomena of morphogenesis." Typical two-dimensional patterns are shown in Figure 9.2.

Turing's work was significantly extended by Gierer and Meinhardt (1972), who provided fundamental principles for patterning via the DDI mechanism. In particular, they showed that, in the case of two chemical components, DDI could occur in only two scenarios: (i) one chemical had to be self-activating and activate the production of the other chemical which, in turn, inhibited the production of the first; (ii) a substrate depletion system in which one chemical (activator) depletes the

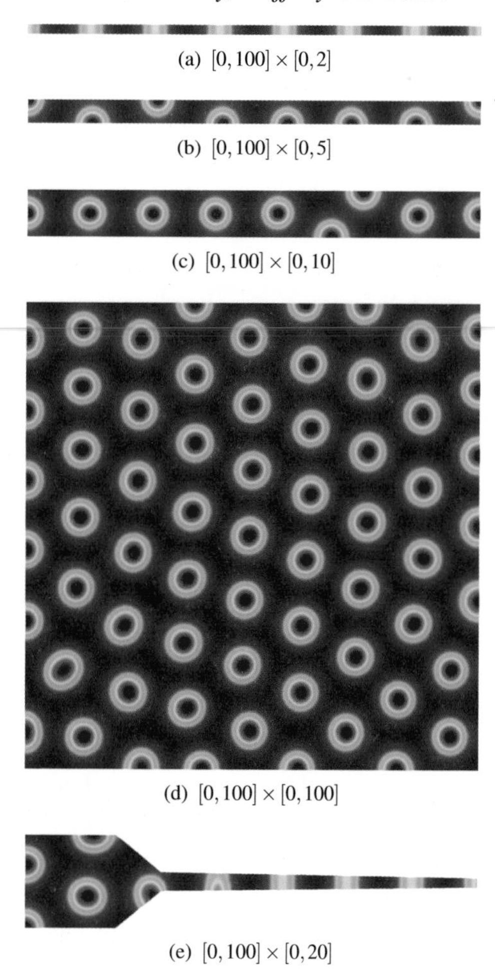

Figure 9.2 Typical patterns produced by a Turing reaction–diffusion model for two chemical reactants. Here we show the concentration of one of the chemicals, the other is out of phase with the solution. Notice that as the domain changes from (a) quasi-one-dimensional to (b)–(d) two-dimensional the patterns become more complex. In (e) the domain geometry is a simplified representation of an animal pelt, showing a transition from spots to stripes. The solution domain is given under each image except (e) where we have solved the simulated system on the tapered domain shown, which is contained within $[0,100] \times [0,20]$. The colour axis runs from 0 (blue) to 4 (red).

other inhibitory chemical (known as a substrate) which, in turn, produces activator. In addition, for both these forms of interactions, patterning requires that the inhibitor diffuses more rapidly than the activator. This is now known as short-range-activation–long-range-inhibition, or local-activation–lateral-inhibition (LALI); see for example Oster (1988).

The LALI mechanism has been applied extensively in explorations of patterning and regeneration for hydra, vein formation in leaves, somite formation in early development, and butterfly wing pigmentation patterns, to name but a few examples. We direct the interested reader to some of the excellent books written on this topic for fuller details and more examples e.g. Meinhardt (1982), Murray (2003) and Meinhardt et al. (2003), and also the very readable article by Kondo and Miura (2010).

We provide an illustration for this counterintuitive notion based on one of Turing's own (slightly imperialistic) analogies involving cannibals and missionaries living on an island. Cannibals self-activate through reproduction, but the act of cannibalism is inhibited by missionaries, who can increase their numbers by converting cannibals to missionaries. Thus one can imagine a stable equilibrium of cannibals and missionaries. However, suppose there is now movement with the missionaries (on bicycles) moving faster than the cannibals. This has the potential to destabilise the equilibrium, with a core of high cannibal activity surrounded by a region of suppressed cannibalism due to missionary presence (Teuscher, 2004). While in real life, human interaction is not as simple and rarely is motion diffusive, such analogies emphasise that Turing's ideas are not restricted to developmental biology and can be found operating in spatial ecology as well as other diverse areas of nature, including chemistry and optics.

9.2 Some developmental applications

Perhaps the most colourful application of these ideas is to animal coat markings. Here, one can exploit the properties of the Laplacian operator in equation (9.1). In many cases the patterns exhibited by the model are characterised by the eigenfunctions of the Laplacian. Thus, for small domains, no pattern will form but, as the domain grows, pattern grows in complexity (see Figure 9.2). One prediction of the model is that on tapering domains, for example a tail, one should see a spot-to-stripe transition, which is indeed observed in many cases. Another prediction is that animals with striped bodies and spotted tails have coat patterns that are not an emergent property of a simple Turing model. Moreover, if an animal is such that it has a plain body colouring then its tail should also be plain according to Turing's mechanism, illustrating an example of a *developmental constraint*. In Figure 9.3 we see that the genet is an excellent example of this, but that the lemur has, unfortunately, no respect for this mathematical theory. In the latter case, a prospective explanation in terms of Turing's mechanism would require an assumption that the parameter values implicit in patterning the body are different to those in the tail, or that the pattern forms due to highly nonlinear interactions.

Another application arises from skeletal patterning in the limb. For example, in the chick limb, at the limb bud stage at which the skeletal pattern is laid down,

(a) (b)

Figure 9.3 Pigmentation patterns observed on (a) the genet, and (b) the ring-tailed lemur.

the limb has a 'paddle-shape' and therefore Turing's theory predicts that patterning complexity should be increasing along the paddle. This is precisely what happens with the humerus–radius/ulna–digits patterning transition, which occurs not only in the chick but is characteristic of many limb architectures. Experiments that reduce the size of the growing limb bud reduce digit number, while those that increase the size of the limb bud result in an increase in digit number. These experimental results are consistent with the theory.

A final arena concerns developmental left–right symmetry breaking. In particular, not everyone has their heart on the left-hand side. A rare genetic disorder, primary cilia dyskinesia (PCD), is associated with a left–right symmetry reversal of the vital organs, such as the heart, in about 50% of cases. This disorder is associated with immotile cilia, which are filamentous cellular protrusions and normally capable of active motion that drives surrounding fluid. Clearly, there is a link between ciliary function and developmental symmetry. The picture emerging in mammals via murine studies is that the initial left–right symmetry breaking event is driven by primary cilia driving flow within the embryonic node, a small covered indentation filled with fluid and located on the embryo surface. The resulting asymmetry of the fluid flow and transport arising from cilia rotating in a single direction then induces differential cellular signalling around the node, though there are numerous competing hypotheses concerning the details of how this signal is transmitted to the surrounding cells. Regardless, a commonly considered, and ultimately testable, hypothesis is that the interaction of the morphogens Nodal and Lefty utilises a Turing-type mechanism to amplify the tiny signal emerging from the nodal flow, instigating the differential development of the embryo's left-hand side and right-hand side (Hamada et al., 2002; Tabin 2006). As such, there is a prospective pathway leading from the suggestion that molecular chirality dictates the direction of embryonic nodal cilia motion to the location of the heart in the mammalian bodyplan (Okada et al., 2005), with Turing's mechanism potentially driving signal amplification.

9.3 Extending Turing

Since the work of Turing there have been a number of models proposed for pattern formation. Some of these consider not only diffusible biochemicals such as morphogens but also cells. The latter can move in response to chemical gradients (a process known as chemotaxis) or deform the extracellular matrix in which they move to set up physical directional cues. In turn these mechanisms can, under certain circumstances, instigate cellular aggregation. It is hypothesised that differentiation then occurs in these aggregates. Other models hypothesise that neurosecretory mechanisms set up patterning (in, for example, molluscs). While these models are all based on different biological hypotheses and lead to systems of equations of different mathematical types, they all give pattern via the LALI mechanism. In fact, their patterns typically emerge as eigenfunctions of the Laplacian in the initial stages of self-organisation and therefore the concept of domain geometry determining pattern complexity, while developed for Turing, applies to all these newer models. This idea was exploited in an evolutionary context by Oster et al. (1988) who noticed that the variations exhibited by salamander (*Ambystoma*) limbs when treated with mitotic inhibitors (which decreased the domain size of the budding limb leading to a loss of digits, precisely as predicted by the LALI mechanism) were very similar to the digit patterns observed in a diggerent salamander (*Proteus*). This is consistent with the observation that *Ambystoma* and *Proteus* share common developmental mechanisms.

One further property of developmental systems is growth. Kondo and Asai (2005) observed that, as the angelfish *Pomocanthus* grows to maturity, its pigmentation pattern of stripes changes. As the stripes move wider apart due to growth, new stripes are inserted so that the stripe wavelength is preserved (Figure 9.4). This is in agreement with the Turing model.

However, it would be highly misleading to suggest that the Turing model suggests that only a few equations uniformly govern the range of developmental patterning that we see. Even observations consistent with Turing's mechanism do not imply that the underlying dynamics is driven by Turing morphogens, and there is no developmental example where the molecular details can be elucidated to the extent that there is unequivocal support for a DDI.

Conversely, the fact that biological complexity surpasses the behaviours of Turing's basic two-component model does not immediately imply that a DDI is not at

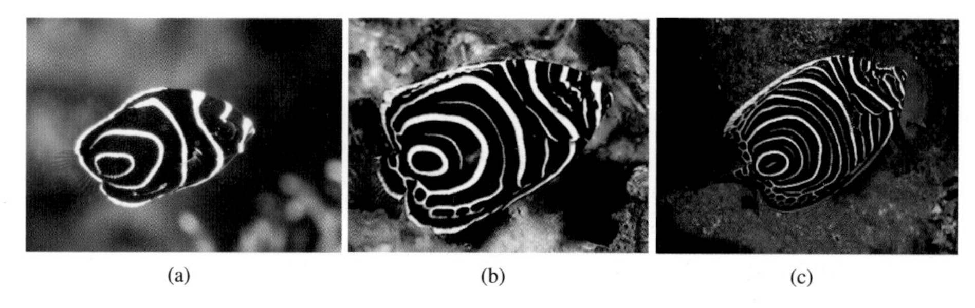

(a) (b) (c)

Figure 9.4 Development of stripes on the marine angelfish *Pomocanthus*. (a) Baby. (b) Juvenile. (c) Mature.

work. For instance, the patterns driven by a Turing mechanism may arise only if certain constraints are satisfied by the model parameters such as the diffusion coefficients, and the production, degradation and interaction rates. These can, in more general settings, also be functions of processes occurring at a lower spatial scale and may even be spatially patterned. This was recognised by Turing, who said, in his original paper, that "most of an organism, most of the time, is developing from one pattern into another, rather than from homogeneity into a pattern." This idea was exploited by Maini et al. (1992) to investigate experiments by Wolpert and Hornbruch (1990) which appeared to contradict the LALI mechanism. In the experiments the anterior portion of the limb bud of a donor chick was grafted onto the anterior part of a host limb bud, so that the resultant double anterior limb bud was the same size as a normal limb bud. The LALI mechanism then predicts that, as the domain size has remained unchanged, the resultant limb should be normal. However, it is observed that two humeri develop. This can be reconciled with the LALI mechanism if one assumes that there is a spatial pattern in one of the parameters in the model upstream of the Turing patterning process, and indeed there is strong evidence to suggest this (Brümmer et al., 1991).

9.4 Critiquing Turing

Such difficulties highlight why Turing's mechanism remains as a tantalising hypothesis rather than an established or refuted mechanism within developmental biology. In particular, while morphogens have since been discovered it is still a matter for strenuous debate as to whether morphogen pairs exist and if they or, more generally, diffusively driven instabilities form pre-patterns as Turing envisioned. There is tentative evidence that the transforming growth factor beta and fibroblast growth factors may play Turing-type roles in limb development (Newman and Bhat, 2007) and that Nodal and Lefty gene products may be a Turing pair (Solnica-Krezel, 2003). In addition, Sick et al. (2006) highlighted that Wnt and Dkk may be a Turing morphogen pair in hair follicle formation in mice, while Garfinkel et al. (2004) provided evidence for a Turing mechanism underlying vascular mesenchymal cell self-organisation during development and identified the morphogens involved. However, the jury is still out. On the other hand, in chemistry, as opposed to developmental biology, it has been unequivocally shown that Turing patterns can arise, in the now famous CIMA (Chloride–Iodide–Malonic-Acid) reaction (Figure 9.5), demonstrating proof of principle. The collection of articles edited by Borckmans et al. (2007) reviews advances in Turing patterns in chemistry as well as other pattern formation phenomena that occur in chemistry.

Figure 9.5 Illustrative examples of chemical Turing patterns for the CIMA (Chloride–Iodide–Malonic-Acid) reaction within a continuously fed open gel reactor. Reprinted with permission from Ouyang and Swinney (1991) Transition to chemical turbulence. *Chaos* 1:411–420. © 1991, American Institute of Physics. The gel is loaded with starch, which changes from yellow to blue if sufficient iodide concentrations establish during the CIMA chemical reaction. The illustrated patterns are essentially stationary states, though some slow movement of interfaces was observed after the pattern had effectively stabilised; the differences between the frames reflect variations in the control parameters of the system, particularly the initial concentrations of the CIMA reactants.

From a theoretical point of view, detailed studies of the Turing model reveal a sensitivity to noise in, for example, the initial fluctuations that drive the instability (Bard and Lauder, 1974) or in the domain geometry on which the reaction is taking place (Bunow et al., 1980). This is obviously a problem in cases where precise patterning is necessary or observed (for example, the number of digits) but not in, for example, the case of some animal coat markings. However, additional aspects of developmental biology, in particular when growth is incorporated into the models, can alleviate initial condition sensitivities (Crampin et al., 1999), though such questions remain unexplored for the prospect of sensitivity to changes in the shape of the domain boundary, or the conditions imposed at these boundaries.

More generally, while genetics alone is insufficient to understand developmental pattern formation, Turing's ideas for morphogen dynamics nonetheless take place within the framework of the cellular molecular biology driving the differential gene expression inherent in the emergence of structure during development. For instance, modern studies reveal that even very simple developmental systems involve complex interactions of numerous receptors and morphogens, while morphogen production requires signal transduction, gene expression and protein production. This complexity is still being unravelled though it is subsumed within simple representations of a very small number of morphogens and their interactions within Turing's framework.

This raises the question of whether Turing's idea is of limited applicability in light of the biological interaction networks emerging in the "omics" era. In particular, extensive changes in predicted system behaviour occur when removing a morphogen or the receptor from a three-component model of hair follicle patterning involving the receptor Edar, connective tissue growth factor and bone morphogenetic protein (Klika et al., 2012). Hence, simply neglecting diffusing elements or receptors to reduce a model of developmental self-organisation to the canonical two-component Turing system is, in general, insufficient and more sophisticated approaches are required.

However, as we have already discussed, numerous developmental systems nonetheless behave in a manner that, at the very least, is suggestive of a Turing system. Hence, even when a literal interpretation of a DDI via pairs of Turing morphogens may ultimately be difficult to justify, there is nonetheless the question of whether Turing morphogens are representations for more complex functional units. These could be a collection of cellular receptors, genes and their products or even whole cells.

Evidence for such a reinterpretation of Turing morphogens is emerging from zebrafish skin patterning studies, which indicate that pigment cells could be functional Turing 'morphogens' (Nakamasu et al., 2009). Such reinterpretations are very much an open area of research and may also alleviate further difficulties with the standard Turing model, for instance the need to have vastly different diffusion rates between the activator and the inhibitor for pattern formation *without* parameter fine-tuning. This approach may also eliminate the need to consider the time delays associated with gene expression, which are troublesome in that incorporating such delays into the classical two-component Turing morphogen model leads to abberrant behaviours (e.g. Seirin-Lee et al., 2010). Even when patterning does occur in the delayed Turing model, there is an extensive patterning lag and a temporal sensitivity that is difficult to reconcile with the highly regulated temporal order observed in many (but not all) developmental phenomena. An alternative reconciliation of such difficulties with empirical observation is that biological interaction networks involving putative Turing morphogen pairs also provide additional stabilising feedback dynamics for reducing temporal sensitivity. Again, this hypothesis remains to be explored.

There are other cases where the application of a Turing-type model, even with a reinterpretation of the interacting elements, cannot be reconciled with experimental data. For example, the model was proposed to account for the stripe-like patterns of the pair-rule genes in the *Drosophila* embryo, but experimental results showed that it is possible to knock out one stripe while keeping the others. This contradicts the Turing model and, in fact, patterning in this case seems to arise from a cascade of interacting gradients in various chemicals (Akam, 1989). Another example is the case of shell pigmentation patterns on molluscs. It has been shown that an astonishingly large variety of these patterns can be produced by Turing-like models on a growing domain (Meinhardt et al., 2003). However, the evidence is increasingly pointing to a neurosecretory mechanism (Boettiger et al., 2009), another of the LALI models.

9.5 The impact of Turing

As can be seen in this brief article, the Turing model has been applied to a wide range of patterning phenomena in developmental biology. The richness in behaviour of this seemingly simple set of equations is remarkable and much wider than we have covered here. This modelling framework has also motivated an enormous amount of mathematical analysis and computational study. Moreover, it has caused a paradigm shift in the thinking of developmental biologists, especially after the short-range-activation–long-range-inhibition idea championed by Meinhardt.

Turing's work has therefore inspired modelling efforts for describing self-organisation built on different biological hypotheses and has provided a framework of patterning principles and developmental constraints consistent with the diffusively driven instability and LALI mechanisms. Indeed, it is ironic to note that this framework, developed during an era of biological list-making, risks being lost in the present era of data generation and collection. In addition, these concepts have also raised more than their fair share of controversy. However, while it is in the nature of modelling that all models are simplifications, the worth of a model in biology must be measured in the number of experiments it motivates that may not have been done otherwise and in the amount it changes the way experimentalists think. By this measure, Turing's 1952 paper is one of the most influential theoretical papers ever written in developmental biology.

Acknowledgements. PKM was partially supported by a Royal Society Wolfson Research Merit Award; TEW would like to thank the EPSRC for support.

References

Akam, A., 1989. Making stripes inelegantly. *Nature* 341:282–283

Alberts, B., Johnson, A., Walter, P., Lewis, J., Raff, M. and Roberts, K., 2002. *Molecular Biology of the Cell*, 5th ed. Garland Science.

Bard, J. and Lauder, I., 1974. How well does Turing's theory of morphogenesis work? *J. Theor. Biol.* 45:501–531.

Borckmans, P., De Kepper, P., Khokhlov, A.R. and Métens, S. (eds.), 2009. *Chemomechanical Instabilities in Responsive Materials*. Springer.

Boettiger, A., Ermentrout, B. and Oster G., 2009. The neural origins of shell structure and pattern in aquatic mollusks. *Proc. Nat. Acad. Sci. USA* 106:6837–6842.

Bunow, B., Kernervez, J.P., Joly, G. and Thomas, D., 1980. Pattern formation by reaction–diffusion instabilities: Application to morphogenesis in *Drosophila*. *J. Theor. Biol.* 84:629–649.

Brümmer, F., Zempel, G., Buhle, P., Stein, J.-C. and Hulser, D.F., 1991. Retinoic acid modulates gap junctional permeability: A comparative study of dye spreading and ionic coupling in cultured cells, *Exper. Cell Res.* 96:158–163.

Crampin, E.J., Gaffney, E.A. and Maini, P.K., 1999. Reaction and diffusion on growing domains: scenarios for robust pattern formation. *Bull. Math. Biol.* 61:1093–1120.

Dillon, R., Maini, P.K. and Othmer, H.G., 1994. Pattern formation in generalised Turing systems: I. Steady-state patterns in systems with mixed boundary conditions. *J. Math. Biol.* 32:345–393.

Garfinkel, A., Tintut, Y., Petrasek, D., Boström, K. and Demer, L.L., 2004. Pattern formation by vascular mesenchymal cells. *Proc. Nat. Acad. Sci. USA* 101:9247–9250.

Gierer, A. and Meinhardt, H., 1972. A theory of biological pattern formation. *Kybernetik* 12:30–39.

Hamada, H., Meno, C., Watanabe, D. and Saijoh, Y., 2002. Establishment of vertebrate left–right asymmetry. *Nature Rev. Genetics* 3:103–113.

Klika, V., Baker, R.E., Headon, D. and Gaffney, E.A. (2012). The influence of receptor-mediated interactions on reaction–diffusion mechanisms of cellular self-organisation. *Bull. Math. Biol.* 74:935–957.

Kondo, S. and Asai, R., 1995. A reaction–diffusion wave on the skin of the marine angelfish Pomacanthus. *Nature* 376:765–768.

Kondo, S. and Miura, T., 2010. Reaction–diffusion model as a framework for understanding biological pattern. *Science* 329:1616–1620.

Maini, P.K., Benson, D.L. and Sherratt, J.A., 1992. Pattern formation in reaction diffusion models with spatially inhomogeneous diffusion coefficients. *IMA J. Math. Appl. Med. Biol.* 9: 197–213.

Meinhardt, H., 1982. *Models of Biological Pattern Formation*. Academic Press.

Meinhardt, H., Prusinkiewicz, P. and Fowler, D., 2003. *The Algorithmic Beauty of Sea Shells*, 3rd ed. Springer.

Murray, J.D., 2003. *Mathematical Biology. II: Spatial Models and Biomedical Applications*. Springer.

Nakamasu, A., Takahashi, G., Kanbe, A. and Kondo, S., 2009. Interactions between zebrafish pigment cells responsible for the generation of Turing patterns. *Proc. Nat. Acad. Sci. USA* 106:8429–8434.

Newman, S.A. and Bhat. R., 2007. Activator–inhibitor dynamics of vertebrate limb pattern formation. *Birth Defects Res.* (Part C) 81:305–319.

Okada, Y., Takeda, S., Tanaka, Y., Belmonte, J.I. and Hirokawa, N., 2005. Mechanism of nodal flow: a conserved symmetry breaking event in left–right axis determination. *Cell* 121:633–644.

Oster, G.F., 1988. Lateral inhibition models of developmental processes. *Math. Biosci.* 90:265–286.

Oster, G.F., Shubin, N., Murray, J.D. and Alberch P., 1988. Evolution and morphogenetic rules. The shape of the vertebrate limb in ontogeny and phylogeny. *Evolution* 42:862–884.

Ouyang, Q. and Swinney, H.L., 1991. Transition to chemical turbulence. *Chaos* 1:411–420.

Seirin-Lee, S., Gaffney, E.A. and Monk, N.A.M., 2010. The influence of gene expression time delays on Gierer–Meinhardt pattern formation systems. *Bull. Math. Biol.* 72:2139–2160.

Sick, S., Reiner, S., Timmer, J. and Schlake, T., 2006. WNT and DKK determine hair follicle spacing through a reaction–diffusion mechanism. *Science* 314:1447–1450.

Solnica-Krezel, L., 2003. Vertebrate development: taming the nodal waves. *Current Biology* 13:R7R9.

Tabin, C.J., 2006. The key to left-right asymmetry. *Cell* 127:27–32.

Teuscher, C., 2004. *Alan Turing: Life and Legacy of a Great Thinker*. Springer.

Thompson, D.W., 1992. *On Growth and Form*. Cambridge University Press.

Turing, A.M., 1952. The chemical basis of morphogenesis. *Phil. Trans. Roy. Soc. London B* 327:37–072.

Widelitz, R.B., Jiang, T.X., Lu, J.F. and Choung, C.M., 2000. Beta-catenin in epithelial morphogenesis: conversion of part of avian foot scales into feather buds with a mutated beta-catenin. *Devel. Biol.* 219: 98–114.

Wolpert, L. and Hornbruch, A. 1990. Double anterior chick limb buds and models for cartilage rudiment specification. *Development* 109:961–966.

10

Walking the Tightrope: The Dilemma of Hierarchical Instabilities in Turing's Morphogenesis

Richard Gordon

Dedicated to Bashir Ahmad and a bright future for physics in his Afghanistan.

Ever since we understood that living organisms are made of fundamentally the same stuff as the rest of the universe, we have been puzzling about what makes life different, and how that difference arose. Alan Turing made a major contribution in putting forth a model for morphogenesis that attempts to bridge the gap between the molecules we are made of and how we look.[1] This is a huge gap to span in trying to solve the problem of how an embryo builds itself. People are about $1.5\,\text{m} = 1.5 \times 10^9\,\text{nm}$ (nanometers) tall and the typical protein molecule is about $30\,\text{nm}$ wide, a ratio of $50\,000\,000 = 5 \times 10^7$ to 1. That protein might contain 1000 amino acids and, if we take into account the relative volumes, not just length, say 70 liters for an adult human and $0.15\,\text{nm}^3$ for an amino acid (Sühnel and Hühne, 2005), this raises the ratio to 5×10^{26} to 1.

When we build a bridge, unless it's over a small ditch, we use many parts and assemble them in sections called spans (Figure 10.1). In biology this has come to be known as 'modularity', and the grand search has been on to discover just what the modules of life are (von Dassow and Munro, 1999; Gilbert and Bolker, 2001; Redies and Puelles, 2001; Newman and Bhat, 2009; Peter and Davidson, 2009; Christensen et al., 2010). At first it was thought that cells represented modules. The 'cell theory' was resisted by a minority of 19th century biologists, who thought that the basic module was the whole organism, not the cell. Certain observations support this 'organismal' theory. First, single cell organisms, such as *Paramecium* and diatoms (Figure 10.2), can have quite complex morphologies; see Gordon (2010); Tiffany et al. (2010); Gordon and Tiffany (2011). Second, some

Published in *The Once and Future Turing*, edited by S. Barry Cooper & Andrew Hodges. Published by Cambridge University Press

[1] I will assume that the reader has read Turing's final paper, 'The chemical basis of morphogenesis' (Turing, 1952) and/or Maini et al. (2015), which brings it up to date. Some prefer to cite this as the Rashevsky–Turing theory because of mathematical and biological similarities in the earlier work of Nicolas Rashevsky (Rosen, 1968; Parisi, 1991; Graham et al., 1993).

Figure 10.1 The Kingston–Rhinecliff Bridge over the Hudson River, New York (Hermeyer and Wantman, 2008), made of many spans, each of which has many smaller parts. Reproduced under the Creative Commons Attribution-Share Alike 3.0 Unported license.

green algae have many nuclei that move in cytoplasmic streaming, unhindered by cell boundaries, yet "...exhibit morphological differentiation into structures that resemble the roots, stems, and leaves of land plants and even have similar functions" (Chisholm et al., 1996); see Cocquyt et al., 2010. Third, we can make 'polyploid' salamanders with up to seven copies of their genome per cell (Fankhauser et al., 1955). The result is an adult with fewer larger cells, which generally reaches the same size (Fankhauser, 1941). So the adult's morphology is not dependent on how many cells it is made of – though its intelligence may be: see Fankhauser et al. (1955). Thus we may have to look for the supposed modular organization of embryos above the cell level.

Turing considered two models, one consisting of a line of cells, another of a 'continuum' line. In the limit of many small cells, they are equivalent. However, the length of the line is important because, along a line, the pattern generated by a Turing model is often a simple wave, of a specific wavelength that depends on the chemical reactions and diffusion rates of hypothesized molecules he called 'morphogens' (Figure 10.3). Thus the number of waves is proportional to the length of the model organism. Attempts to overcome this limitation have included adding growth to the model (Crampin et al., 1999) or making the concentration of one morphogen depend on the size of the organism (Ishihara and Kaneko, 2006; cf. Gordon, 1966). These modifications are needed because, for many species, shape is reasonably independent of size.

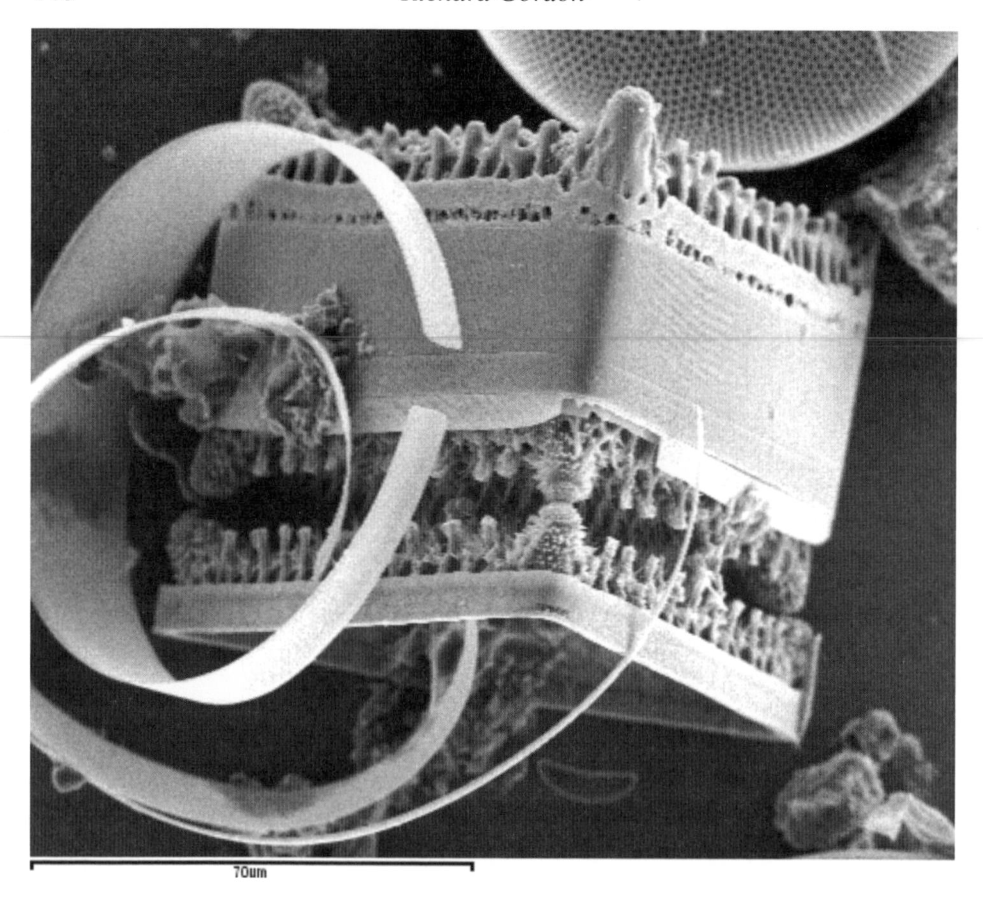

Figure 10.2 An ornate, single cell, box-like diatom *Triceratium favus*. This diatom makes clonal chains, and is here seen attached to part of an adjacent cell. The organic components have been removed, leaving only the silica shells. The curved girdle bands are from another species of cylindrically shaped diatom. Reproduced with permission of Mary Ann Tiffany, who took this SEM (scanning electron micrograph).

It used to be thought (in the time before the mammalian egg was discovered) that we were all just blow up dolls, called homunculi, and that the sperm was thus a miniature human that only needed to be inflated. Actually, this idea originated as an unproven hypothesis and a spoof that has been propagated in textbooks since the 1700s to contrast preformationism with epigenesis (the idea that the embryo somehow forms itself, and is not preformed). Apparently few, except the duped, ever took it seriously (Cobb, 2006).

Turing introduced the concept that morphogenesis results from processes that 'break symmetry'. You can watch such a process yourself. Draw a thin string of honey across a dish, and it will break up into drops (Figure 10.4); see Rayleigh

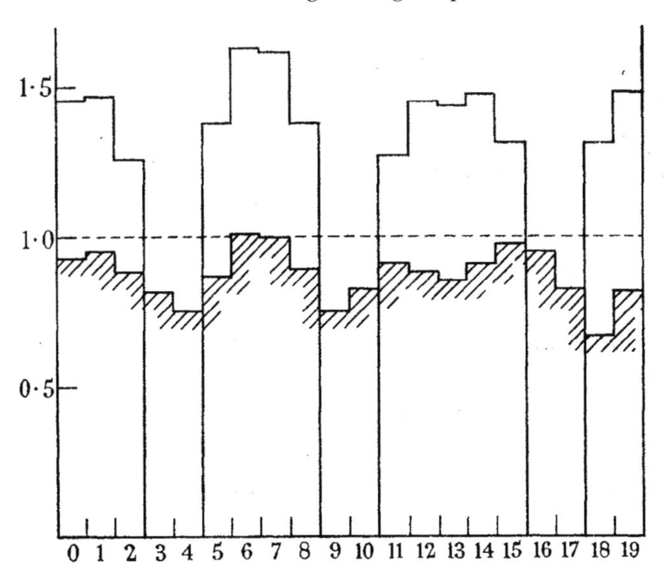

Figure 10.3 Turing's own illustration of a homogeneous equilibrium (broken line) along a line of 20 cells, which he arranged in a ring (with left and right edges coinciding) to simplify the calculation (Turing, 1952). He assumed some initial small random departures from uniformity and showed the concentration of one of the morphogens versus cell number at two times: the incipient random pattern (hatched) with four peaks and the final pattern (solid lines). The pattern thus settled down to three peaks and three deep valleys. Any deviation from the broken equilibrium line will produce a similar result; therefore this homogeneous equilibrium is referred to as 'unstable'. Reproduced with permission of the Royal Society of London.

(1892). This process is driven by surface tension, and comes about because the surface area of the drops is smaller than the surface area of the cylinder of fluid from which they form. However, a cylinder has greater symmetry than a row of drops: it has rotational symmetry about a line and translational symmetry along the line, while a row of drops has translational symmetry only for multiples of the specific distance between the drops. Thus drop formation partly breaks the initial symmetry.

We all start as a single drop of mist with spherical symmetry (Evsikov et al., 1994), a fertilized human egg only 70 micrometers in diameter (Figure 10.5). The problem of embryogenesis is how that drop breaks its symmetry to form one of us. We are different front to back, head to toe, inside to outside, and left to right. We are not even bilaterally symmetric, as we know from our heart being on one side, and most of us are generally left- or right-handed rather than ambidextrous. Our brain has an asymmetric twist to it (Hellige, 1993). A great Photoshop experiment is to take face-on pictures of people you know and split them down the middle, making two new pictures with the left half and its mirror reflection, and the same

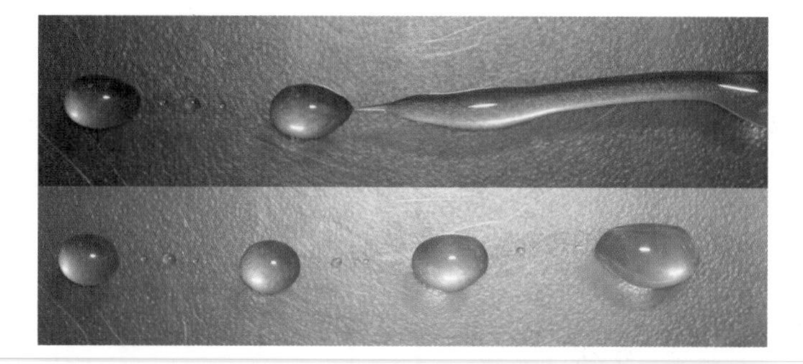

Figure 10.4 A strand of honey breaking into drops. With thanks to Susan Crawford-Young.

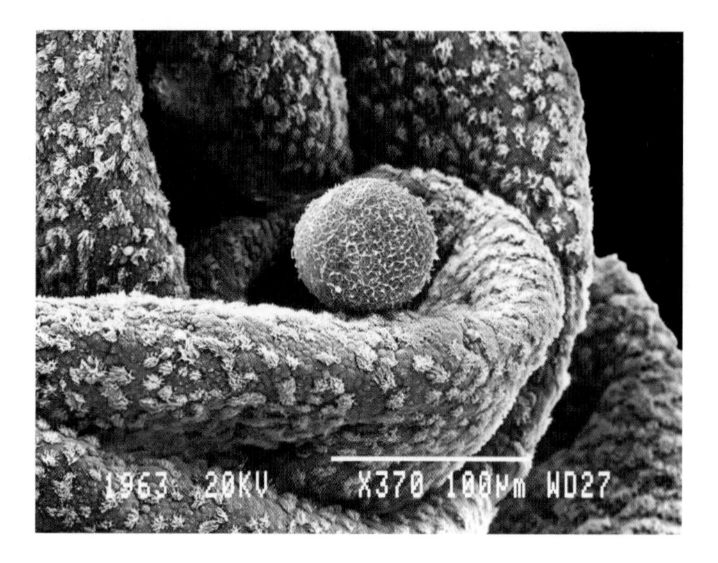

Figure 10.5 Colored scanning electron micrograph (SEM) of a human egg in the fallopian tube. When fused with a sperm, the egg becomes a one-celled embryo called a zygote. From Nikas (2011), reproduced with kind permission of Yorgos Nikas, Athens Innovative Microscopy (http://www.aim.cat).

for the right half. For most people one gets two faces that are quite from the original (Figure 10.6).

If we probe a little deeper, we learn that we are made of up to 7000 different kinds of cells in the course of development (Bard et al., 1998a). But we all started as one cell (Figure 10.5), when fertilized, a union of egg and sperm (zygote), obviously of just one kind. So much more is going on in embryogenesis than lots of cell division and growth in mass. The cells are also becoming different from one another, a process of symmetry breaking: somehow two 'daughter' cells that result from the splitting of their 'mother' cell in two sometimes become different from one another, and perhaps also different from the mother cell. This process is

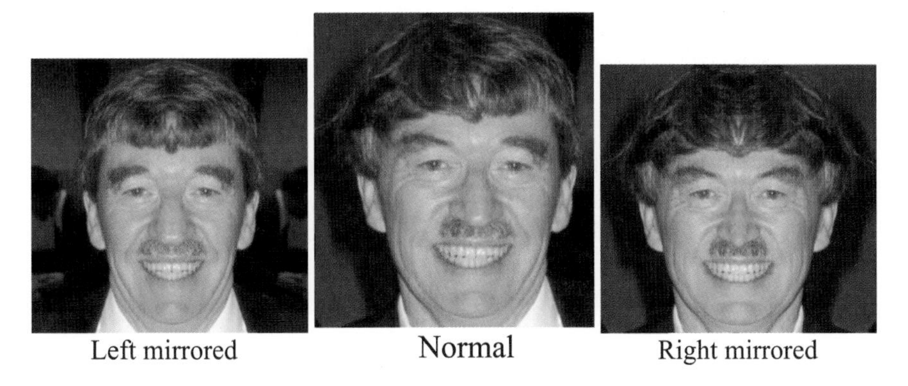

Figure 10.6 The two sides of my friend and colleague, David Hoult (Tomanek et al., 2000), reproduced with his permission.

called ‘differentiation’. In a way, differentiation is one of the three major outstanding biological problems of our day, the other two being the origin of life (Gordon and Hoover, 2007; Gordon, 2008; Damer et al., 2012; Sharov and Gordon, 2013) and consciousness (Tuszynski and Gordon, 2012). As it is less exciting and occurs daily before our eyes – as in plants flowering, babies being born and wounds healing – differentiation gets much less press. But for the same reason it is a much more accessible question than the other two.

So, can Turing’s model handle differentiation? In a way it can, and Turing showed how just two cells, stuck side by side, containing the same morphogens at nearly identical concentrations, can end up having different concentrations of each morphogen: “This breakdown of symmetry or homogeneity may be illustrated by the case of a pair of cells originally having the same, or very nearly the same, contents” (Turing, 1952). It would seem that he found the essence of the problem of cell differentiation, but let’s try it again. To make this clear, I’ll use letters to represent cells of a given kind. Thus we start with one cell, A, and let us suppose it divides to produce two daughter cells, B and C, that may each be different from A:

$$A \Longrightarrow BC.$$

In the next step B divides into cells D and E, and C divides into cells F and G:

$$BC \Longrightarrow DEFG.$$

But we have a problem here. Unless cells B and C can somehow influence one another, we could (by symmetry) just as well get

$$BC \Longrightarrow EDFG,$$
$$BC \Longrightarrow DEGF,$$
$$BC \Longrightarrow EDGF.$$

Furthermore (again by symmetry) cell A could just as well have produced

$$A \Longrightarrow CB$$

because BC can differ from CB if there is a left/right polarity to each cell. We thus see that there are even more possibilities:

$$\begin{aligned} CB &\Longrightarrow FGDE, \\ CB &\Longrightarrow FGED, \\ CB &\Longrightarrow GFDE, \\ CB &\Longrightarrow GFED. \end{aligned}$$

Thus after two cell divisions (assumed synchronous) we have eight distinct possible arrangements of the four different kinds of cells. The problem gets exponentially worse with each step of cell division and its symmetry breaking. This shows that there is a symmetry to symmetry breaking and, if we are to have a specific organism result, we need to figure out how to break this 'higher order' combinatoric 'metasymmetry' (Gordon, 2006). "...when multiple, hierarchical steps of symmetry breaking are involved, the combinatorics could lead to many alternative, morphologically untenable organisms" (Gordon, 1999). Most attempts to fit Turing's model to real cell differentiation, including his, have dealt with only one step in the process.

The number of different cell types in the course of mouse development has been estimated to be as high as 7000 with eight hierarchical levels (Bard et al., 1998a,b). Since $2^{12} < 7000 < 2^{13}$, the number of bifurcations in the cell lineage tree producing a mouse is around 12 to 13 on average, suggesting four or five more levels are to be found. Many kinds of cells look alike, and require further investigation to tell them apart. If half these cell types are present in the adult mouse, and we arranged one cell of each kind along a line, there would be 3500! (factorial) different arrangements *a priori*, or well over 10^{10000}. In three dimensions the possibilities are even greater. This is so even without considering that there are many cells of each kind.

These are enormous combinatorics that are somehow overcome in normal embryo development. The problem was formulated in the 1890s by Hans Driesch: "How do cells differentiate in the right place, at the right time, into the right kinds?" (see Reid, 1985 and also Gordon, 1999). It is not yet solved, though. This 'miracle', that each of us is not a jumble of cell types, has been missed by creationists and the Intelligent Design advocates, who make much of such probability arguments (Gordon, 2008). The problem is real, because teratomas and teratocarcinomas do indeed contain "a disorganized mixture of various tissues" from all three embryonic germ layers (Bulić-Jakuš et al., 2006).

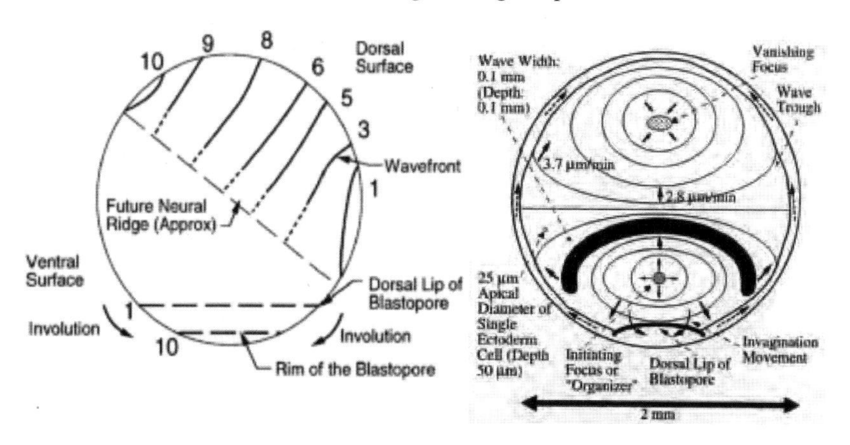

Figure 10.7 The ectoderm contraction wave on the embryo of a salamander, the axolotl *Ambystoma mexicanum*: side view (Brodland et al., 1994) and top view (Gordon et al., 1994) Reproduced with slight modifications with permission of John Wiley and Sons and Elsevier Books, respectively. The ectoderm expansion wave (not shown), which traverses the ventral (bottom) surface, starts as the ectoderm contraction wave ends. Positions of the bottom of the furrow-shaped contraction wave are shown at one hour intervals. The thick bar shows the width of the strongly contracted band of cells, which remain contracted for about 10 minutes.

The answer may lie in finding the proper modules for embryonic development, above the single cell level.

One clue may be that, in most organisms we meet face to face, cells differentiate in groups, not one at a time. Let's call these groups 'embryonic tissues'. What defines tissues has been somewhat of a mystery. In 1985 I had a wonderful eureka experience (Appendix 3 in Gordon, 1999) in which I realized that this problem had a possible solution in the mechanics of the cytoskeleton at the apical (outer) end of the cells in salamander embryos, and that differentiation could spread in response to physical waves traversing parts of the embryo. After I predicted their existence (Gordon and Brodland, 1987), the first such wave was discovered by Natalie K. Björklund-Gordon in 1990 (Brodland et al., 1994; Gordon, 1999). The 'ectoderm contraction wave' leaves the tissue called neuroepithelium in its wake, the tissue that later forms the brain and spinal cord (Figure 10.7). We found many other waves, and called them 'differentiation waves' (Gordon et al., 1994). Only half of the ectoderm is traversed by the ectoderm contraction wave. The other half is traversed by the 'ectoderm expansion wave', and in the process it becomes another tissue: epidermis (later skin).

We thus have a potentially higher-order module than the cell: a pair of differentiation waves traversing complementary parts of a tissue, dividing it into two new tissues, regardless of the number of cells in the original tissue. But at this point we can only speculate on what sets up the conditions for a wave to propagate, what determines whether it is an expansion or contraction wave, what determines where it starts, what determines its trajectory, and what determines how and where it stops (Gordon, 1999). To start finding answers to these questions, we need detailed observations of the waves and the cells through which they propagate, which we hope to organize in an online four-dimensional dataset called Google Embryo (Gordon, 2009; Gordon and Westfall, 2009).

Turing pointed out that, with three morphogens, his equations had solutions consisting of travelling waves. Thus it is possible that differentiation waves propagate as Turing waves (Gordon, 1999). Certainly, we can anticipate at least three interacting components: microfilaments made of the protein actin, microtubules made of proteins called tubulins, and a calcium ionic current (Jaffe, 1999). The first two contribute a mechanical component, of which Turing was quite aware:

> ...the description of the state consists of two parts, the mechanical and the chemical. The mechanical part of the state describes the positions, masses, velocities and elastic properties of the cells, and the forces between them ... The interdependence of the chemical and mechanical data adds enormously to the difficulty, and attention will therefore be confined, so far as is possible, to cases where these can be separated ... In this paper it is proposed to give attention rather to cases where the mechanical aspect can be ignored and the chemical aspect is the most significant. These cases promise greater interest, for the characteristic action of the genes themselves is presumably chemical. (Turing, 1952) .

(Compare Howard et al., 2011.) We now know that one characteristic action of genes is to produce mechanochemically active proteins such as actin and tubulins, and with the use of computers such multiphysics problems can now be tackled. But the hard work is still in the future (Fleury and Gordon, 2012).

The philosopher Daniel Dennett imagined embryonic development to be analogous to construction cranes that build a tall building, but with each crane also constructing such a crane on itself (Dennett, 1995; Gordon, 1999). Just as in the problem of how high can a child stack blocks (Figure 10.8) – see Iwasaki and Honda (2000), Blanchard and Hongler (2002), Tokumitsu and Honda (2005) – or a house of cards, or how far one can walk a tightrope, one would anticipate that soon the whole superstructure would just come tumbling down. Yet this simplistic analogy to cranes, one of our current civil engineering technologies, is about as advanced as our current thinking gets on how embryos build themselves.

In some sense, every organism that develops from a single cell is undergoing an act of creation. What symmetry breaking has taught us is that, contrary to the opinion of some philosophers, we can indeed get something from nothing, in this case structure from uniformity. But we have to face an obstacle in our theories, namely that, having done one step of symmetry breaking, how do we explain the next step, and the next, etc.? This cannot be answered simply by invoking chaos,

Figure 10.8 Child stacking blocks: my grandson Nick, with permission of his mother, Nola Sandor-Collins, photographed by his aunt, Mary Ann Marquez.

fractals or branching processes, because those all generate much the same structure at all levels.

If we take the idea that the basic modules of embryogenesis are differentiation waves, and that expansion/contraction pairs of waves consecutively split each tissue into two new tissues (Björklund and Gordon, 1994) then embryogenesis can be thought of in terms of a 'differentiation tree'. What is unique about this tree is that every branch is different, representing a different cell type (Figure 10.9). A plausible DNA mechanism for duplicating and grafting branches of the differentiation tree onto itself (Gordon, 1999) creates novelty, because the two branches can diverge by mutation during subsequent evolution. This leaves the descendents with new tissues and thus new possibilities for morphology and behavior (West-Eberhard, 2002). Our predecessors, the bacteria, can sometimes produce three or four cell types, such as vegetative cells, heterocysts, spores and hormogonia in cyanobacteria – formerly known as blue green algae – (Meeks et al., 2002), or the three spatially separated motile, matrix-producing, and sporulating cells of cloned *Bacillus subtilis* colonies (Lopez et al., 2009). An important event in the subsequent evolution of eukaryotes (cells with nuclei, such as ours, that are symbiotic associations of bacteria; see Watson, 2006) was the invention of what I call 'continuing differentiation' (Gordon, 1999): the ability to keep producing new cell types. This may be one of the major acts of creation in evolution.

Continuing differentiation may have occurred first in organisms like nematode worms, where every cell type is represented by usually just one cell (Gordon, 1999), and most cell divisions (called 'asymmetric') give rise to two new kinds of cells, just as in our alphabet strings (though arranged in 3D). How asymmetric divisions get oriented in space is an active research enquiry (Hwang and Rose, 2010; Zernicka-Goetz, 2011). For 'mosaic' organisms such as nematodes, the cell lineage tree (the 'family tree' showing which cells came from which, starting from

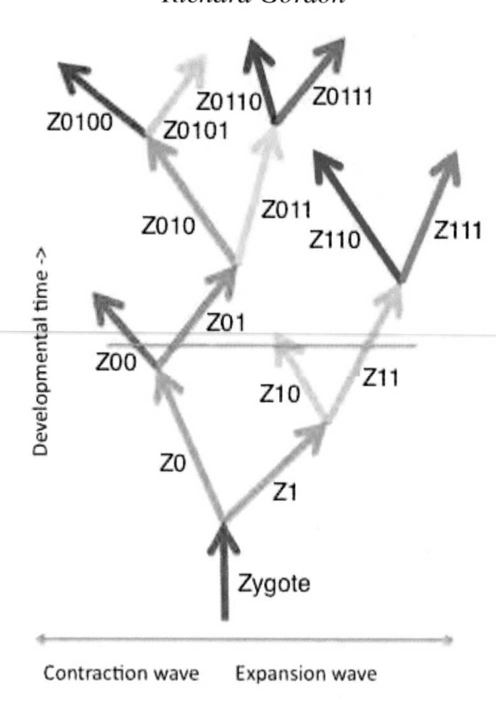

Figure 10.9 A schematic differentiation tree. Each branch represents the splitting of a transient embryonic tissue into two new tissues. Those tissues existing at any given moment can be determined by drawing a horizontal line, as shown. Terminal arrows could represent cell types that die off, or that go into a holding pattern (stem cells; Gordon, 2006, 2011), or the arrows could be extended vertically to indicate they are terminally differentiated cell types found in the adult. Next to each tissue is its binary differentiation code, the sequence of contraction (**0**) and expansion (**1**) waves its ancestor cells have participated in, to be read left to right. From Gordon N.K. and Gordon R. (2015), reproduced with permission of World Scientific Publishing Company.

the fertilized one-cell embryo) is also the differentiation tree, when rearranged by cell size (Alicea et al., 2014). The nematode *Caenorhabditis elegans* ends up one millimeter long with 900 cell types, or what we might call 900 mostly single-cell tissues. This suggests that in organisms that led to us, another major evolution event occurred: the ability for cells to differentiate in groups via differentiation waves that traverse many cells, producing the modules of which we are constructed.

One problem with differentiation trees is the question of how a cell 'knows' what kind of cell it is, especially given that all the cells contain DNA with the same nucleotide sequence? Because the answer does not lie in the DNA sequence, we call such a problem 'epigenetic' (Gilbert and Epel, 2009). If we let **1** stand for an expansion wave and **0** for a contraction wave, and write the history of the waves in which a cell and its cell ancestors have participated, from left to right, then we can specify each cell type after the fertilized egg by a binary number, as shown in Figure 10.9. We called this the 'differentiation code' (Björklund and Gordon, 1994). How the code is stored in each cell and read out are unsolved problems. Perhaps Turing's code-cracking work (Maurer, 2015) will help.

In summary Turing, in 'The chemical basis of morphogenesis', laid the groundwork for what we now call embryo physics (Beloussov and Gordon, 2006) in which we try to reverse engineer the embryo (Gordon and Melvin, 2003) to figure out how it manages to build itself while breaking symmetries and metasymmetries at every level, with enough consistency to produce a viable you and me. We don't know yet how this is done. However, we can now discern six major events in the evolution of bacteria to people:

(1) asymmetric cell division, resulting in differentiated cells, i.e., two cells of different kinds;
(2) the symbiotic union of certain bacteria to produce eukaryotic cells;
(3) the invention of continuing differentiation;
(4) the breaking of metasymmetries to produce fairly unique arrangements of the many resulting cell types;
(5) the invention of an epigenetic differentiation code;
(6) differentiation waves, which allow the single-cell tissues resulting from asymmetric cell divisions to expand to multicellular tissues.

Many future Turings will be needed to crack these mysteries.

Acknowledgements. Thanks to Stephen A. Krawetz, Lu Kai, Stephen P. McGrew, Yorkos Nikas and Jack Rudloe for careful readings and insightful comments.

References

Alicea, B., McGrew, S., Gordon, R., Larson, S., Warrington, T. and Watts, M. (2014). DevoWorm: differentiation waves and computation in *C. elegans* embryogenesis. Available at `http://www.biorxiv.org/content/early/2014/10/03/009993` [Accessed March 26, 2015].

Bard, J.B.L., Baldock, R.A. and Davidson, D.R. (1998a). Elucidating the genetic networks of development: a bioinformatics approach. *Genome Res.*, **8** (9), 859–63.

Bard, J.B.L., Kaufman, M.H., Dubreuil, C., Brune, R.M., Burger, A., Baldock, R.A. and Davidson, D.R. (1998b). An internet-accessible database of mouse developmental anatomy based on a systematic nomenclature. *Mech. Dev.*, **74** (1/2), 111–20.

Beloussov, L.V. and Gordon, R. (2006). Preface. Morphodynamics: Bridging the gap between the genome and embryo physics. *Int. J. Dev. Biol.*, **50** (2/3), 79–80.

Björklund, N.K. and Gordon, R. (1994). Surface contraction and expansion waves correlated with differentiation in axolotl embryos. I. Prolegomenon and differentiation during the plunge through the blastopore, as shown by the fate map. *Computers & Chemistry*, **18** (3), 333–45.

Blanchard, P. and Hongler, M.-O. (2002). How many blocks can children pile up? Some analytical results. *J. Phys. Soc. Japan*, **71** (1), 9–11.

Brodland, G.W., Gordon, R., Scott, M.J., Björklund, N.K., Luchka, K.B., Martin, C.C., Matuga, C., Globus, M., Vethamany-Globus, S. and Shu, D. (1994). Furrowing surface contraction wave coincident with primary neural induction in amphibian embryos. *J. Morphol.*, **219** (2), 131–42.

Bulić-Jakuš, F., Ulamec, M., Vlahović, M., Sinčić, N., Katušić, A., Jurić-Lekić, G., Šerman, L., Krušlin, B. and Belicza, M. (2006). Of mice and men: teratomas and teratocarcinomas. *Collegium Antropologicum*, **30** (4), 921–4.

Chisholm, J.R.M., Dauga, C., Ageron, E., Grimont, P.A.D. and Jaubert, J.M. (1996). 'Roots' in mixotrophic algae. *Nature*, **381** (6581); erratum, **382** (6583) 565.

Christensen, D.J., Campbell, J. and Stoy, K. (2010). Anatomy-based organization of morphology and control in self-reconfigurable modular robots. *Neural Comput. Appl.*, **19** (6), 787–805.

Cobb, M. (2006). *Generation: The Seventeenth-Century Scientists Who Unraveled the Secrets of Sex, Life, and Growth.* Bloomsbury.

Cocquyt, E., Verbruggen, H., Leliaert, F. and De Clerck, O. (2010). Evolution and cytological diversification of the green seaweeds (Ulvophyceae). *Mol. Biol. Evol.*, **27** (9), 2052–61.

Crampin, E.J., Gaffney, E.A. and Maini, P.K. (1999). Reaction and diffusion on growing domains: scenarios for robust pattern formation. *Bull. Math. Biol.*, **61** (6), 1093–120.

Damer, B., Newman, P., Norkus, R., Gordon, R. and Barbalet, T. (2012). Cyberbiogenesis and the EvoGrid: a twenty-first century grand challenge. In *Genesis – In the Beginning: Precursors of Life, Chemical Models and Early Biological Evolution*. J. Seckbach (ed.). Springer, pp. 267–88.

Dennett, D.C. (1995). *Darwin's Dangerous Idea: Evolution and the Meanings of Life.* Simon & Schuster.

Evsikov, S.V., Morozova, L.M. and Solomko, A.P. (1994). Role of ooplasmic segregation in mammalian development. *Roux's Arch. Dev. Biol.*, **203**, 199–204.

Fankhauser, G. (1941). Cell size, organ and body size in triploid newts (*Triturus viridescens*). *J. Morphol.*, **68**, 161–77.

Fankhauser, G., Vernon, J.A., Frank, W.H. and Slack, W.V. (1955). Effect of size and number of brain cells on learning in larvae of the salamander, *Triturus viridescens*. *Science*, **122** (3172), 692–3.

Fleury, V. and Gordon, R. (2012). Coupling of growth, differentiation and morphogenesis: an integrated approach to design in embryogenesis. In *Origin(s) of Design in Nature: A Fresh, Interdisciplinary Look at How Design Emerges in Complex Systems, Especially Life*. L. Swan, R. Gordon and J. Seckbach (eds.). Springer, pp. 385–428.

Gilbert, S.F. and Bolker, J.A. (2001). Homologies of process and modular elements of embryonic construction. *J. Exp. Zool.*, **291** (1), 1–12.

Gilbert, S.F. and Epel, D. (2009). *Ecological Developmental Biology: Integrating Epigenetics, Medicine, and Evolution*. Sinauer Associates.

Gordon, N.K. and Gordon, R. (2015). *Embryogenesis Explained.* World Scientific.

Gordon, R. (1966). On stochastic growth and form. *Proc. Natl. Acad. Sci. USA*, **56** (5), 1497–504.

Gordon, R. (1999). *The Hierarchical Genome and Differentiation Waves: Novel Unification of Development, Genetics and Evolution*. World Scientific and Imperial College Press.

Gordon, R. (2006). Mechanics in embryogenesis and embryonics: prime mover or epiphenomenon? *Int. J. Dev. Biol.*, **50** (2/3), 245–53.

Gordon, R. (2008). Hoyle's tornado origin of artificial life, a computer programming challenge. In *Divine Action and Natural Selection: Science, Faith and Evolution*, J. Seckbach and R. Gordon (eds.). World Scientific, pp. 354–67.

Gordon, R. (2009). Google Embryo for building quantitative understanding of an embryo as it builds itself: II. Progress toward an embryo surface microscope. *Biological Theory: Integrating Development, Evolution, and Cognition*, **4** (4), 396–412.

Gordon, R. (2010). Diatoms and nanotechnology: early history and imagined future as seen through patents. In *The Diatoms: Applications for the Environmental and Earth Sciences*, J.P. Smol and E.F. Stoermer (eds.). Cambridge University Press, pp. 585–602.

Gordon, R. (2011). Epilogue: the diseased breast lobe in the context of X-chromosome inactivation and differentiation waves. In *Breast Cancer: A Lobar Disease*, T. Tot (ed.). Springer, pp. 205–10.

Gordon, R., Björklund, N.K. and Nieuwkoop, P.D. (1994). Dialogue on embryonic induction and differentiation waves. *Int. Rev. Cytology*, **150**, 373–420.

Gordon, R. and Brodland, G.W. (1987). The cytoskeletal mechanics of brain morphogenesis. Cell state splitters cause primary neural induction. *Cell Biophysics*, **11**, 177–238.

Gordon, R. and Hoover, R.B. (2007). Could there have been a single origin of life in a Big Bang universe? *Proc. SPIE*, **6694**, doi:10.1117/12.737041.

Gordon, R. and Melvin, C.A. (2003). Reverse engineering the embryo: a graduate course in developmental biology for engineering students at the University of Manitoba, Canada. *Int. J. Dev. Biol.*, **47** (2/3), 183–7.

Gordon, R. and Tiffany, M.A. (2011). Possible buckling phenomena in diatom morphogenesis. In *The Diatom World*, J. Seckbach and J.P. Kociolek (eds.). , Springer, pp. 245–72.

Gordon, R. and Westfall, J.E. (2009). Google Embryo for building quantitative understanding of an embryo as it builds itself: I. Lessons from Ganymede and Google Earth. *Biological Theory: Integrating Development, Evolution, and Cognition*, **4** (4), 390–5, with supplementary appendix.

Graham, J.H., Freeman, D.C. and Emlen, J.M. (1993). Antisymmetry, directional asymmetry, and dynamic morphogenesis. *Genetica*, **89** (1/3), 121–37.

Hellige, J.B. (1993). *Hemispheric Asymmetry, What's Right and What's Left*. Harvard University Press.

Hermeyer, D. and Wantman, S. (2008). Kingston–Rhinecliff bridge. Available at `http://en.wikipedia.org/wiki/File:Kingston-Rhinecliff_Bridge2.JPG` [Accessed July 13, 2011].

Howard, J., Grill, S.W. and Bois, J.S. (2011). Turing's next steps: the mechanochemical basis of morphogenesis. *Nature Rev. Molec. Cell Biol.*, **12** (6), 392–8.

Hwang, S.Y. and Rose, L.S. (2010). Control of asymmetric cell division in early *C. elegans* embryogenesis: teaming-up translational repression and protein degradation. *BMB Rep.*, **43**, (2), 69–78.

Ishihara, S. and Kaneko, K. 2006. Turing pattern with proportion preservation. *J. Theor. Biol.*, **238** (3), 683–93.

Iwasaki, S. and Honda, K. (2000). How many blocks can children pile up? – Scaling and universality for a simple play. *J. Phys. Soc. Japan*, **69**, (6), 1579–81.

Jaffe, L.F. (1999). Organization of early development by calcium patterns. *BioEssays*, **21** (8), 657–67.

Lopez, D., Vlamakis, H. and Kolter, R. (2009). Generation of multiple cell types in *Bacillus subtilis*. *FEMS Microbiol. Rev.*, **33** (1), 152–63.

Maini, P.K., Woolley, T., Gaffney, E. and Baker, R. (2015). Turing's theory of developmental pattern formation. In *The Once and Future Turing – Computing the World*. S. Barry Cooper and A. Hodges (eds). Cambridge University Press, pp. 137–49.

Maurer, U. (2015). Cryptography and computation after Turing. In *The Once and Future Turing – Computing the World*. S. Barry Cooper and A. Hodges (eds). Cambridge University Press, pp. 54–78.

Meeks, J.C., Campbell, E.L., Summers, M.L. and Wong, F.C. (2002). Cellular differentiation in the cyanobacterium *Nostoc punctiforme*. *Arch Microbiol.*, **178** (6), 395–403.

Newman, S.A. and Bhat, R. (2009). Dynamical patterning modules: a 'pattern language' for development and evolution of multicellular form. *Int. J. Dev. Biol.*, **53** (5/6), 693–705.

Nikas, Y. (2011). Human egg in the fallopian tube [#470]. Available at `http://www.eikonika.net/v2/photo_info.php?photo_id=470` [Accessed July 25, 2011].

Parisi, J. (1991). Global symmetry aspects of a compartmentalized reaction–diffusion system. *Comp. and Math. Appl.*, **22** (12), 23–31.

Peter, I.S. and Davidson, E.H. (2009). Modularity and design principles in the sea urchin embryo gene regulatory network. *FEBS Lett.*, **583** (24), 3948–58.

Rayleigh, Lord (1892). On the instability of a cylinder of viscous liquid under capillary force. *Phil. Mag.*, **34**, 145–54.

Redies, C. and Puelles, L. (2001). Modularity in vertebrate brain development and evolution. *BioEssays*, **23** (12), 1100–11.

Reid, R.G.B. (1985). *Evolutionary Theory: The Unfinished Synthesis*. Croom Helm.

Rosen, R. (1968). Turing's morphogens, two-factor systems and active transport. *Bull. Math. Biophys.*, **30** (3), 493–499.

Sharov, A.A. and Gordon, R. (2013). Life before Earth. Available: `http://arxiv.org/abs/1304.3381`.

Sühnel, J. and Hühne, R. (2005). The Amino Acid Repository. Available at `http://www.imb-jena.de/IMAGE_AA.html` [Accessed July 13, 2011].

Tiffany, M.A., Gordon, R. and Gebeshuber, I.C. (2010). *Hyalodiscopsis plana*, a sublittoral centric marine diatom, and its potential for nanotechnology as a natural zipper-like nanoclasp. *Polish Botanical J.*, **55** (1), 27–41.

Tokumitsu, N. and Honda, K. (2005). Crossover scaling in piling block games. *J. Phys. Soc. Japan*, **74** (6), 1873–4.

Tomanek, B., Hoult, D.I., Chen, X. and Gordon, R. (2000). A probe with chest shielding for improved breast MR imaging. *Mag. Res. Med.*, **43** (6), 917–20.

Turing, A.M. (1952). The chemical basis of morphogenesis. *Phil. Trans. Roy. Soc. London*, **B237**, 37–72.

Tuszynski, J.A. and Gordon, R. (2012). A mean field Ising model for cortical rotation in amphibian one-cell stage embryos. *BioSystems*, **109** (3), 381–9.

von Dassow, G. and Munro, E. (1999). Modularity in animal development and evolution: elements of a conceptual framework for EvoDevo. *J. Exp. Zool.*, **285** (4), 307–25.

Watson, R.A. (2006). *Compositional Evolution: The Impact of Sex, Symbiosis, and Modularity on the Gradualist Framework of Evolution*. MIT Press.

West-Eberhard, M.J. (2002). *Developmental Plasticity and Evolution*. Oxford University Press.

Zernicka-Goetz, M. (2011). Proclaiming fate in the early mouse embryo. *Nat. Cell Biol.*, **13** (2), 112–14.

Part Four: Biology, Mind, and the Outer Reaches of Quantum Computation

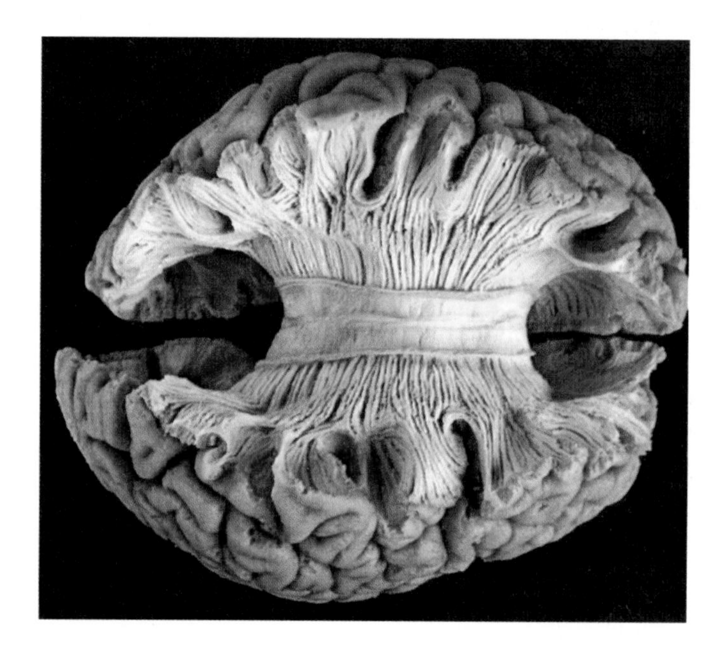

This section focuses on the issue which is both most central and most mysterious: how does the concept of computability relate to the physical world and, in particular, to its quantum-mechanical nature? Ueli Maurer's article in Part 1 refers to the development of quantum computing since 1980s, with its striking results for the analysis of computational complexity. But such quantum computation does not change the absolute boundary of computability. The discussion in this section attacks deeper issues, taking its lead from Turing's analysis of classical computing in 1936. In particular, the question of the reduction (or measurement) process in quantum mechanics, comes to the fore.

Turing, influenced by Eddington, seems to have thought about the meaning of quantum mechanics for the brain as early as 1932. Indeed it is even possible that this line of thought propelled him towards the analysis of mental operations that went into his 1936 paper. Yet quantum mechanics is thereafter absent from his arguments until the 1950s. It is not possible to know what view he had arrived at, but from his remarks on what is now known as the Quantum Zeno Paradox, he must have considered the interpretation of reduction process to be an important problem. Since the 1980s, this whole question has been greatly influenced by the arguments of Roger Penrose. His theme is that what Turing in 1950 called the 'Mathematical Objection to Machine Intelligence', an objection which in 1950 Turing dismissed, is a serious objection indeed. Moreover it must involve the deeper understanding of physical law, involving a new understanding of the reduction process bringing in gravity, i.e the macroscopic curvature of space-time.

It is a striking fact that 80 years after von Neumann's mathematical axiomatisation of quantum mechanics, which Turing read immediately after its publication in English in 1932, there is still no agreed picture of what it means. There is no clear picture of what a wave function is, how random outcomes arise, or how the macroscopic world (notably the brain) relates to the quantum mechanical description of its constituent particles. The two articles in this section, by Stuart Kauffman and Scott Aaronson, address these fundamental questions. Whilst they have a similar point of departure as Penrose, in that they question the nature of the reduction process, they put forward interpretations which do not involve gravity. Scott Aaronson explicitly refers to Turing's early and late thoughts and says "The reason I'm writing this essay is that I think I now understand what Turing could have meant by these remarks."

Both of these provocative articles share the spirit of scientific adventure, that led Turing to explain during the course of his prohibitively technical 1939 paper exploring *Systems of logic based on ordinals*, that:

> Mathematical reasoning may be regarded ... as the exercise of a combination of ...intuition and ingenuity ... In pre-Gödel times it was thought by some that all the intuitive judgments of mathematics could be replaced by a finite number of ... rules. The necessity for intuition would then be entirely eliminated. In our discussions, however, we have gone to the opposite extreme and eliminated not intuition but ingenuity, and this in spite of the fact that our aim has been in much the same direction.

One may detect here an implicit concern with the computational nature of the human brain. What is it that is computationally special about 'intuition' that it might take one up the hierarchy of theories beyond the classically computable? Did Turing *really* mean to engage with a modelling question still debated in today's world?

One aspect of Turing's thinking which is so special, is its responsiveness to the actuality of the world he passed through in his short life. The experience of embodied computation at Bletchley Park leads into the brilliant closing years of his scientific career in which he faces up to the challenges of brain and morphogenic functionality. His allegiance to logical structure shows in his 'bottom-up' work on connectionist models, and his well-known comment that "We are not interested in the fact that the brain has the consistency of cold porridge."

But his influential 'top-down' Turing Test for machine intelligence, and comprehensive success with certain morphogenic examples, shows a clear respect for embodied computation, and its potentially transgressive computational nature. His Universal Computer may have underpinned the functionalist paradigm underlying today's computer dominated world, but throughout his life he was alive to the essential nature of embodied contexts – the quantum physics (he attended Dirac's lectures in Cambridge), the biology (his mother's sketch of him engrossed in inspecting the daisy during hockey at school), the preoccupation with brain/mentality – from which grew troubling challenges.

Scott Aaronson's beautifully written extended essay describes his early impression that "Turing [had] finally unmasked the pretensions of anyone who claimed there was anything more to mind, brain, or the physical world than the unfolding of an immense computation." After which he goes on to confess discovering "that Turing's own views about these issues, as revealed in his lectures as well as private correspondence, were much more complicated than my early caricature." Aaronson's discussion of free will is very much in the Turing computational tradition, describing how "a physical system will be 'free' if and only if it's unpredictable in a sufficiently strong sense."

Stuart Kauffman also explores the physics with a view to transcending the classical Turing model. His Turing interpretation and departure point is:

> Turing, in the Turing Test, or 'Imitation Game', soon turned to the question of whether the human mind was itself a Turing machine. He thought, after careful consideration, that the answer was "Yes". He did, however, retain doubts, partially reflected in his use of humans, not algorithms, as the judges in the Turing Test. Turing scholars rightly admire his capacity to doubt himself.

This is a crucial point, of course – and a key observation that future researchers will doubtless need to take on-board.

11

Answering Descartes: Beyond Turing

Stuart Kauffman

Introduction

The first half of the twentieth Century was filled with a stunning group of scientists, Einstein, Bohr, von Neumann and others. Alan Turing ranks near the top of this group. I am honored to contribute to this volume commemorating his work. How much do we owe one mind? His was a pivotal role in the cracking of the Nazi war code, which profoundly aided the defeat of Nazism. His invention of the Turing machine has revolutionized modern society, from universal Turing machines to all digital computers and the IT revolution. His model of morphogenesis, the first example of a 'dissipative structure', to use Prigogine's phrase, is one I have myself used as a developmental biologist.

I rightly praise Turing, but seek in this chapter to go beyond him. The core issue is the human mind. Two lines of thought, one stemming from Turing himself, the other from none other than Bertrand Russell, have led to a dominant view that the human mind arises as some kind of vast network of logic gates, or classical physic-sconsciousness neurons', to use F. Crick's phrase in *The Astonishing Hypothesis* (Crick, 1994), connected in the 10^{11} neurons of the human brain.

I think this view could well be right, but is more likely to be wrong. My aim in this chapter is to sketch the lines of thought that lead to the standard view in computer science and much of neurobiology, and note some of the philosophic claims for and doubts about the claim; but most importantly I wish to explore the emerging behavior of open quantum systems in an environment, their new physics, and, centrally, our capacity to construct what I will call non-algorithmic, non-determinate yet non-random Trans-Turing Systems. As we shall see, Trans-Turing Systems are not determinate, for they inherit the indeterminism of their open quantum system aspects, yet they are non-random owing to their classical aspects. They are new to us and may move us decisively beyond the beauty, and limitations, of Turing's justly famous, but purely classical-physics, discrete-time and discrete-state machine.

Beyond the above, I shall make one truly radical proposal that I believe grows out of "sum over all possible histories" formulation of quantum mechanics (Feynman, 1948). This formulation is fully accepted as an equivalent formulation of quantum mechanics. I will show that Feynman's formulation evades Aristotle's Law of the Excluded Middle, while classical-physics and, *a fortiori*, algorithmic discrete-state, discrete-time Turing machines, obey the Law of the Excluded Middle. Following the philosopher Pierce (1960) – who pointed out that 'Possibles' evade the Law of the Excluded Middle, while 'Actuals' and 'Probables' obey that law – and Whitehead (1978), I shall propose for our consideration a new dualism, *res potentia* and *res extensa*, the realms of the ontologically real Possible and ontologically real Actual, linked, hence truly united, by quantum measurement. In contrast, the dualism of Descartes, *res cogitans*, thinking stuff, and *res extensa*, his mechanistic world philosophy, have never been united. I believe *res potentia* may be a consistent and new interpretation of 'closed' quantum systems prior to measurement. These ideas and other much less radical ones resting on open quantum systems lead to new and testable hypotheses in molecular biology, cellular biology, and neurobiology, and, hopefully, a new line of ideas in the philosophy of mind including the following proposals: how mind acts acausally on brain; an ontologically responsible free will; what consciousness *is*; the experimentally testable loci of qualia as associated with quantum measurement itself; the irreducibility of both qualia and quantum measurement; the unity of consciousness, i.e. the 'qualia binding problem' and its cognate 'frame problem' in computer science. From these, technological advances in a number of directions may flow.

Mind as machine

As already noted, there are two strands – from Turing and the Turing machine, and from Bertrand Russell – that both lead to the view of the mind as a classical physics 'computing machine'. The strand from Turing is well known. It begins with the Turing machine, the very definition of algorithmic behavior. To recall, a Turing machine consists, in general, of an infinite tape divided into squares. On each square, one from a finite number of more than one symbol, say 0 and 1, is written. A reading head begins poised over one square. The head contains two sets of rules. The first rule prescribes the following actions. If situated over a tape square with a given symbol written on it, the head will stay where it is, or move one step to the left or right, or erase the symbol on the square below it and write a symbol from the defined alphabet of discretely different symbols on that square.

The second rule specifies that, under the above conditions, the reading head will change from one of a finite number of discrete internal states to some internal state. Thereafter, the system iterates. There is, in addition the crucial 'halting state'.

Turing showed that any recursive computation that could be carried out could be carried out by a universal Turing machine. From this followed wonderful theorems about the formal undecidability of the 'Halting problem', the demonstration that most irrational numbers were not computable, and other remarkable advances.

The feature of the Turing machine that I wish to emphasize is that it is **absolutely definite** or **determinate**. Given the symbols written on the tape, and the rules in the reading head, its behavior at each step is fully determined. This determinate behavior is essential to the algorithmic character of the Turing machine. Because it is determinate, the Turing machine is bound by classical physics. However, Turing machines are discrete-state and discrete-time systems, while classical physics more generally is based on continuous variables and continuous time and is also deterministic and can, since Poincaré, exhibit deterministic chaos.

Computer scientists often distinguish between algorithms that may halt with an answer, and those that are 'processes', such as Holland's genetic algorithm (Holland, 1975) which just continues or halts at some defined success criterion.

Turing (1952), in the Turing Test, or 'Imitation Game', soon turned to the question of whether the human mind was itself a Turing machine. He thought, after careful consideration, that the answer was *Yes*. He did, however, retain doubts, partially reflected in his use of humans, not algorithms, as the judges in the Turing Test. Turing scholars rightly admire his capacity to doubt himself.

Russell and onwards to mind as machine

At the turn of the twentieth century, Bertrand Russell, having just published with Whitehead the *Principia Mathematica*, turned to the problem of maximally reliable knowledge of the 'external world'. We could be wrong, he reasoned, that there actually was a chair in the room. But we could hardly be wrong that "We seemed to be seeing a chair". That is, statements about our experiences, say visual, were less corrigible, or error-prone, than our statements about the external world. Russell and his contemporaries, including the young Ludwig Wittgenstein, hoped to build up knowledge of the external world from experience itself.

Pause and look at the room or world around you. You experience a 'whole' visual field called in neuroscience the 'unity of consciousness'. This unity will be central to my interests. However, Russell threw away the unity of consciousness in his very first philosophic move. He invented, whole cloth, 'sense data', such as 'red here' or the musical note 'A-flat now' (Russell, 1912). That is, Russell shattered the unity of consciousness into bits, soon to be related to computational 'bits'. The next step

(Russell, 1912) was to invent 'sense data statements', e.g.

It is true for Kauffman that A-flat now.

Why did Russell make this move? Because his *Principia* hoped to construct the entire mathematical world from first-order predicate calculus. Then the hope was that the statement, "There is a chair in the room", could be translated into a **logically equivalent statement** comprised of a finite list of true or false sense data statements and quantifiers such as 'There Exists', and 'For All'. If the move worked, knowledge of the external world would be set on a firm foundation.

The discussion took perhaps 40 years but the move, culminating in *Tractatus Logico-Philosophicus*, (Wittgenstein, 1921), did not work. The statement, "There is a chair in the room" could not be translated into a logically equivalent set of sense data statements in first-order predicate calculus. Philosophers gave up on the idea that there was a 'basement' language from which all other knowledge of the world, captured in propositions, could be formulated.

Later, in his transforming opus *Philosophical Investigations*, Wittgenstein (1953) pointed out that there was no basement language. Rather, language about legal proceedings could not be translated into logically equivalent sets of statements about ordinary human actions. Each 'level' constituted a 'language game', not reducible to a lower level. Thus: "Kauffman is guilty of murder." requires for its understanding a co-defined set of concepts such as 'trial', 'jury', 'legally admissible evidence', 'legally competent to stand trial' . . . that cannot be translated or 'reduced' into sets of statements about ordinary human actions.

This step is critical, for it says that there is no logical procedure, certainly no first-order logic, to get from a lower-level language game – here normal human action – to a higher-level language game – here legal language. But then there is no first-order logic 'algorithmic procedure' to get from the lower to the higher language. Yet we learn legal language. This is one line of argument suggesting that the human mind is not merely algorithmic.

Despite some philosophers giving up on a basement language, the early cyberneticians, W. McCulloch and W. Pitts, in 1943 published a seminal paper that would lead to the contemporary theory of neural networks and 'connectionism'.

McCulloch and Pitts showed that in a network of on/off formal neurons, constructed in a feed forward network, with *N* formal neurons per row and *M* rows, and in which the input row 'neurons' could be placed in any arbitrary combination of '1' and '0' states, the network, with arbitrary threshold Boolean functions such as AND, OR, and NOT, could compute any logical function on the 'states' of the input neurons.

Implicitly, they identified the '1' or '0' state of a formal neuron with the truth or falseness of a Russellian sense data statement, such as, "For Kauffman, 'A-flat now' is true", which might be encoded by a '1' on the first neuron in the input layer to the feedforward network of formal neurons.

More generally McCulloch and Pitts (1943) considered networks with feedback loops. They entitled their paper *A logical calculus of the ideas immanent in nervous activity*.

In this step, McCulloch and Pitts set the stage for the now generally accepted view in computer science, neurobiology, and much of the philosophy of mind, that an 'idea' in the mind was logically identical to the on or/off states of a set of formal neurons, or in contemporary neurobiology, with the axonal firing or not of members of a set of 'consciousness neurons'.

Note that McCulloch and Pitts chose the terms, 'immanent in nervous activity'. In some magical way, the sense data features, or sense experiences, i.e. 'qualia', are 'slipped' into the '1' and '0' behaviors of the formal neural net.

Note further that this conceptual move:

(1) assumes that there is a basement language, captured in the '1' and '0' states of the formal neurons;
(2) has, with Russell, thrown away the Unity of Consciousness and will have to reconstruct it.

In contemporary neurobiology this issue has returned as the famous 'binding problem', i.e., how does the firing of unconnected 'consciousness neurons' become bound into a unity of consciousness, or more simple examples such as this, from *The Astonishing Hypothesis*, Crick (1994):

> Suppose I see a yellow triangle and a blue square.
> Suppose 'yellow', 'triangle', 'blue' and 'square' are, in fact, processed in different, unconnected, areas of the brain.
> How do 'yellow' and 'triangle' become bound together, while 'blue' and 'square' become bound together?

Following the logic of McCulloch and Pitts, the early hope was brain 'grandmother cells' that fired if and only if you saw a combination of features, sense data, that corresponded your grandmother. Now reconsider the number of relational features of your visual field. How many grandmother cells would be required, each to encode by firing 'if and only if' presented with one of each of the possible combinations of up to, say 30 features at a time, out of say 10,000 features you can discriminate? The answer is

$$\binom{10000}{30} = \frac{(10000!)}{(99770!) \times (30!)},$$

a vast number. Crick (1994) concluded that the idea does not work, it would take more than the 10^{11} neurons to encode all the sets of relational features you see. One current hope is a 40 hertz oscillation in the brain. The idea is that if 'yellow' and 'triangle' neurons fire at the same phase of the oscillation they will be bound, and if 'blue' and 'square' fire at a different phase, they too will be bound. Well, maybe, but how do we squeeze maybe trillions of combinations of relational degrees of freedom into different phases of a 40 hertz oscillation? I find it implausible. While detailed work on binding is beyond the scope of this chapter, in general the issue remains: binding anatomically unconnected classical-physics neurons and their presumed qualia, or experiences.

Note that this binding problem arises, descending from Russell, with the idea of sense data and sense data statements, true or false, as a **digital and propositional encoding** of our experience of the world in our unity of consciousness. Below I will offer an unexpected **analog and non-propositional** encoding which may solve the binding problem. But there is another deeper issue: McCulloch and Pitts, and all later neural network theory, cannot meet Wittgenstein's language-game argument that there is no 'basement language' and that the learning of higher language games cannot be based on algorithmic procedures from that basement language.

Despite the warning of Wittgenstein connectionism has flowered, much along the ideas above but with important improvements such as 'back propagation', (Werbos, 1994), and 'Hopfield networks' (Hopfield, 1953), with attractors encoding classes or memories and content-addressable memory. These are now the basis of voice recognitions systems around the world. But the language-game problem remains unsolved, so on this ground mind seems not to be algorithmic.

I point to another important line of evidence that the mind is not algorithmic. I ask you to name all the possible uses of screwdrivers: screwing in screws, opening paint cans, tied to the end of a stick to spear fish, rented to locals to spear fish and you take 5% of the catch ... Is there a statable list of the possible uses of a screwdriver for all possible purposes? I think not. How would we construct such a list, know we had completed the list, or at least made it 'infinite but recursively enumerable'? More precisely, the set of uses of a screwdriver alone or with other objects or processes is indefinite and also unorderable. But this means that there is **no algorithmic** procedure to effectively enumerate all possible uses of a screwdriver. This is the famous 'frame' problem of computer science, yet to be solved algorithmically. Again: there **is no bounded or recursively orderable set of functionalities of human artefacts for all possible purposes**, so no algorithm can list such uses yet we literally discover and invent them all the time in the evolution of the econosphere. We routinely solve the frame problem. If so, the human mind is not always algorithmic.

I note that Penrose (1989, 1994) also argues that the human mind is not always algorithmic on the basis of its capacity to prove incompleteness theorems such as

Gödel's theorem and the Halting Problem. I join Penrose, who precedes me, but on different grounds, in thinking that the mind is not algorithmic and I join him in thinking that quantum mechanics is related to consciousness.

Mind, consciousness, and the mind as machine

Two major positions can be taken with respect to mind as a classical-physics, and further, a discrete-space, discrete-time, and discrete-state algorithmic computational machine with inputs from a discrete-space, discrete-time and discrete-state environment. First, we are not conscious at all but are zombies. This view was discussed by Dennett (1991), and is, in part, a sophisticated form of logical behaviorism making use of an extensively developed computer science framework. A contrary argument was made by Searle (1997) in his debated but famous Chinese Room argument. This claims to show that mind is not a Turing machine, which is merely syntactic in its manipulation of symbols having no semantics and hence cannot experience the meanings of words.

In one form or another, the view of the mind–brain system as a network of classical-physics neurons, with continuous variables and continuous time and interacting in classical-physics causal ways via action potentials, vast networks with classical-physics inputs and outputs, is the dominant view today. Edelman (1992), Crick (1994), Searle (1997), and most working neuroscientists hold this view. According to Searle, functionalists such as H. Putnam and D. Lewis are 'property dualists' who see mental terms such as 'believe' as constituted by a classical-physics causal network, whether made of neurons or beer cans. Searle (1997) asserted that functionalists do not mean by mental terms the actual experience of, for example, pain. These two paragraphs cannot characterize the vast scholarly work touched on above, yet those efforts have neither answered Descartes, introduced below, nor found a home for consciousness itself.

Then, whence consciousness, experience, qualia? A popular view is that at some level of complexity of a network of logic gates, whether electronic, or water bowls pouring into one another above and below a 0/1 threshold, or classical-physics continuous-time and continuous-state neurons, consciousness will 'emerge'. It is commonly pointed out that a single H_2O molecule is not wet, but a sufficient collection of them is. So too can consciousness emerge.

Perhaps consciousness can so emerge, but here is the first deep problem. If the emergent consciousness is a classical-physics 'process', for example an electromagnetic field as some argue, then it is a deterministic classical-physics system. Consider Newton's three laws of motion and and his law of universal gravitation, and billiard balls moving on a table. The boundary conditions of the table and the current positions, momenta and diameters of the balls completely determine the entire future trajectory, perhaps deterministically chaotic, of the balls.

But if the mind–brain is a deterministic machine, we can have no ontologically real and responsible free will. I walk down the street and kill a little old lady with a frying pan, yet I am not responsible. I was physically determined to whack her. Even in the face of deterministic chaos I have no ontologically real responsible free will, merely perhaps the epistemic illusion of one.

Thus, the familiar view, derived from Turing and Russell, may be right; consciousness may be a classical-physics 'something', but we buy it at the price of no ontologically real responsible free will.

This is a high price to pay. To avoid it, I will offer below a set of ideas that appear to afford us, among other things, an ontologically responsible free will.

There is another huge set of problems, derived from Descartes (1637) in his *Discourse on Method*. As mentioned earlier, Descartes postulated a famous dualism: *res cogitans*, thinking stuff, and *res extensa*, his mechanical world view, which later led to Newton and celestial mechanics and thence to classical physics.

But the problem immediately arose of how *res cogitans* is connected to *res extensa*. Descartes proposed the pineal gland. The idea does not work.

Given Newton, here is the issue. If the brain is a deterministic dynamical system, like billiard balls on a table, then the current state of the brain is entirely sufficient for the next state of the brain. Then there is **nothing for mind to do. Worse, there is no way for mind – experiences – to act on brain**! What should mind do, somehow magically to cause the billiard balls to swerve despite the sufficiency of Newton's laws?

This central problem arises due to the **causal closure of classical physics**. It is due to causal closure that we claim that Newton's laws (in differential form), plus the initial positions and momenta and diameters of the billiard balls and boundary conditions, will, once integrated, be entirely sufficient to yield the entire future trajectory of the balls on the table.

Thus, the Turing model of the machine mind leaves us with no free will, and mind, experiences or qualia, if they can arise at all, are unable to affect the classical-physics machine aspect of the mind–brain system. We retreat to mind as a mere epiphenomenon, of no effect in our actions as humans, or as a 'compatibilism' which rejoices that at least as deterministic systems we can train one another to be moral machines.

In truth, we have been stuck with this cycle of problems since Descartes. Turing machine minds are frozen in the same way. If the central problem above is due to the causal closure of classical physics, then I believe we must forsake the limitations of classical physics and purely classical-physics 'consciousness neurons' for a view that embraces the non-deterministic behavior arising from quantum mechanics.

I turn now to such a radically different approach to the mind-body problem. It will take us through open quantum systems, the 'poised realm' between open quantum and classicality for all practical purposes (FAPP) to non-determinate, hence non-algorithmic, yet non-random trans-Turing systems, beyond Turing, to my tentative postulate about a new dualism, ontologically real *res potentia*, the realm of the Possible, and *res extensa*, the realm of the Actual, linked – hence united – by quantum measurement. This postulate is also an interpretation concerning what the unmeasured Schrödinger wave is 'about', where we have had no idea since the Schrödinger equation (Schrödinger, 1926). The postulate of *res potentia* leads to a resulting idea of consciousness as a participation in *res potentia*, i.e. in ontologically real possibilities and strengthens the independent hypothesis that qualia, i.e. conscious experiences, are associated with quantum measurement. Most of what I shall say is independent of an ontologically real *res potentia*. But there is more: we escape the digital 'propositional' model of mind with the realization that a quantum wave process in a potential well knows in an analog – not a propositional or digital – way its potential-well boundary condition or 'context', as part of solving the binding problem. I will link this analog 'knowing of qualia' to quantum entanglement among many synapses in the brain as the candidate loci of quantum behavior, and the **quantum measurement** of those entangled degrees of freedom to achieve non-local EPR high correlations (Aspect et al., 1982) hence 'binding' of vastly many qualia, one per measurement, to solve the binding problem and achieve the unity of consciousness.

Answering Descartes

Quantum biology emerged with the discovery that chlorophyl wrapped by its chromophore-bearing antenna protein can be quantum coherent for 700 femtoseconds or more (Ishizaki et al., 2010). I believe, however, that quantum coherence may be only a small part of the quantum effects in biology. We biologists may find ourselves learning and collaborating with quantum physicists and quantum chemists in ways we cannot yet tell. This part of the chapter is an attempt to peer into this new territory.

Closed quantum systems and the two-slit experiment

Many readers will be familiar with the famous two-slit experiment; see for example Feynman et al. (1964). A photon gun emits photons, say one per minute, at a screen with two slits close together; behind the screen is a photodetector, say a film emulsion. If either slit is covered, one obtains a bright spot on the photodetector behind the open slit. Stunningly, if both slits are open, one obtains the famous bars of light, dark, light, dark..., an interference pattern. No classical objects, e.g. classical particles, can yield this result. It is a hallmark of quantum mechanics (QM).

A classical analogy helps us understand the underlying time-dependent Schrödinger equation of QM. Imagine a sea wall with two gaps, and a beach beyond. Let a series of plane waves approach the wall. As it passes through the gaps, each wave yields two semicircular wave patterns that approach the beach. If these semicircular patterns overlap at the beach, there will be points on the beach where the crests of the two wave patterns coincide, yielding a higher wave crest. Similarly there will be points where the troughs of two waves coincide yielding lower troughs. But there will also be points on the beach where the peak of one wave coincides with the trough of another wave and the two will cancel entirely.

The Schrödinger time-dependent linear wave equation produces similar waves. Where peaks and peaks coincide, or troughs and troughs coincide, one obtains a bright bar of photons in 'constructive interference'. Where peaks meet troughs, they cancel yielding dark bars in 'destructive interference' and hence the interference pattern. An 'action' variable in the equation keeps track of the phases in time and space of the Schrödinger waves. Quantum 'weirdness' arises due to the linearity of the equation, for sums and differences of solutions are also solutions. This linearity permits the famous Schrödinger's Cat paradox in which a cat in a box, prior to measurement, is **simultaneously** both dead and alive.

It is notable that, at first no one knew what was 'waving' in the Schrödinger wave equation. However, the axiomatization of quantum mechanics (von Neumann, 1932), included this propagating Schrödinger wave and the mysterious quantum measurement process. Here each wave has an amplitude. The **square** of the modulus of an amplitude, called the Born rule (Born, 1989), yields the **probability** that that amplitude will be measured in von Neumann's Process 1, or 'R' process with its controversial 'collapse of the wave function' of many amplitudes to only one, which can become classical as in the spot each photon makes on the screen of the two-slit experiment. In general, there is, to the best of my knowledge, no agreed derivation of quantum measurement from within QM.

Open quantum systems

The emergence of the classical world from the propagation of the Schrödinger wave equation is a deep mystery. One of the current best hypotheses requires distinguishing a quantum 'system' from its 'environment' yielding an 'open quantum system' and its 'environment'. The key idea is that phase information within the open quantum system can be lost, acausally, to the quantum environment. This process is called 'decoherence' (Zurek, 2002). Then, within the system, the 'action' gradually loses information about where the peaks and valleys of the Schrödinger wave 'are', so constructive and destructive interference cannot happen, nor can interference patterns. This interference hallmark of quantum effects is gradually lost and classicality is approached arbitrarily closely, reaching classicality FAPP.

Decoherence is well established experimentally. It disrupts quantum coherent qubit behavior in quantum computers.

Critically, decoherence is yielding new physics. First, decoherence takes time. A typical time scale is a femtosecond. During that time phase information is being lost from the quantum system. The Schrödinger wave equation is time-reversible. But decoherence is a dissipative process, so it is not fully describable by the Schrödinger equation. New physics is expected and found.

I give three examples of this new physics. First, we are all familiar with radioactive decay half-life, which is due to the closed quantum system Poisson-distributed decay of the radioactive nucleus whose integral is the familiar halflife of exponential decay. In the confirmed quantum anti-Zeno effect, the decay is not exponential (Fischer et al., 2001). New physics.

Second, of particular interest to biologists, decoherence can alter the rate of chemical reactions (Prezhdo, 2000). Decoherence happens in cells. What are the implications for molecular, cellular, neural, biomedical, drug, and other behaviors? We don't yet know.

An essential feature of decoherence is that the weird superposition states, the cat simultaneously dead and alive, decohere very rapidly, leaving what are called one or more 'pure states', if more than one, this is called a mixed state. Thus the cat is either dead or alive, but not simultaneously both. We don't know which until quantum measurement happens[1].

Third, recoherence, including to a new superposition state, is possible for open quantum systems.

(i) Several papers by Paz and Roncaglia (2009), Cormack and Paz (2010), Cai et al. (2008), and Briegel and Popescu (2008) show that a quantum entangled state can decohere to classicality, FAPP, and recohere again.
(ii) Imposition of a classical field can induce recoherence (Mainos, 2000).
(iii) The quantum error correction theorem (Shor, 1995), states that if in a quantum computer some qubits are partially decoherent then measurement can be undertaken, and information injected, correcting the qubits back to full coherence.

In summary, and stunningly, for open quantum systems it is just becoming known

[1] Seth Lloyd, Miles Blencowe, personal communications.

that both decoherence to classicality FAPP and its **reverse**, recoherence, perhaps to a new quantum coherent superposition state, can occur. Then, in principle, quantum degrees of freedom, including biomolecules, can 'hover' between open quantum behavior and classicality, FAPP. It is right to stress, as above, that this may have very large implications for the actions of molecules in cells, and drug discovery, design, and action. After all, we treat biomolecules as classical. We may be wrong.

The poised realm

Gabor Vattay, Samuli Niiranen, and I have proposed 'The poised realm' between fully coherent quantum behavior in **open** quantum systems and classicality, FAPP. Picture a two-dimensional coordinate X, Y system. The Y-axis rises from the origin, where there is open quantum coherent behavior, via decoherence to classicality FAPP up the Y-axis, and via recoherence down the Y-axis to open quantum coherent behavior. The X-axis is new, comprising 'order', 'criticality' and 'chaos'. The two axes box in the poised realm. The X-axis, order, criticality, and chaos is well defined in the classical limit, and now is being extended to embrace partially open quantum behavior in the presence of different extents of decoherence and recoherence.

The motion along the X-axis from the origin, characterized classically by a frictionless pendulum, can be obtained in at least two ways. The first uses the 'Hamiltonian" of the classical system. A pendulum is perfectly ordered. If released from different initial heights, the frictionless pendulum describes roughly circular orbits in a coordinate space of position and velocity. These circular orbits are parallel, hence neither converge nor diverge. Mathematically, this lack of divergence or convergence is described by a zero-valued Lyapunov exponent. As one moves out along the X-axis, the Hamiltonian of the system changes. In the ordered regime, the Lyapunov exponent remains constant at zero. But when the Hamiltonian is deformed enough, at 'criticality' the Lyapunov exponent becomes slightly positive, and the onset of divergence of flows in state space constitutes chaos. As the Hamiltonian is modified further, the Lyapunov exponent becomes more positive. This kink at 'criticality' is a 'second-order phase transition', and is well established (Mackay, 1993). A second means of moving out along the X-axis consists in using a 'kicked quantum rotor'. A quantum rotor is a one-dimensional hoop of states around which a quantized electric charge rotates. It can be kicked by a laser with, say, intensity K. As K increases, Vattay[2] has shown that at first there are many amplitudes propagating, then few, then a single amplitude transforms to 'classical' diffusive behavior in momentum space. This classicality is reversible if K is de-

[2] G. Vattay 'Transition from quantum to classical behavior in kicked quantum rotors as kicking intensity increases'. Personal communication.

creased or the Hamiltonian is changed. Thus, classicality, presumably FAPP, can be **reversibly** achieved up the Y-axis or out along the X-axis.

The non-algorithmic, non-determinate, yet non-random Trans-Turing System

I recall here the fully algorithmic Turing machine described above. Several points, sketched above, are essential:

(i) all contemporary computers are based on the Turing machine;
(ii) the Turing machine is **completely definite**; it is the perfect instantiation, restricted to discrete space, time, and state, of classical physics and Descartes' *res extensa* machine world view;
(iii) this definite behavior of a Turing machine defines algorithmic behavior;
(iv) critically, a major contemporary view in neuroscience and computer science and much of the philosophy of mind is that the mind–brain system *must be* algorithmic – some huge system of interconnected logic gates or, more broadly, of continuous-time and continuous-state classical-physics neurons firing.

I now describe non-algorithmic, non-determinate, but also non-random Trans-Turing Systems (TTS). None has been constructed. I believe they are constructible. Moreover, the mind–brain system may be not only a vast *non*-algorithmic, non-determinate system, in contrast with classical physics in general, but also a non-random TTS. More broadly, classical physics is state determined. The mind-brain system may be partially open quantum and poised realm, hence, whether it works via decoherence to classicality FAPP or via quantum measurement, the mind–brain system may not be a state-determined system.

The central ideas are simple. A TTS 'lives in' the poised realm and perhaps involves quantum measurement in the poised realm.

- First, there are quantum degrees of freedom propagating in short-lived superposition states that decay rapidly due to decoherence. But these short-lived superposition states undergo constructive and destructive interference and will be one basis for non-determinacy in the TTS when coupled to decoherence to classicality FAPP, or quantum measurement. Thus a TTS is not algorithmic, not determinate, and not state determined, in contrast with a Turing machine.
- Second, either via decoherence or motion out along the X-axis or both, quantum degrees of freedom become classical FAPP or, via quantum measurement, become classical *simpliciter*. Both decoherence and measurement are acausal and yield the non-determinant behavior of the TTS.
- Third, there are, in addition, coupled classical degrees of freedom in the TTS.

- Fourth, when quantum degrees of freedom and either superposition states or pure states become classical FAPP, or are measured, that **alters** in different specific ways the effects of the now classical degrees of freedom on one another, and thus alters the non-random collective dynamics of the coupled classical degrees of freedom. In turn this altered non-random classical behavior alters non-randomly the behavior of the remaining quantum degrees of freedom.
- Fifth, in turn this non-random alteration of the behavior of the remaining quantum degrees of freedom alters non-randomly whichever of the open quantum degrees of freedom decohere or move out along the X-axis to classicality FAPP. In particular, higher quantum amplitudes tend to decohere with higher probability. So non-randomly altered quantum behavior, including altered constructive and destructive interference, alters non-randomly which amplitudes become higher, and thus alters non-randomly which amplitudes decohere to classicality FAPP.
- Sixth, in turn classical FAPP degrees of freedom can recohere, for example, if they are driven by a coherent electromagnetic field whose intensity and period distribution can be tuned non-randomly, thereby injecting information. The recoherent degrees may achieve a new controlled superposition state, thereby altering non-randomly the constructive, destructive, and pure-state behaviors among themselves and other quantum amplitudes, and thereby non-randomly affecting which amplitudes achieve higher amplitudes and tend to decohere and also non-randomly altering the behaviors of the coupled classical degrees of freedom in the TTS.

These six ideas are the building blocks of a Trans-Turing System.

Part of a TTS has been realized in a computation in de la Lande et al. (2011). The authors considered a quantum system of many nuclei and many electrons. This system has two potential wells, with corresponding states, say A and B. The vertical Y-axis represents energy. The X-axis represents a chemical reaction coordinate. The two potential wells overlap at some point in the X–Y plane, in what is called the 'seam region'. At this seam the nuclei are in a superposition of states, simultaneously A and Not A, B and Not B. Via gradual decoherence the nuclei fall into one of the minima of the wells, and become classical FAPP. But in turn this alters the effect of the now classical FAPP nuclei on the electron cloud which does not rapidly decohere. Thus, if the nuclei are now in the well corresponding to A, say, the electrons behave differently than if the now classical nuclei FAPP are in the other well and the electrons behave differently if the nuclei are still a superposition in the seam region.

This model is the first instantiation in quantum chemistry of which I know in which some quantum degrees of freedom, here the nuclei in a superposition of A and Not A and simultaneously B and Not B, decohere to classicality FAPP, to a well corresponding to *A* or to *B*, and thereby alter the behavior of the remaining quantum degrees of freedom, the non-decohering electrons.

A more refined calculation would allow the many nuclei in this system to decohere in some sequential order. As they do, the newly classical FAPP nuclei will yield a sequential alteration in the behavior of at least the electrons and probably the remaining open quantum system superposition nuclei, as well as the other now classical FAPP nuclei. That research lies in the future as does the study of such a system if the classical FAPP nuclei can be made to recohere to some perhaps new superposition state, perhaps by an external field, perhaps by interactions of many such subsystems within a molecule.

The essential points about a TTS are:

(1) Its behavior is not Turing-definite, because of the constructive and destructive interference of superposition states, followed by their falling to a classical FAPP state where high amplitudes preferentially decohere, and where the remaining quantum pure states will also decohere probabilistically or by quantum measurement. Further, the total constructive and destructive interference behavior, and further controlled recoherence behavior, alters non-randomly which states achieve high amplitude and so decohere preferentially to classicality FAPP with what probability, or are quantum measured, by the Born rule, with what probability. The ongoing behavior is not definite, hence is **not** algorithmic. The behavior is not state determined. The behavior is also non-random.

(2) The above behavior is, as noted, not 'quantum random', as in the case of radioactive decay, for a further reason: the classical degrees of freedom have their own Hamiltonian, and hence non-random dynamics, which may, in addition, affect non-randomly the behaviors of the quantum degrees of freedom, and so which quantum amplitudes, via constructive and destructive interference, become high amplitudes and preferentially decohere or are preferentially measured via the Born rule. The behavior is both non-deterministic, and non-random.

(3) The TTS may receive quantum, open quantum, poised realm, and classical inputs and output open quantum, poised realm, and classical output behaviors. So it is a non-algorithmic, non-deterministic via decoherence to classicality FAPP or quantum measurement, yet non-random information processing system. Consequently if a TTS, as a single or coupled system is constructible, perhaps in liposomes or nano-devices, we have a new non-algorithmic, non-state-determined, and not random 'device' – that is, unlike a Turing machine or logic gate or deterministic classical physical system – to consider for the mind–brain system. We are no longer almost forced to the conclusion that mind–brain

must be described by classical physics, definite, and either discrete-time and discrete-state logic gates or continuous-time continuous-variable deterministic 'consciousness' neurons, coupled into a huge network. TTS may also take us far beyond the Turing machine technologically.

A responsible free will

As noted above, the view that consciousness emerges from a vast network of classical-physics logic gates or classical-physics neurons may be entirely correct. However, it has a big price: that we are deterministic so have no ontologically real responsible free will. Such a system could exhibit chaotic behavior, yielding the 'illusion' of free will, but such a free will would not be ontologically real, for the classical-physics neural system would remain deterministic.

But there is another horn to this free-will dilemma if we seek an ontologically real and responsible free will and then try to use standard quantum randomness. I have a radioactive nucleus in my brain, I walk down the street, the nucleus randomly decays, and I kill a little old man so my 'free will' is ontologically real due to quantum indeterminism. But killing the old man is not my fault, just random quantum chance! I have no *responsible* free will if we use quantum randomness.

But a TTS is both *not* deterministic, hence not algorithmic, and *not* quantum random, it is something entirely new. I hope this can break the horns of the standard responsible free-will dilemma and allow for an ontologically real and responsible free will. I believe more is needed, building upon the idea of Ross Ashby's famous homeostat, (1952), with its subset of "essential [classical-physics] variables" that must be kept in bounds, to provide an internal "goal state" for the total system, to begin to yield a non-random but non-deterministic responsible free will.

Ashby's starting sketch, even if right, is inadequate. There is no mention of some analog or actuality of sensory inputs, or motor outputs or the capacity of coupled TTSs, or set of entangled TTSs, joined to the classical aspects of the brain, presumably classical-physics neural networks, to classify its environments and act appropriately given goals and subgoals. Below, in proposing the testable hypothesis that qualia are associated with quantum measurement, it seems that 'experiences' have, as a natural dual, the rudiments of an 'I', that which experiences. 'Agency', on this view, is an elaboration of these rudiments in the total mind–brain system. I note that Penrose (1989, 1994), seeks a non-deterministic, non-algorithmic, yet non-random behavior of consciousness via a modified non-deterministic, so non-algorithmic, yet non-random quantum measurement process, i.e. 'objective reduction', which might be associated with quantum gravity. Unlike Penrose, who may surely be right, I seek the same non-deterministic, so non-algorithmic, yet non-random behavior in TTSs operating in the poised realm.

Answering Descartes: how can mind act on brain

Owing to the causal closure inherent in classical physics, we have remained frozen with the Cartesian problem for 350 years. Mind has nothing to do and no way to do it. I believe that open quantum systems and the mind–brain system as one or trillions of interlocked TTSs may afford an answer to Cartesian dualism, for it breaks the causal closure of classical physics. Decoherence is an *acausal process*. Thus if the mind–brain system lies in the poised realm, decoherence of 'mind' to classicality FAPP allows 'mind' to have *acausal consequences for brain*, without acting causally on brain. We have indeed escaped the causal closure of classical physics.

But we want 'mind' to do this many times in our lives. TTSs, living in the poised realm, where recoherence, perhaps to new superposition states, allows 'mind' to repeatedly decohere and to have acausal consequences for material brain.

Quantum measurement can occur in TTSs. But quantum measurement, von Neumann's process 1 or 'R' process, is also acausal, and furthermoe allows 'mind' to have acausal consequences for brain. Moreover, even if von Neumann's process 1 or 'R' depend upon the Born rule and the square of the amplitudes to achieve the probability of its acausal measurement, the ongoing behavior of TTSs modifies *non-randomly* which amplitudes are propagating and which achieve high amplitudes and tend to decohere or be measured, so the total behavior is non-random. Once measured, a classical degree of freedom can also flower into quantum behavior again, allowing repeated acausal mind–brain action. The non-random but non-determinate total behavior may support a responsible free will.

Res potentia *and* res extensia *linked by quantum measurement*

I now come to the most radical proposition in this chapter. It could be wrong yet the remainder of this chapter would stay largely intact. I am, with proper hesitation, about to propose a new dualism: *res potentia*, the realm of the ontologically real Possible; and *res extensa*, the realm of the ontologically real Actual, linked – hence united – by quantum measurement. The very basis of this is quantum mechanics itself.

I turn first to the late nineteenth century American philosopher C.S. Pierce. He noted that Actuals and Possibles obey Aristotle's Law of the Excluded Middle. Here it is:

The table is or is not in the room.

There is nothing 'in the middle'. Hence the statement,

"The table simultaneously is and is **not** in the room"

is a contradiction in terms, always false. Now consider:

> "The probability of 5234 heads out of 10,000 fair coin flips is simultaneously 0.245 and is not 0.245".

It too is a contradiction in terms, always false. Classical physics obeys Aristotle's Law of the Excluded Middle. But, said Pierce, "Possibles evade the law of the Excluded Middle: A is possibly true and A is possibly not true" is *not* a contradiction.

Now consider the "sum over all possible histories" formulation of quantum mechanics (Feynman, 1948). A photon on its way through the two slits mentioned earlier, simultaneously takes all possible pathways through the two slits to the photoreceptor. It follows that the single photon 'simultaneously possibly *does* and possibily *does not* pass through the left slit'. This is *not a contradiction.*

The critical implication is that Feynman's formulation of quantum mechanics *evades* Aristotle's Law of the Excluded Middle. Therefore, I claim, Feynman's formulation of quantum mechanics is fully interpretable in terms of ontological *real* possibles, *res potentia.* The unmeasured Schrödinger wave concerns *res potentia. Res potentia* proposes an answer to what the unmeasured Schrödinger wave is 'about'.

This is a huge step, not to be taken lightly. I note that Aristotle himself toyed with the reality of 'potentia'. And British philosopher Whitehead (1978) proposed ontologically real Possibles which gave rise to ontologically real Actuals which gave rise to ontologically real Possibles: $P \rightarrow A \rightarrow P \rightarrow A$.

The idea may be radical, and may be right, but I am not the first to propose it. We will find evidence consistent with the reality of *res potentia* below in the Conway–Kochen Strong Free Will Theorem. Further, quantum physicists are very close to the concept of *res potentia.* I quote Zeh (2007):

> In classical physics you can and do assume that only one of the possibilities is real (that is why you call them possibilities). It is your knowledge that was incomplete before the observation. Mere possibilities cannot interfere with one another to give effects in reality. In particular, if you would use the dynamical laws to trace back in time the improved information about the real state, you would also get improved knowledge about the past. This is different in quantum theory (for pure states): in order to obtain the correct state in the past (that may have been recorded in a previous measurement), you need all apparent 'possibilities' (all components of the wave function – including the non-observed ones). So they must have equally been real.

Clearly Zeh is saying that possibilities ... must have been equally real. *Res potentia* removes the quotes from 'possibilities' to propose an ontologically real *res potentia.*

Moreover, a founder of quantum mechanics, Werner Heisenberg, often spoke of *potentia* sometimes as "probabilities' (Heisenberg, 1955), sometimes as 'possibilities' (Heisenberg, 1958), as a separate ontologically real realm along with ontologically real Actuals. I am following Heisenberg with my *res potentia* as a realm of ontologically real Possibles.

See Epperson (2004) for a cogent discussion of many quantum authors and an ontological dualism based on real Actuals and real Probables that *do* obey the Law of the Excluded Middle.

I stress again that unlike Descartes' *res cogitans* and *res extensa*, never united, *res potentia* and *res extensia* truly *are* united by quantum measurement.

What is *consciousness?*

Philosopher of the mind Jerry Foder (2000) quipped that "Not only have we no idea what consciousness 'is', we have no idea what it would be like to have an idea what consciousness 'is' ".

To my surprise, *res potentia* leads to an obvious idea about what consciousness *is*. Consciousness is a participation in the 'possible', an ontologically real *res potentia*. I offer three pieces of evidence:

(1) Where exactly is the possibility I will skate across town reading the New York Times and not be hit by a car? I think we all feel that the 'possibility' itself is not spatially locatable; it is not spatially extended.
(2) Now consider your experienced visual field. Where exactly is your experienced field located? I think we all sense that our experienced field is not located spatially; it is not spatially extended.
(3) Shortly I will propose that qualia are associated with quantum measurement and furthermore hypothesize that entanglement of many quantum degrees of freedom, perhaps among neurotransmitter receptor molecules in anatomically unconnected synapses in the brain, may, by each being quantum measured, yield causally non-local Einstein–Podolsky–Rosen (EPR) high correlations of now bound qualia (Aspect et al., 1982) to solve the 'qualia binding problem' in neurobiology. Non-local correlations are 'non-local' because they are beyond speed of light signaling and 'instantaneous', hence also 'non-spatial'.

This non-spatial character of possibilities, experience and non-local EPR quantum measurements may be happenstance or a clue. Taking this parallel as a clue may lead us forward in new ways.

Qualia may be associated with quantum measurement

Where is it natural to locate experience itself, the blueness of blue, the taste of wine, qualia? I suggest qualia are associated with quantum measurement, i.e. Possible 'becomes' Actual or

$$\text{Possible} \longrightarrow \text{Actual}.$$

As we shall see, this leads to testable consequences. It is not a bald hypothesis standing alone for, as just noted, I will propose below that quantum entanglement among many quantum degrees of freedom in anatomically unconnected synapses, and non-local EPR correlations achieved by a set of quantum measurements of these entangled degrees of freedom may help solve the 'qualia binding problem' and the unity of consciousness issue. Thus solving the binding problem may require the hypothesis that quantum measurement is associated with qualia. The hypothesis should be testable in the brain. More, entanglement to solve the binding problem is testable. I note that the physicist Henry Stapp (2007) has different but somewhat related ideas; see also Penrose (1994).

On the hypothesis that qualia are associated with quantum measurement, successive qualia from successive measurement events can be non-randomly related. Quantum measurement is, in itself, random via the Born rule above and, for example, von Neumann's measurement process. In a total TTS, however, whose classical degrees of freedom are behaving dynamically and non-randomly and thereby affecting non-randomly the remaining quantum degrees of freedom and their superpositions, successive quantum measurement of those quantum degrees of freedom, each random as a measurement, will be non-randomly affected by the behavior of the classical degrees of freedom. Thus successive qualia can be non-randomly related.

A critical feature of quantum measurement, my physicist friends assure me, is that it has never been derived from within quantum mechanics. Granted *res potentia*, such a derivation may be disallowed. The "X is Possible" of *res potentia* does *not* imply "X is Actual" of *res extensa*. Our difficulties with such a derivation since 1927 may be ontological, not technical or mathematical. If *res potentia* is ontologically real, the same ontological issue may bear on our failure to unite general relativity and quantum mechanics: the "X is Possible" of unmeasured quantum mechanics does *not* imply the "X is Actual" of general relativity.

On *res potentia*, a second feature of measurement becomes equally important. What if the 'becomes' of Possible 'becomes' Actual? What is the status of '$\longrightarrow$' in $P \longrightarrow A$? It is not a classical becoming like water freezing, nor the unitary propagation of the Schrödinger wave. As a 'becoming' it seem not to be an **existing state** at all. Qualia are a 'becoming' not an 'existence'. Nor can '$\longrightarrow$' be a mathematizable deductive entailing process, for if it were, it would enable deduction from 'X is Possible' to 'X is Actual', which is invalid if *res potentia* is real. Then there is no mechanism for the quantum measurement captured in von Neumann's ad hoc Process 1 or 'R' process.

The above paragraph depends upon the ontological reality of *res potentia*. But the proposal of a real *res potentia* ties to the Strong Free Will Theorem, (Conway and Kochen, 2009), which states that if physicists have free will so do electrons, that the world is non-determinate, that there can be no mechanism for quantum measurement, and that a relevant property does not exist before its measurement. This theorem rests on free will for the physicist. But above I have argued that TTSs in the poised realm, *without replying on an ontologically real res potentia*, may afford an ontologically responsible free will. Responsible free will may well require qualia, i.e. experience, which I propose is associated with quantum measurement. This again is a proposal that does not require the reality of *Res potentia*. But a responsible free will supports the claims of the Strong Free Will Theorem. Conversely, this theorem states that, given the free-willed physicist, the world is non-determinate. This is consistent with the hypothesis of the reality of *res potentia*. More, by this theorem, if qualia are associated with quantum measurement, there is no mechanism for that measurement. But measurements yield classical degrees of freedom that, as such, can have classical causal effects on the classical world. Mind and qualia, can, via acausal measurement, act causally on the world classically. Perhaps, as I propose below, neurotransmitter receptors are the loci of quantum measurement and qualia. Then the classical variable consequences could alter post synaptic voltage gate behaviors leading to neural firing or not. In turn qualia themselves emerge as irreducible.

The vice of this view is that it hides the mystery of qualia in the mystery of measurement. The virtue of this 'hiding' is that it may explain, at last, why we cannot isolate or pin down an irreducible character of qualia. The philosopher Chalmers (1996) also proposed on independent grounds that qualia are irreducible.

The hypothesis that qualia are associated with quantum measurement leaves open the question of whether measurement is merely necessary for qualia, or more strikingly, whether it is also sufficient. I note just below the start of means to test whether quantum measurement is necessary for qualia. I have no idea how to test whether quantum measurements are sufficient for qualia, with the stunning implications that would have for a kind of protoqualia wherever quantum measurement occurs in the universe.

I stress that this hypothesis does not say what qualia *are*.

The hypothesis that qualia are associated with quantum measurement is testable. Anesthetics bind to hydrophobic pockets in neurotransmitter receptors in the synapses (Hameroff, 1958). If they freeze the receptors so that they cannot quantum measure, no more qualia can arise. Moreover, *Drosophila* can be anesthetized by ether. Consider selecting for ease of being anesthetized and seek the molecular components involved in easy anesthetization. The normal, or wild type, versions of these molecules may be involved in consciousness and their quantum and quantum measurement properties studied. If wild-type proteins can carry out quantum measurement, but mutant ones cannot, we have initial evidence that qualia may indeed be associated with quantum measurement.

If I assume qualia are associated only with measurement, a potential role for unmeasured quantum behavior in the mind–brain system could be unconscious mental processing, which may have classical consequences via decoherence to classicality FAPP, without measurement. Possibly this bears on results demonstrating neural activities 200 or more milliseconds before conscious awareness of a decision to act (Libet, 1990). This too should be testable.

Standing the brain on its head

I begin with a stunning fact. The box jelly fish, with only a loose neural net and no evolved brain, but eyes that have evolved to see shape and color, swims at five knots adroitly avoiding obstacles (Morris, 2010). An evolved brain is not needed for these feats. Also, choanoflagellates, single cell precursors to animals, have many molecular components of synapses (King et al., 2003).

Some readers of this chapter may not know the neuroanatomy of the human brain or much about its physiology. For those, and in brief, we have about 10^{11} neurons, each with an average of 6000 synapses. Cell bodies have descending axons which may or may not branch, but each of which ends on synapses, associated with synaptic spikes on dendrites, in arborizations which lead into cell bodies. When an action depolarization potential travels down an axon to a synapse, presynaptic vesicles release one of a set of neurotransmitters, such as GABA, which crosses the synaptic cleft to the adjacent dendrite of the post-synaptic neuron and binds to post-synaptic neurotransmitter receptors which are often in clusters of many proteins. In turn, often this leads to the opening of an ion channel, a transient flow of ions, and a very short-term depolarization or hyper-polarization (excitatory or inhibitory respectively) of the tiny local patch of dendritic transmembrane potential. These local changes flow to the cell body and are summed. If the resulting transmembrane potential at the cell body is more than -20 mV, an action potential is initiated and travels down the axon. Most neurobiologists think classical-physics action potentials in neurons carry a 'neuronal code' underlying consciousness.

Crick (1994) noted in a throwaway line, that vast amounts of information about tiny time-space alterations in dendritic transmembrane potentials and behaviors of synaptic molecules are thrown away in neural classical-physics action potentials. What if we consider 'standing the brain on its head', and supposing that this vast amount of information in and around synapses and local dendritic regions is the 'business end' of the brainsensory-motor system. This does not vitiate at all the huge amount of work on neural circuitry and classical action potentials and information processing by classical neural action in the brain. However, it does raise the possibility that the 'neural correlates of consciousness' may lie in synaptic and local dendritic, possibly poised-realm behavior, perhaps in quantum measurement processes.

I note that Beck and Eccles (1994) considered quantum processes in synapses.

Quantum entanglement, Niirnan's idea, and the binding problem

Consider, says Crick, a yellow triangle and blue square. Let 'yellow', 'triangle', 'blue', and 'square' be processed in different, anatomically unconnected brain areas. How are they bound into yellow triangle and blue square? This, as I have described, is the binding problem. Crick focuses his hope on squeezing perhaps millions of distinct sets of features to be bound into different phases of the 40 hertz oscillation.

The first idea I propose is to use quantum entanglement to link quantum processes in different, anatomically unconnected, synapses in order to start to solve the binding problem. Entanglement occurs if a quantum degree of freedom, say a photon, decays into two lower energy photons that go off in opposite directions, even becoming so far apart that even light cannot travel between them in the *interval between quantum measurements of the two entangled photons*. Quantum mechanics says, and it is confirmed again and again, that the *two quantum measurements* will be highly correlated, even though no light or information can have traveled between the two sites. This is 'EPR non-local correlation' (Aspect et al., 1982). I stress that in the entangled state, the two photons remain in a *single* quantum state.

I want to try to use quantum entanglement among many synaptic degrees of freedom to try to solve the binding problem. Hence, as I have emphasized, it is very attractive to me that these quantum correlations require quantum *measurement* of the entangled degrees of freedom, and I have already supposed that quantum measurement itself is associated with qualia. Then these many entangled degrees of freedom in a *single* quantum state when measured yield qualia that *are* bound. The hypothesis that qualia are associated with quantum measurement does not stand alone, it may afford a part of an answer to the unity of consciousness.

Clearly, such entanglement may require long-range entanglement among anatomically unconnected synapses and neurons in the brain connecting the 'right' set of, say, synaptic molecules. How and whether this may be accomplished is, at present, uncertain, but see below. Samuli Niianen had a lovely idea: "If you measure the position and momentum of a single classical gas particle in a box, do you know about the shape of the box?" No, you do not. "But", he continued, "a quantum wave process in a potential well that serves as its boundary condition 'knows' about the shape of that potential well, for example in its measured energy spectrum!". He is right.

Think of music in a room and of trying to describe the air pressure waves using bits. Now think of 1000 differently shaped drum heads carefully placed in the room. Their patterns of vibrations, i.e. the eigenfunction spectra of the drums bound to drumheads, 'know' the music in the room, and do so in an analog, embodied way, not in a digital or propositional way. A telephone is not digital either. This leads to the idea that the brain's sensory system and indeed the whole brain, may tune the synaptic or local dendritic transmembrane potentials in tiny time-space regions of the brain, such as synapses and adjacent local dendritic membranes; thus they jointly 'cover', like many tuned antennas, the visual scene so that quantum processes in those potential wells, when entangled and measured, yield both a unity of consciousness and solve the binding problem in an analog rather than a digital way.

Two final points. It now appears that increasing the number of entangled degrees of freedom *increases* the quantum EPR correlations. This increase is the opposite of the curse of dimensionality. It helps the binding problem. Second, local actions can alter which quantum degrees of freedom are entangled, perhaps offering an account of the serially shifting focus of attention, and they might entangle the 'correct' set of quantum degrees of freedom for each focus of attention (Science News, 2010).

Can all this be correct? I certainly do not know. But the ideas seem coherent and testable, and jointly seem to offer a new purchase on manifold problems.

Programming Trans-Turing Systems

We have known about Turing Machines since the mid-1930s and programming the von Neumann architecture for over 50 years. We have no experience with TTSs. But we face a problem: how would we achieve a TTS that 'does something we want'?

There seem to be two approaches. Simulate the TTS on a digital computer and evolve a population of TTSs, i.e. of interacting entangled, measured, TTSs, to yield the behavior desired. This is analogous to the genetic algorithm of Holland (1975). But real quantum behavior cannot be achieved on a Turing machine.

Another approach which may be worth considering is creating self-reproducing molecular systems, perhaps autocatalytic sets of polymers in dividing liposomes, supplied with energy by pyrophospate or in other ways and capable of open-ended evolution. Recent work shows that:

(1) collectively autocatalytic sets arise as the diversity of polymers in a reaction set is increased (Kauffman, 1986; Hordijk et al., 2010);

(2) such systems can undergo open ended evolution (Fernando et al., 2011);
(3) liposomes can grow and divide (Luisi et al., 2004);
(4) a collectively autocatalytic set in a reproducing container can yield synchronization of the reproduction of each (Filsetti et al, 2010): experimental collectively autocatalytic sets have been constructed (Wagner and Ashkenasy, 2009); libraries of stochastic DNA, RNA, peptides, polypeptides and proteins can be made (Kauffman, 1993), so we can test for the emergence of collectively autocatalytic sets.

It is an exciting prospect that work on the origin of life and work on TTSs may come together. Moreover, Darwinian pre-adaptations among such co-evolving protocells generate new, un-pre-statable, biological functions that maintain one or more such protocells and hence solve the frame problem (Kauffman, 2008). Co-evolving TTSs in protocells may well solve the frame problem too. Work with minimal cells as vehicles for TTS evolution and co-evolution may be possible (Hotz, 2010). In addition nanotechnology, perhaps with populations of nano-devices that can be subjected to Holland's genetic algorithm, may prove useful.

Conclusions

I have argued that classical-physics Turing machines as models of the mind are possible, but leave us at best with no free will, and an epiphenomenal consciousness. I believe that we can begin to go beyond Turing, to create non-algorithmic, non-determinate, and non-randomly behaving Trans-Turing Systems, living in the poised realm, perhaps in self-reproducing protocells, or perhaps as nano-devices, open to evolution or co-evolution to achieve useful ends. I propose tentative answers to Descartes about mind and body. Many ideas in this chapter are new or even radical science. They may portend transformations in quantum physics, quantum chemistry, a new poised realm behavior of biomolecules hovering between quantum and classical behaviors, a new approach to neurobiology, the philosophy of mind, and the radical possibility of *res potentia* with consciousness as a participation in the Possible, qualia as irreducible and associated with quantum measurement which also may be irreducible, and entanglement and quantum measurement to achieve a unity of consciousness. I hope these concepts point the way forward for us all.

Postscript

In the main text of this chapter, I have explored the possibility that the qualia binding problem of neurobiology might be solvable via the entanglement of the many relevant quantum variables in anatomically unconnected neurons or their synapses via the non-local correlations among such variables seen in quantum measurements of them. One possibility is that infrared photons which can penetrate the brain at 300K, corresponding roughly to their typical energies, might be the quantum variables which entangle or couple the relevant quantum variables in anatomically disconnected neurons or synapses.

I have not, in the main text, specified a possible mechanism that could conceivably accomplish this but do so here, as a further working hypothesis. The absorption and emission lines of atoms are very sharp. In molecules those lines are a bit less sharp. But in the presence of decoherence, and increasing with the degree of decoherence, those absorption and emission lines broaden significantly. Thus in a neuron or synapse, a relevant quantum variable which is decoherent is able to absorb wavelengths that lie outside the smaller range of wavelengths that the same quantum variable can absorb if coherent.

The above fact suggests a possible means for a quantum variable, *A*, in a neuron or synapse, to communicate with a specific subset of one or more relevant quantum variables, subset *B*, in anatomically unconnected neurons or synapses. Let *A* emit one or a shower of photons whose wavelengths are above or below the narrow range of the quantum variables in subset *B* when those variables are coherent, but are within the wider range of the quantum variables in set *B* when these variables are sufficiently decoherent. Then the decoherent variables in subset *B* can receive infrared photons from *A*, while the relevant quantum variables in other neurons or synapses, outside the subset of neurons, *B*, in the rest of the brain cannot receive photons from *A*.

Therefore, by time-varying levels of decoherence and recoherence or measurement, changing sets of emitting and absorbing quantum variables in anatomically unconnected neurons or synapses in the brain can be specifically coupled to one another by infrared photons. This provides a tentative mechanism for coupling changing relevant quantum variables in anatomically unconnected neurons or synapses, whose quantum measurements, I hypothesize, are associated with qualia, and perhaps solving the unity of consciousness, qualia binding problem. In addition, by shifting the entangled subsets of quantum variables, where each subset generates a different 'unity of consciousness", the system may also be shifting its 'focus of attention'.

Finally, there is new evidence supporting the reality of the *X*- and *Y*-axes of the poised realm. In work led by quantum physicist Gabor Vattay, he, Samuli Niiranen and I have posted an article entitled 'Quantum biology on the edge of quantum chaos'[3]. This article shows that the rate of decoherence varies dramatically on the

[3] Link: `http://arxiv.org/abs/1202.6433`.

X-axis expressing the order, criticality, chaos, of the poised realm, from a very slow power law in the ordered and critical regimes, to a very fast exponential decay of coherence in the chaotic regime. The experimentally measured rated of decoherence of a light harvesting molecule is indeed a power law well fitted by our predicted power law slope. Tentative evidence suggests this light harvesting molecule is critical on the X-axis. In many physical systems, decoherence is known to be exponential. The variation in decoherence on the X-axis supports the hypothesis that the X-axis is real. More, Vattay, at a CERN-hosted *Origin of Life* meeting in 2013, presented evidence that different organic molecules are at different locations, ordered, critical and chaotic, on the X-axis, again supporting the reality of the X-axis. With respect to the reality of the Y-axis, varying from quantum coherent at the origin, then decohering to classicality FAPP up the Y-axis, and recohering to quantum coherent down the Y-axis, decoherence is fully established. Now later experimental evidence, (Tiwarji et al., 2013) demonstrates recoherence experimentally. This supports the claim that variables can live reversibly in the poised realm, hovering reversibly between quantum and classical FAPP behavior.

Acknowledgements The author was partially supported as a Finnish Distinguished Professor at the Tampere University of Technology by the TEKES Foundation. Some of this material was published on `npr.org/blogs/13.7`.

References

R. Ashby (1952). *Design for a Brain: The Origin of Adaptive Behavior*. Chapman & Hall.

A. Aspect, J. Dalibard, and G. Rogers (1982). Experimental test of Bell's inequalities using time-varying analyses. *Phys. Rev. Lett.*, **44**, 1804–1807.

F. Beck and J.C. Eccles (1994). Quantum processes in the brain. A scientific basis for consciousness. In *Neural Basis of Consciousciousness*, edited by Naoyuki Osaka, Benjamins.

Max Born (1989). *Atomic Physics*, 8th edition, Dover.

H.J. Briegel and S. Popescu (2008). Entanglement and intra-molecular cooling in biological systems – a quantum thermodynamic perspective. ArXiv:0806.4552v1 [quant-ph] (accessed 27 June 2008).

J. Cai, S. Popescu, and H.J. Briegel (2008). Dynamic entanglement in oscillating molecules. ArXiv:0809.4906v1 [quant-ph] (accessed 29 September 2008).

David J. Chalmers (1996). *The Conscious Mind: In Search of a Fundamental Theory*. Oxford University Press.

J.H. Conway and S. Kochen (2009). The strong free will theorem. *Notices of the AMS*, **56** (2), 226–232.

C. Cormack and J.P. Paz (2010). Observing different phases for the dynamics of entanglement in an ion trap. *Phys. Rev. A*. **81**, 022306.

Francis Crick (1994). *The Astonishing Hypothesis: The Scientific Search For the Soul*, Simon and Schuster.

A. de la Lande, J. Rezac, B. Levy, B. Sanders, and D.R. Salahub (2011). Transmission coefficients for chemical reactions with multiple states: the role of quantum decoherence. *J. Am. Chem. Soc.*, **13**, 3883–3894.

Daniel C. Dennett (1991). *Consciousness Explained.* Little Brown.

Rene Descartes (1637). *Discourse on Method.* Open Court Publishing, reprint 1962.

Gerald Edelman (1992). *Bright Air, Brilliant Fire: On the Matter of Mind.* Basic Books.

Michael Epperson (2004). *Quantum Mechanics and the Philosophy of Alfred North Whitehead*, Fordham University Press.

C. Fernando, V. Vasas, M. Santos, S. Kauffman, and E. Szathmary (2011). Spontaneous formation and evolution of autocatalytic sets within compartments. Submitted.

R.P. Feynman (1948). The space–time formulation of nonrelativistic quantum mechanics. *Rev. Mod. Phys.*, **20**, 367–387.

R.P. Feynman, R. Leighton, and M. Sands (1964). *The Feynman Lectures on Physics, Volume 3*, Addison–Wesley.

A. Filsetti, R. Serra, T. Carletti, M. Villiani, and I. Poli (2010). Non-linear protocell models: synchronization and chaos. *Eur. J. Phys. B*, **77**, 249–256.

M.C. Fischer, B. Butierrez-Medina, and M.G. Raizen (2001). Observation of the quantum Zeno and anti-Zeno effects in an unstable system. *Phys. Rev. Lett.*. **87** (4), 040402–1–040402-4.

Jerry Fodor (2000). *The Mind Doesn't Work That Way: The Scope and Limits of Computational Psychology*. MIT Press.

S.R. Hameroff (2006). The entwined mysteries of anesthesia and consciousness: is there a common underlying mechanism? *Anesthesiology*, **105**, 400–412.

Werner Heisenberg (1955). The development of the interpretation of the quantum theory. In *Neils Bohr and the Development of Physics*, edited by Wolfgang Pauli. McGraw–Hill.

Werner Heisenberg (1958). *Physics and Philosophy: The Revolution in Modern Science.* Harper and Row.

John H. Holland (1975). *Adaptation in Natural and Artificial Systems.* University of Michigan Press.

J.J. Hopfield. (1982). Neural networks and physical systems with emergent collective computational abilities, *PNAS*, **79** (8), 2554–2558.

W. Hordijk, J. Hein, and M. Steel (2010). Autocatalytic sets and the origin of life. *Entropy*, **12** (7), 1733–1742.

R.L. Hotz (2010). Scientists create synthetic organism. *Wall Street Journal*, 21 May.

A. Ishizaki, T.R. Calhoun, G.S. Schlou-Cophen, and G.R. Fleming (2010). Quantum coherence and its interplay with protein environments in photosynthetic electronic energy transfer. *Phys. Chem. Chem. Phys.*, **12** (27), 7319–7337.

S.A. Kauffman (1986). Autocatalytic sets of proteins. *J. Theor. Bio.*, **119** 1–24.

Stuart Kauffman (1993). *Origins of Order: Self-organization and Selection in Evolution.* Oxford University Press.

Stuart Kauffman (2008). *Reinventing the Sacred.* Basic Books.

N. King, C.T. Hittinger, and S.B. Carroll (2003). Evolution of key cell signalling and adhestion proteins predates animal origins. *Science*, **301** (5631), 361–363.

B. Libet (1990). Cerebral processes that distinguish conscious experience from unconscious mental functions. In *The Principles of Design and Operation of the Brain*, edited by J.C. Eccles and O.D. Creutzfeldt. Springer.

P.L. Luisi, P. Stano, S. Rasi, and F. Mavelli (2004). A possible route to prebiotic vesicle reproduction. *Artifical Life*, **10**, 297–308.

R.S. Mackay (1993). *Renormalization in Area-Preserving Maps*. World Scientific.

C. Mainos (2000). Laser induced coherence in ultraphoton excitation of individual molecules. *Phys. Rev. A*, **61**, 063410–6.

W. McCulloch and W. Pitts (1943). A logical calculus of the ideas immanent in nervous activity. *Bull. Math. Biophys.*, **7**, 115–133.

S.C. Morris (2010). In *Atoms and Eden: Conversations on Religion and Science*, edited by Steve Paulson. Oxford University Press.

J.P. Paz, and A.J. Roncaglia. (2009). Entanglement dynamics during decoherence. *Quantum Inf. Process*, **8**, 535–548.

Roger Penrose (1989). *The Emperor's New Mind: Concerning Computers, Minds and the Laws of Physics*. Oxford University Press.

Roger Penrose (1994). *Shadows of the Mind: A Search for the Missing Science of Consciousness*. Oxford University Press.

C.S. Pierce (1960). *Collected Papers* Volumes I and II, edited by Charles Hartshorne and Paul Weiss. Harvard University Press.

O.V. Prezhdo (2000). Quantum anti-Zeno acceleration of a chemical reaction. *Phys. Rev. Lett.*, **85**, 4413–4417.

Bertrand Russell (1912). *The Problems of Philosophy*. Williams and Norgate.

E. Schrödinger (1926). An undulatory theory of the mechanics of atoms and molecules. *Phys. Rev.*, **28** (6), 1049–1070.

Science News 2010. Inducing Entanglement. 20 November issue.

John R. Searle (1997). The mystery of consciousness. In *The New York Review of Books*.

P.W. Shor. (1995). Scheme for reducing decoherence in quantum computer memory. *Phys. Rev. A*, **52**, 2493–2496.

Henry P. Stapp (2007). *Mindful Universe: Quantum Mechanics and the Participating Observer*. Springer.

V. Tiwarji, W.K. Peters, and D.M. Jonas (2013). Electronic resonance with anticorrelated pigment vibrations drives photosynthetic energy transfer outside the adiabatic framework. *PNAS*, **10** (4), 1203–1208.

Alan Turing (1952). Can automatic calculating machines be said to think? Transcript of a discussion broadcast on BBC Third Programme. Reproduced in *The Essential Turing: The Ideas that Gave Birth to the Computer Age*, edited by Jack B. Copeland. Oxford University Press (2004).

John von Neumann (1932). *Mathematical Foundations of Quantum Mechanics*, translated by R.T. Beyer (1996 edition). Princeton University Press.

N. Wagner and G. Ashkenasy (2009). Symmetry and order in systems chemistry. *J. Chem. Phys.*, **130**, 164907–164911.

Paul J. Werbos (1994). *The Roots of Backpropagation. From Ordered Derivatives to Neural Networks and Political Forecasting*. Wiley.

Alfred North Whitehead (1978). *Process and Reality: An Essay in Cosmology*, corrected edition, edited by D. Griffin and D. Sherburne. Free Press.

Ludwig Wittgenstein (1921). Logisch-Philosophische Abhandlung. *Annalen der Naturphilosophische*, **14**.
Ludwig Wittgenstein (1953). *Philosophical Investigations*. Blackwell.
D. Zeh (2007). http://www.fqxi.org/community/forum/topic/39 (third comment from top).
W.H. Zurek. (2002). Decoherence and the transition from quantum to classical – revisited. *Los Alamos Science*, **27**.

12

The Ghost in the Quantum Turing Machine

Scott Aaronson

Abstract

In honor of Alan Turing's hundredth birthday, I unwisely set out some thoughts about one of Turing's obsessions throughout his life, the question of physics and free will. I focus relatively narrowly on a notion that I call 'Knightian freedom': a certain kind of in-principle physical unpredictability that goes beyond probabilistic unpredictability. Other, more metaphysical, aspects of free will I regard as possibly outside the scope of science.

I examine a viewpoint, suggested independently by Carl Hoefer, Cristi Stoica, and even Turing himself, that tries to find scope for 'freedom' in the universe's boundary conditions rather than in the dynamical laws. Taking this viewpoint seriously leads to many interesting conceptual problems. I investigate how far one can go toward solving those problems and, along the way, I encounter (among other things) the no-cloning theorem, the measurement problem, decoherence, chaos, the arrow of time, the holographic principle, Newcomb's paradox, Boltzmann brains, algorithmic information theory, and the common prior assumption. I also compare the viewpoint explored here with the more radical speculations of Roger Penrose.

The result of all this is an unusual perspective on time, quantum mechanics, and causation, of which I myself remain skeptical but which has several appealing features. Among other things, it suggests interesting empirical questions in neuroscience, physics, and cosmology; and it takes a millennia-old philosophical debate into some underexplored territory.

This material is based upon work supported by the National Science Foundation under Grant No. 0844626. Also supported by a DARPA YFA grant, an NSF STC grant, a TIBCO Chair, and a Sloan Fellowship.
Published in *The Once and Future Turing*, edited by S. Barry Cooper & Andrew Hodges. Published by

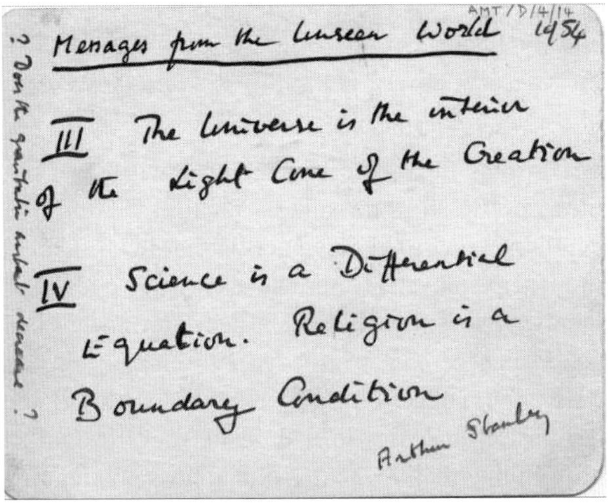

Messages from the Unseen World

The Universe is the interior of the Light Cone of the Creation

Science is a Differential Equation. Religion is a Boundary Condition

Figure 12.1 Postcard from Alan M. Turing to Robin Gandy, March 1954 (reprinted in Hodges (2012)). Reproduced from The Turing Digital Archive with permission of C.E.M. Yates.

12.1 Introduction

When I was a teenager, Alan Turing was at the top of my pantheon of scientific heroes, above even Darwin, Ramanujan, Einstein, and Feynman. Some of the reasons were obvious: the founding of computer science, the proof of the unsolvability of the Entscheidungsproblem, the breaking of the Nazi Enigma code, the unapologetic nerdiness and the near-martyrdom for human rights. But beyond the facts of his biography, I idolized Turing as an 'über-reductionist': the scientist who had gone further than anyone before him to reveal the mechanistic nature of reality. Through his discovery of computational universality, as well as the Turing Test criterion for intelligence, Turing finally unmasked the pretensions of anyone who claimed there was anything more to mind, brain, or the physical world than the unfolding of an immense computation. After Turing, it seemed to me, one could assert with confidence that all our hopes, fears, sensations, and choices were just evanescent patterns in some sort of *cellular automaton*; that is, a huge array of bits, different in detail but not in essence from Conway's famous *Game of Life*,[1] getting updated in time by simple, local, mechanistic rules.

So it's striking that Turing's own views about these issues, as revealed in his lectures as well as private correspondence, were much more complicated than my early caricature. As a teenager, Turing devoured the popular books of Sir Arthur Eddington, who was one of the first (though not, of course, the last!) to speculate about the implications of the then-ongoing quantum revolution in physics for ancient questions about mind and free will. Later, as a prize from his high school in 1932, Turing selected John von Neumann's just-published *Mathematische Grundlagen der Quantenmechanik* (von Neumann, 1932): a treatise on quantum mechanics famous for its mathematical rigor, but also for its perspective that the collapse of the wavefunction ultimately involves the experimenter's mental state. As detailed by Turing biographer Andrew Hodges (2012), these early readings had a major impact on Turing's intellectual preoccupations throughout his life and probably even influenced his 1936 work on the theory of computing.

Turing also had a more personal reason for worrying about these 'deep' questions. In 1930, Christopher Morcom – Turing's teenage best friend, scientific peer, and (probably) unrequited love – died from tuberculosis, sending a grief-stricken

[1] Invented by the mathematician John Conway in 1970, the Game of Life involves a large two-dimensional array of pixels, with each pixel either 'live' or 'dead.' At each (discrete) time step, the pixels get updated via a deterministic rule: each live pixel 'dies' if less than 2 or more than 3 of its 8 neighbors are alive, and each dead pixel 'comes alive' if exactly 3 of its 8 neighbors are alive. *Life* is famous for the complicated, unpredictable patterns that typically arise from a simple starting configuration and repeated application of the rules. Conway (see Levy, 1992) has expressed certainty that, on a large enough *Life* board, living beings would arise, who would then start squabbling over territory and writing learned PhD theses! Note that, with an *exponentially*-large *Life*board (and, say, a uniformly random initial configuration), Conway's claim is vacuously true, in the sense that one could find essentially any regularity one wanted just by chance. But one assumes that Conway meant something stronger.

Turing into long ruminations about the nature of personal identity and consciousness. Let me quote from a remarkable disquisition, entitled *Nature of Spirit*, that the 19-year-old Turing sent in 1932 to Christopher Morcom's mother.

> It used to be supposed in Science that if everything was known about the Universe at any particular moment then we can predict what it will be through all the future. This idea was really due to the great success of astronomical prediction. More modern science however has come to the conclusion that when we are dealing with atoms and electrons we are quite unable to know the exact state of them; our instruments being made of atoms and electrons themselves. The conception then of being able to know the exact state of the universe then really must break down on the small scale. This means then that the theory which held that as eclipses etc. are predestined so were all our actions breaks down too. We have a will which is able to determine the action of the atoms probably in a small portion of the brain, or possibly all over it. The rest of the body acts so as to amplify this. (Quoted in Hodges, 2012)

The rest of Turing's letter discusses the prospects for the survival of the 'spirit' after death, a topic with obvious relevance to Turing at that time. In later years, Turing would eschew that sort of mysticism. Yet even in a 1951 radio address *defending* the possibility of human-level artificial intelligence, Turing still brought up Eddington, and the possible limits on prediction of human brains imposed by the uncertainty principle:

> If it is accepted that real brains, as found in animals, and in particular in men, are a sort of machine it will follow that our digital computer suitably programmed, will behave like a brain. [But the argument for this conclusion] involves several assumptions which can quite reasonably be challenged. [It is] necessary that this machine should be of the sort whose behaviour is in principle predictable by calculation. We certainly do not know how any such calculation should be done, and it was even argued by Sir Arthur Eddington that on account of the indeterminacy principle in quantum mechanics no such prediction is even theoretically possible.[2] (Reprinted in Shieber, 2004)

Finally, two years after his sentencing for 'homosexual indecency,' and a few months before his tragic death by self-poisoning, Turing wrote the striking aphorisms that appears in Figure 12.1: "The universe is the interior of the light-cone of the Creation. Science is a differential equation. Religion is a boundary condition."

The reason I'm writing this essay is that I *think* I now understand what Turing could have meant by these remarks. Building on ideas of Hoefer (2002), Stoica

[2] As Hodges (personal communication) points out, it's interesting to contrast these remarks with a view Turing had expressed just a year earlier, in *Computing Machinery and Intelligence* (Turing, 1950): "It is true that a discrete-state machine must be different from a continuous machine. But if we adhere to the conditions of the imitation game, the interrogator will not be able to take any advantage of this difference." Note that there's no actual contradiction between this statement and the one about the uncertainty principle, especially if we distinguish (as I will) between simulating a *particular* brain and simulating *some* brain-like entity able to pass the Turing test. However, I'm not aware of any place where Turing explicitly makes that distinction.

(2008), and others, I'll examine a perspective – which I call the 'freebit perspective,' for reasons to be explained later – that locates a nontrivial sort of freedom in the universe's boundary conditions even while embracing the mechanical nature of the time-evolution laws. We'll find that a central question, for this perspective, is how well complicated biological systems like human brains can *actually* be predicted: not by hypothetical Laplace demons but by prediction devices compatible with the laws of physics. It's in the discussion of this predictability question (and *only* there) that quantum mechanics enters the story.

Of course, the idea that quantum mechanics might have *something* to do with free will is not new; neither are the problems with that idea or the controversy surrounding it. While I chose Turing's postcard for the opening text of this essay, I also could have chosen a striking claim by Niels Bohr, from a 1932 lecture about the implications of Heisenberg's uncertainty principle:

> [W]e should doubtless kill an animal if we tried to carry the investigation of its organs so far that we could tell the part played by the single atoms in vital functions. In every experiment on living organisms there must remain some uncertainty as regards the physical conditions to which they are subjected, and the idea suggests itself that the minimal freedom we must allow the organism will be just large enough to permit it, so to say, to hide its ultimate secrets from us. (Reprinted in Bohr, 2010)

Or this, from the physicist Arthur Compton:

> A set of known physical conditions is not adequate to specify precisely what a forthcoming event will be. These conditions, insofar as they can be known, define instead a range of possible events from among which some particular event will occur. When one exercises freedom, by his act of choice he is himself adding a factor not supplied by the physical conditions and is thus himself determining what will occur. That he does so is known only to the person himself. From the outside one can see in his act only the working of physical law. (Compton, 1957)

I want to know:

> *Were Bohr and Compton right or weren't they? Does quantum mechanics (specifically, say, the no-cloning theorem or the uncertainty principle) put interesting limits on an external agent's ability to scan, copy, and predict human brains and other complicated biological systems, or doesn't it?*

Of course, one needs to spell out carefully what one means by 'interesting limits,' an 'external agent,' the 'ability to scan, copy, and predict,' and so forth.[3] But once that's done, I regard the above as an unsolved scientific question, and a big one. Many people seem to think the answer is obvious (though they disagree on what it is!), or else they reject the question as meaningless, unanswerable, or irrelevant. In this essay I'll argue strongly for a different perspective: that we can easily

[3] My own attempt to do so is in Appendix 12A.

imagine worlds consistent with quantum mechanics (and all other known physics and biology) where the answer to the question is yes, and other such worlds where the answer is no. And we don't yet know which kind we live in. The most we can say is that, like P versus NP or the nature of quantum gravity, the question is well beyond our *current* ability to answer.

Furthermore, the two kinds of world lead, not merely to different philosophical stances but to different visions of the remote future. Will our descendants all choose to upload themselves into a digital hive-mind, after a 'technological singularity' that makes such things possible? Will they then multiply themselves into trillions of perfect computer-simulated replicas, living in various simulated worlds of their own invention, inside of which there might be further simulated worlds with still more replicated minds? What will it be *like* to exist in so many manifestations: will each copy have its own awareness, or will they comprise a single awareness that experiences trillions of times more than we do? Supposing all this to be possible, is there any reason why our descendants might want to hold back on it?

Now, if it turned out that Bohr and Compton were wrong – that human brains were as probabilistically predictable by external agents as ordinary digital computers equipped with random-number generators – then the freebit picture that I explore in this essay would be falsified, to whatever extent it says anything interesting. It should go without saying that I see the freebit picture's vulnerability to future empirical findings as a feature rather than a bug.

In summary, I'll make no claim to show here that the freebit picture is *true*. I'll confine myself to two weaker claims:

(1) That the picture is *sensible* (or rather, not obviously much crazier than the alternatives): many considerations one might think would immediately make a hash of this picture, fail to do so for interesting reasons.

(2) That the picture is *falsifiable*: there are open empirical questions that need to turn out one way rather than another for this picture to stand even a *chance* of working.

I ask others to take this essay as I do: as an exercise in what physicists call model building. I want to see *how far I can get* in thinking about time, causation, predictability, and quantum mechanics in a certain unusual but apparently consistent way. 'Resolving' the millennia-old free will debate isn't even on the table! The most I can hope for, if I'm lucky, is to construct a model whose strengths and weaknesses help to move the debate slightly forward.

12.1.1 'Free will' versus 'freedom'

There's one terminological issue that experience has shown I need to dispense with before anything else. In this essay, I'll sharply distinguish between 'free will' and another concept, which I'll call 'freedom,' and will mostly concentrate on the latter.

By 'free will,' I'll mean a metaphysical attribute that I hold to be largely outside the scope of science – and which I can't even *define* clearly, except to say that, if there's an otherwise-undefinable thing that people have tried to get at for centuries with the phrase 'free will,' then free will is that thing! More seriously, as many philosophers have pointed out, 'free will' seems to combine two distinct ideas: first, that your choices are 'free' from any kind of external constraint; and second, that your choices are not arbitrary or capricious, but are 'willed by you.' The second idea – that of being 'willed by you' – is the one I consider outside the scope of science, for the simple reason that no matter what the empirical facts were, a skeptic could always deny that a given decision was 'really' yours, and hold the true decider to have been God, the universe, an impersonating demon, etc. I see no way to formulate, in terms of observable concepts, what it would even mean for such a skeptic to be right or wrong.

But crucially, the situation seems different if we set aside the 'will' part of free will, and consider only the 'free' part. Throughout, I'll use the term *freedom*, or *Knightian freedom*, to mean a certain strong kind of physical unpredictability: a lack of determination, even probabilistic determination, by knowable external factors. That is, a physical system will be 'free' if and only if it's unpredictable in a sufficiently strong sense, and 'freedom' will simply be that property possessed by free systems. A system that's not 'free' will be called 'mechanistic.'

Many issues arise when we try to make the above notions more precise. For one thing, we need a definition of unpredictability that does *not* encompass the 'merely probabilistic' unpredictability of (say) a photon or a radioactive atom – since, as I'll discuss in Section 12.3, I accept the often-made point that *that* kind of unpredictability has nothing to do with what most people would call 'freedom,' and is fully compatible with a system's being 'mechanistic.' Instead, we'll want what economists call 'Knightian' unpredictability, meaning unpredictability that we lack a reliable way even to quantify using probability distributions. Ideally, our criteria for Knightian unpredictability will be so stringent that they won't encompass systems like the Earth's weather – for which, despite the presence of chaos, we arguably *can* give very well-calibrated probabilistic forecasts.

A second issue is that unpredictability seems observer-relative: a system that's unpredictable to one observer might be perfectly predictable to another. This is the reason why, throughout this essay, I'll be interested less in particular methods of prediction than in the *best* predictions that could ever be made, consistent both with the laws of physics and with the need not to destroy the system being studied.

This brings us to a third issue: it's not obvious what should count as 'destroying' the system, or which interventions a would-be predictor should or shouldn't be allowed to perform. For example, in order to ease prediction by a human brain, should a predictor first be allowed to replace each neuron by a 'functionally-equivalent' microchip? How would we decide whether the microchip *was* functionally equivalent to the neuron?

I'll offer detailed thoughts about these issues in Appendix 12A. For now, though, the point I want to make is that, once we *do* address these issues, it seems to me that 'freedom' – in the sense of Knightian unpredictability – is perfectly within the scope of science. We're no longer talking about ethics, metaphysics, or the use of language: only about whether such-and-such a physical system is or isn't predictable in the relevant way! A similar point was recently made forcefully by the philosopher Mark Balaguer (2009), in his interesting book *Free Will as an Open Scientific Problem.* (However, while I strongly agree with Balaguer's basic thesis, as discussed above I reject any connection between freedom and 'merely probabilistic' unpredictability, whereas Balaguer seems to accept such a connection.)

Surprisingly, my experience has been that many scientifically minded people will happily accept that humans plausibly *are* physically unpredictable in the relevant sense. Or at least, they'll accept my own position, that whether humans *are or aren't* so predictable is an interesting empirical question whose answer is neither obvious nor known. The point where many people object to my approach is not the one I would have expected. Again and again, people have conceded that chaos, the no-cloning theorem, or some other phenomenon might make human brains physically unpredictable – indeed, they've seemed oddly indifferent to the question of whether they do or don't! But they never fail to add: "even if so, who cares? we're just talking about unpredictability! that obviously has nothing to do with *free will*!"

For my part, I grant that free will can't be *identified* with unpredictability, without doing violence to the usual meanings of those concepts. Indeed, it's precisely because I grant this that I write, throughout the essay, about 'freedom' (or 'Knightian freedom') rather than 'free will.' I insist, however, that unpredictability has *something* to do with free will – in roughly the same sense that verbal intelligence has *something* to do with consciousness, or optical physics has *something* to do with subjectively-perceived colors. That is, some people might see unpredictability as a pale empirical shadow of the 'true' metaphysical quality, free will, that we really want to understand. But the great lesson of the scientific revolution, going back to Galileo, is that understanding the 'empirical shadow' of something is *vastly* better than not understanding the thing at all! Furthermore, the former might already be an immense undertaking, as understanding human intelligence and the physical universe turned out to be (even setting aside the 'mysteries' of consciousness and metaphysics). Indeed I submit that, for the past four centuries, 'start with the shadow' has been a spectacularly fruitful approach to unravelling the mysteries of the universe: one that's succeeded where greedy attempts to go behind the shadow have failed. If one likes, the goal of this essay is to explore what happens when one applies a 'start with the shadow' approach to the free-will debate.

Personally, I'd go even further than claiming a vague connection between unpredictability and free will. Just as displaying intelligent behavior (by passing the Turing Test or some other means) might be thought a *necessary condition* for consciousness if not a sufficient one, so I tend to see Knightian unpredictability as a necessary condition for free will. In other words, if a system were completely predictable (even probabilistically) by an outside entity – not merely in principle but in practice – then I find it hard to understand why we'd still want to ascribe 'free will' to the system. Why not admit that we now fully understand what makes this system tick?

However, I'm aware that many people sharply reject the idea that unpredictability is a necessary condition for free will. Even if a computer in another room perfectly predicted all of their actions, days in advance, these people would still call their actions 'free,' so long as 'they themselves chose' the actions that the computer also predicted for them. In Section 12.2.5, I'll explore some difficulties to which this position leads when carried to science-fiction conclusions. For now, though, it's not important to dispute the point. I'll happily settle for the weaker claim that unpredictability has *something* to do with free will, just as intelligence has something to do with consciousness. More precisely: in both cases, even when people *think* they're asking purely philosophical questions about the latter concept, much of what they want to know often turns out to hinge on 'grubby empirical questions' about the former concept![4] So if the philosophical questions seem too ethereally inaccessible, then we might as well focus for a while on the scientific ones.

12.1.2 Note on the chapter title

The term 'ghost in the machine' was introduced by Gilbert Ryle (2008). His purpose was to ridicule the notion of a 'mind-substance': a mysterious entity that exists outside of space and ordinary physical causation; has no size, weight, or other

[4] A perfect example of this phenomenon is provided by the countless people who claim that even if a computer program passed the Turing test, it still wouldn't be conscious – and then, without batting an eye, defend that claim using arguments that presuppose that the program *couldn't* pass the Turing test after all! ("Sure, the program might solve math problems, but it could never write love poetry," etc. etc.) The temptation to hitch metaphysical claims to empirical ones, without even realizing the chasm one is crossing, seems incredibly strong.

material properties; is knowable by its possessor with absolute certainty (while the minds of others are *not* thus knowable); and somehow receives signals from the brain and influences the brain's activity, even though it's nowhere to be found *in* the brain. Meanwhile, a *quantum Turing machine*, defined by Deutsch (1985b) (see also Bernstein and Vazirani, 1997), is a Turing machine able to exploit the principle of quantum superposition. As far as anyone knows today (Aaronson, 2005), our universe seems to be efficiently simulable by – or even 'isomorphic to' – a quantum Turing machine, which would take as input the universe's quantum initial state (say, at the Big Bang), and then run the evolution equations forward.

12.1.3 Level of this essay

Most of this essay should be accessible to any educated reader. In a few sections, though, I assume familiarity with basic concepts from quantum mechanics, or (less often) from relativity, thermodynamics, Bayesian probability, or theoretical computer science. When I *do* review concepts from those fields, I usually focus only on the ones most relevant to whatever point I'm making. (To do otherwise would make the essay even more absurdly long than it already is!)

In the main text, I've tried to keep the discussion extremely informal. I've found that, with a contentious subject like free will, mathematical rigor (or the pretense of it) can easily obfuscate more than it clarifies. However, for interested readers, I did put some more technical material into appendices: a suggested formalization of 'Knightian freedom' in Appendix 12A; some observations about prediction, Kolmogorov complexity, and the universal prior in Appendix 12B; and a suggested formalization of the notion of 'freebits' in Appendix 12C.

Readers seeking an accessible introduction to some of the established theories invoked in this essay might enjoy my recent book *Quantum Computing Since Democritus* (Aaronson, 2013).[5]

12.2 FAQ

In discussing a millennia-old conundrum like free will, a central difficulty is that *almost everyone already knows what he or she thinks* – even if the certainties that one person brings to the discussion are completely at odds with someone else's. One practical consequence is that, no matter how I organize this essay, I'm bound to make a large fraction of readers impatient; some will accuse me of dodging the

[5] For the general reader, other good background reading for this essay might include *From Eternity to Here* by Sean Carroll (2010), *The Beginning of Infinity* by David Deutsch (2011), *The Emperor's New Mind* by Roger Penrose (1989), or *Free Will as an Open Scientific Problem* by Mark Balaguer (2009). Obviously, none of these authors necessarily endorses much of what I say (or vice versa)! What the books have in common is simply that they explain one or more ideas invoked in this essay in much more detail than I do.

real issues by dwelling on irrelevancies. So without further ado, I'll now offer a Frequently Asked Questions list. In the 12 questions below, I'll engage with determinists, compatibilists, and others who might have strong *a priori* reasons to be leery of my whole project. I'll try to clarify my areas of agreement and disagreement, and hopefully convince the skeptics to read further. Then, after developing my own ideas in Sections 12.3 and 12.4, I'll come back and address still further objections in Section 12.5.

12.2.1 Narrow scientism

For thousands of years, the free-will debate has encompassed moral, legal, phenomenological, and even theological questions. You seem to want to sweep all of that away, and focus exclusively on what would some would consider a narrow scientific issue having to do with physical predictability. Isn't that presumptuous?

On the contrary, it seems presumptuous *not* to limit my scope! Since it's far beyond my aims and abilities to address all aspects of the free-will debate, as discussed in Section 12.1.1 I decided to focus on one issue: the physical and technological questions surrounding how well human and animal brains can ever be predicted, in principle, by external entities that also want to keep those brains alive. I focus on this for several reasons: because it seems underexplored; because I might have something to say about it; and because even if what I say is wrong, the predictability issue has the appealing property that *progress* on it seems possible. Indeed, even if one granted – which I don't – that the predictability issue has nothing to do with the 'true' mystery of free will, I'd still care about the former at least as much as I cared about the latter!

However, in the interest of laying all my cards on the table, let me offer some brief remarks on the moral, legal, phenomenological, and theological aspects of free will.

On the moral and legal aspects, my own view is summed up beautifully by this Ambrose Bierce poem (Bierce, 2009):

> There's no free will, says the philosopher
> To hang is most unjust.
> There's no free will, assent the officers
> We hang because we must.

For the foreseeable future, I can't see that the legal or practical implications of the free-will debate are nearly as great as many commentators have made them out to be, for the simple reason that (as Bierce points out) any implications would apply 'symmetrically,' to accused and accuser alike.

But I would go further: I've found many discussions about free will and legal responsibility to be downright *patronizing*. The subtext of such discussions usually seems to be:

> We, the educated, upper-class people having this conversation, *should* accept that the entire concept of 'should' is quaint and misguided, when it comes to the uneducated, lower-class sorts of people who commit crimes. Those poor dears' upbringing, brain chemistry, and so forth absolve them of any real responsibility for their crimes: the notion that they had the 'free will' to choose otherwise is just naïve. My friends and I are *right* because we accept that enlightened stance, while other educated people are *wrong* because they fail accept it. For us educated people, of course, the relevance of the categories 'right' and 'wrong' requires no justification or argument.

Or conversely:

> Whatever the truth, we educated people *should* maintain that all people are responsible for their choices – since otherwise, we'd have no basis to punish the criminals and degenerates in our midst, and civilization would collapse. For *us*, of course, the meaningfulness of the word 'should' in the previous sentence is not merely a useful fiction, but is clear as day.

On the phenomenological aspects of free will: if someone claimed to know, from introspection, either that free will exists or that it doesn't exist, then of course I could never refute that person to his or her satisfaction. But precisely because one can't decide between conflicting introspective reports, in this essay I'll be exclusively interested in what can be learned from scientific observation and argument. Appeals to inner experience – including my own and the reader's – will be out of bounds. Likewise, while it might be impossible to avoid grazing the 'mystery of consciousness' in a discussion of human predictability, I'll do my best to avoid addressing that mystery head-on.

On the theological aspects of free will: probably the most relevant thing to say is that, even if there existed an omniscient God who knew all of our future choices, that fact wouldn't concern us in this essay, *unless* God's knowledge could somehow be made manifest in the physical world, and used to *predict* our choices. In that case, however, we'd no longer be talking about 'theological' aspects of free will, but simply again about scientific aspects.

12.2.2 Bait-and-switch

Despite everything you said in Section 12.1.1, I'm still not convinced that we can learn anything about free will from an analysis of unpredictability. Isn't that a shameless 'bait-and-switch'?

Yes, but it's a shameless bait-and-switch with a distinguished history! I claim that, *whenever* it's been possible to make definite progress on ancient philosophical problems, such progress has almost always involved a similar 'bait-and-switch.' In other words: one replaces an unanswerable philosophical riddle Q by a 'merely' scientific or mathematical question Q', which captures *part* of what people have wanted to know when they've asked Q. Then, with luck, one solves Q'.

Of course, even if Q' is solved, centuries later philosophers might still be debating the exact relation between Q and Q'! And further exploration might lead to *other* scientific or mathematical questions – Q'', Q''', and so on – which capture aspects of Q that Q' left untouched. But from my perspective, this process of 'breaking off' answerable parts of unanswerable riddles, and then trying to answer those parts is the closest thing to philosophical progress that there is.

Successful examples of this breaking-off process fill intellectual history: the use of calculus to treat infinite series; the link between mental activity and nerve impulses; natural selection; set theory and first-order logic; special relativity; Gödel's theorem; game theory; information theory; computability and complexity theory; the Bell inequality; the theory of common knowledge; Bayesian causal networks. Each of these advances addressed questions that could rightly have been called 'philosophical' before the advance was made. And after each advance, there was *still* plenty for philosophers to debate about truth and provability and infinity, space and time and causality, probability and information and life and mind. But crucially, it seems to me that the technical advances transformed the philosophical discussion as philosophical discussion *itself* rarely transforms it! And therefore, if such advances don't count as 'philosophical progress,' then it's not clear that anything should.

Appropriately for this essay, perhaps the *best* precedent for my bait-and-switch is the Turing test. Turing began his famous paper 'Computing Machinery and Intelligence' (Turing, 1950) with the words:

> I propose to consider the question, "Can machines think?"

But after a few pages of ground-clearing, he wrote:

> The original question, "Can machines think?" I believe to be too meaningless to deserve discussion.

So with legendary abruptness, Turing simply *replaced* the original question by a different one: "Are there imaginable digital computers which would do well in the imitation game" – i.e., which would successfully fool human interrogators in a teletype conversation into *thinking* they were human? Though some writers would later accuse Turing of conflating intelligence with the 'mere simulation' of it, Turing was perfectly clear about what he was doing:

> I shall replace the question by another, which is closely related to it and is expressed in relatively unambiguous words ... We cannot altogether abandon the original form

of the problem, for opinions will differ as to the appropriateness of the substitution and we must at least listen to what has to be said in this connexion.

The claim is not that the new question, about the imitation game, is *identical* to the original question about machine intelligence. The claim, rather, is that the new question is a worthy candidate for what we *should* have asked or *meant* to have asked, if our goal was to learn something new rather than endlessly debating definitions. In math and science, the process of revising one's original question is often the core of a research project, with the actual answering of the revised question being the relatively easy part!

A good replacement question Q' should satisfy two properties:

(a) Q' should capture some *aspect* of the original question Q – so that an answer to Q' would be *hard to ignore* in any subsequent discussion of Q;
(b) Q' should be precise enough that one can see what it would mean to make *progress* on Q': what experiments one would need to do, what theorems one would need to prove, etc.

The Turing test, I think, captured people's imaginations precisely because it succeeded so well at (a) and (b). Let me put it this way: if a digital computer were built that aced the imitation game, then *it's hard to see what more science could possibly say* in support of machine intelligence being possible. Conversely, if digital computers were proved unable to win the imitation game, then it's hard to see what more science could say in support of machine intelligence *not* being possible. Either way, though, we're no longer 'slashing air,' trying to pin down the true meanings of words like 'machine' and 'think': we've hit the relatively solid ground of a science and engineering problem. Now if we want to go further we need to *dig* (that is, do research in cognitive science, machine learning, etc). This digging might take centuries of backbreaking work; we have no idea if we'll ever reach the bottom. But at least it's something humans know how to do and have done before. Just as important, diggers (unlike air-slashers) tend to uncover countless treasures besides the ones they were looking for.

By analogy, in this essay I advocate replacing the question of whether humans have free will by the question of how accurately their choices can be predicted, in principle, by external agents compatible with the laws of physics. And while I don't pretend that the 'replacement' question is identical to the original, I do claim the following: if humans turned out to be arbitrarily predictable in the relevant sense, then *it's hard to see what more science could possibly say in support of 'free will being a chimera.'* Conversely, if a fundamental reason were discovered why the appropriate 'prediction game' *couldn't* be won, then it's hard to see what more science could say in support of 'free will being real.'

Either way, I'll try to sketch the research program that confronts us if we take the question seriously: a program that spans neuroscience, chemistry, physics, and even cosmology. Not surprisingly, much of this program consists of problems that scientists in the relevant fields are already working on, or longstanding puzzles of which they're well aware. But there are also questions – for example, about the 'past macroscopic determinants' of the quantum states occurring in nature – which as far as I know *haven't* been asked in the form they take here.

12.2.3 Compatibilism

Like many scientifically-minded people, I'm a *compatibilist*: someone who believes free will can exist even in a mechanistic universe. For me, "free will is as real as baseball," as the physicist Sean Carroll memorably put it.[6] That is, the human capacity to weigh options and make a decision 'exists' in the same sense as Sweden, caramel corn, anger, or other complicated notions that might interest us but that no one expects to play a role in the fundamental laws of the universe. As for the fundamental laws, I believe them to be completely mechanistic and impersonal: as far as they know or care, a human brain is just one more evanescent pattern of computation, along with sunspots and hurricanes. Do you dispute any of that? What, if anything, can a compatibilist take from your essay?

I have a lot of sympathy for compatibilism – certainly more than for an incurious mysticism that doesn't even try to reconcile itself with a scientific worldview. So I hope compatibilists will find much of what I have say 'compatible' with their own views!

Let me first clear up a terminological confusion. Compatibilism is often defined as the belief that free will is compatible with *determinism*. But as far as I can see, the question of determinism versus indeterminism has almost nothing to do with what compatibilists actually believe. After all, most compatibilists happily accept quantum mechanics, with its strong indeterminist implications (see the question in Section 12.2.7), but regard it as having almost no bearing on their position. No doubt some compatibilists find it important to stress that *even if classical physics had been right*, there still would have been no difficulty for free will. But it seems to me that one can be a 'compatibilist' even while denying that point, or remaining agnostic about it. In this essay, I'll simply define 'compatibilism' to be the belief that free will is compatible with a *broadly mechanistic worldview* – that is, with a universe governed by impersonal mathematical laws of *some* kind. Whether it's important that those laws be probabilistic (or chaotic, or computationally universal, or whatever else), I'll regard as internal disputes within compatibilism.

[6] See blogs.discovermagazine.com/cosmicvariance/2011/07/13/free-will-is-as-real-as-baseball/ .

I can now come to the question: is my perspective compatible with compatibilism? Alas, at the risk of sounding lawyerly, I can't answer without a further distinction! Let's define *strong compatibilism* to mean the belief that the statement 'Alice has free will' is compatible with the actual physical existence of a machine that predicts all of Alice's future choices – a machine whose predictions Alice herself can read and verify after the fact (where by 'predict,' we mean 'in roughly the same sense that quantum mechanics predicts the behavior of a radioactive atom': that is, by giving arbitrarily accurate probabilities, in cases where deterministic prediction is physically impossible). By contrast, let's define *weak compatibilism* to mean the belief that 'Alice has free will' is compatible with Alice living in a mechanistic, law-governed, universe – but *not* necessarily with her living in a universe where the prediction machine can be built.

Then my perspective is compatible with weak compatibilism but incompatible with strong compatibilism. My perspective *embraces* the mechanical nature of the universe's time-evolution laws, and in that sense is proudly 'compatibilist.' On the other hand, I care whether our choices can *actually* be mechanically predicted – not by hypothetical Laplace demons but by physical machines. I'm troubled if they are, and I take seriously the possibility that they aren't (e.g., because of chaotic amplification of unknowable details of the initial conditions).

12.2.4 Quantum flapdoodle

The usual motivation for mentioning quantum mechanics and mind in the same breath has been satirized as 'quantum mechanics is mysterious, the mind is also mysterious, ergo they must be related somehow'! Aren't you worried that, merely by writing an essay that *seems* to take such a connection seriously, you'll fan the flames of pseudoscience? That any subtleties and caveats in your position will quickly get lost?

Yes! Even though I can only take responsibility for what I write, not for what various Internet commentators, etc. might mistakenly *think* I wrote, it would be distressing to see this essay twisted to support credulous doctrines that I abhor. So for the record, let me state the following:

(a) I don't think quantum mechanics, or anything else, lets us 'bend the universe to our will,' except through interacting with our external environments in the ordinary causal ways. Nor do I think that quantum mechanics says 'everything is holistically connected to everything else' (whatever that means). Proponents of these ideas usually invoke the phenomenon of *quantum entanglement* between particles, which can persist no matter how far apart the particles are. But contrary to a widespread misunderstanding encouraged by generations of 'quantum mystics,' it's an elementary fact that entanglement does *not* allow instantaneous communication. More precisely, quantum mechanics is 'local' in the following sense: if Alice and Bob share a pair of entangled particles, then nothing that Alice does to her particle only can affect the probability of any outcome of any measurement that Bob performs on his particle only.[7] Because of the famous Bell inequality, it's crucial that we don't interpret the concept of 'locality' to mean *more* than that! But quantum mechanics' revision to our concept of locality is so subtle that neither scientists, mystics, nor anyone else anticipated it beforehand.

(b) I don't think quantum mechanics has vindicated Eastern religious teachings, any more than (say) Big Bang cosmology has vindicated the Genesis account of creation. In both cases, while there are interesting parallels, I find it dishonest to seek out only the points of convergence while ignoring the many inconvenient parts that don't fit! Personally, I'd say that the quantum picture of the world – as a complex unit vector $|\psi\rangle$ evolving linearly in a Hilbert space – is not a close match to *any* pre-twentieth-century conception of reality.

(c) I don't think quantum mechanics has overthrown Enlightenment ideals of science and rationality. On the one hand, quantum mechanics does overthrow the 'naïve realist' vision of particles with unambiguous trajectories through space, and it does raise profound conceptual problems that will concern us later on. On the other hand, the point is still to describe a physical world external to our minds by positing a 'state' for that world, giving precise mathematical rules for the evolution of the state and testing the results against observation. Compared with classical physics, the reliance on mathematics has only increased; while the Enlightenment ideal of describing Nature as we find it to be, rather than as intuition says it 'must' be, is emphatically upheld.

(d) I don't think the human brain is a quantum computer in any interesting sense. As I explained in Aaronson (2013), at least three considerations lead me to this opinion. First, it would be nearly miraculous if complicated entangled states – which today, can generally survive for at most a few seconds in near-absolute-zero laboratory conditions – could last for any appreciable time in the hot, wet environment of the brain. (Many researchers have made some version of that argument, but see Tegmark (1999) for perhaps the most detailed version.) Second, the sorts of tasks quantum computers are known to be good at (for example, factoring large integers and simulating quantum systems) seem like a terrible fit to the sorts of tasks that *humans* seem be good at, or that could have plausibly had survival value on the African savannah! Third, and most importantly, I don't see anything that the brain being a quantum computer

[7] Assuming, of course, that we don't *condition* on Alice's knowledge – something that could change Bob's probabilities even in the case of mere classical correlation between the particles.

would plausibly help to *explain*. For example, why would a conscious quantum computer be any less mysterious than a conscious *classical* computer? My conclusion is that, *if* quantum effects play any role in the brain, then such effects are almost certainly short-lived and microscopic.[8] At the 'macro' level of most interest to neuroscience, the evidence is overwhelming that the brain's computation and information storage are classical. (See Section 12.6 for further discussion of these issues in the context of Roger Penrose's views.)

(e) I don't think consciousness is in any sense necessary to bring about the 'reduction of the wavefunction' in quantum measurement. And I say that, despite freely confessing to unease with all existing accounts of quantum measurement! My position is that, to whatever extent the reduction of the wavefunction is a real process at all (as opposed to an artefact of observers' limited perspectives, as in the many-worlds interpretation), it must be a process that can occur even in interstellar space, with no conscious observers anywhere around. For otherwise, we're forced to the absurd conclusion that the universe's quantum state evolved linearly via the Schrödinger equation for billions of years, *until* the first observers arose (who: humans? monkeys? aliens?) and looked around them – at which instant the state suddenly and violently collapsed!

If one likes, whatever I *do* say about quantum mechanics and mind in this essay will be said in the teeth of the above points. In other words, I'll regard points (a)–(e) as sufficiently well established to serve as useful *constraints*, which a new proposal ought to satisfy as a prerequisite to being taken seriously.

12.2.5 Brain-uploading: who cares?

Suppose it were possible to 'upload' a human brain to a computer, and thereafter predict the brain with unlimited accuracy. Who cares? Why should anyone even worry that that would create a problem for free will or personal identity?

For me, the problem comes from the observation that it seems impossible to give any operational difference between a perfect predictor of your actions and a second copy or instantiation of yourself. If there are two entities, both of which respond to every situation exactly as 'you' would, then by what right can we declare that only one such entity is the 'real' you, and the other is just a predictor, simulation, or model? But having multiple copies of you in the same universe seems to open a Pandora's box of science-fiction paradoxes. Furthermore, these paradoxes aren't

[8] This is not to say, of course, that the brain's activity might not *amplify* such effects to the macroscopic, classical, scale – a possibility that will certainly concern us later on.

'merely metaphysical': they concern *how you should do science* knowing there might be clones of yourself, and which predictions and decisions you should make.

Since this point is important, let me give some examples. Planning a dangerous mountain-climbing trip? Before you go, make a backup of yourself – or two or three – so that if tragedy should strike, you can restore from backup and then continue life as if you'd never left. Want to visit Mars? Don't bother with the perilous months-long journey through space; just use a brain-scanning machine to 'fax yourself' there as pure information, whereupon another machine on Mars will construct a new body for you, functionally identical to the original.

Admittedly, some awkward questions arise. For example, after you've been faxed to Mars, what should be done with the 'original' copy of you left on Earth? Should it be destroyed with a quick, painless gunshot to the head? Would *you* agree to be 'faxed' to Mars, knowing that that's what would be done to the original? Alternatively, if the original were left alive, then what makes you sure you would 'wake up' as the copy on Mars? At best, wouldn't you have 50/50 odds of still finding yourself on Earth? Could that problem be 'solved' by putting a thousand copies of you on Mars, while leaving only one copy on Earth? Likewise, suppose you return unharmed from your mountain-climbing trip, and decide that the backup copies you made before you left are now an expensive nuisance. If you destroy them, are you guilty of murder? Or is it more like suicide? Or neither?

Here's a 'purer' example of such a puzzle, which I've adapted from the philosopher Nick Bostrom (2002). Suppose an experimenter flips a fair coin while you lie anesthetized in a white, windowless, hospital room. If the coin lands heads, then she'll create a thousand copies of you, place them in a thousand identical rooms, and wake each one up. If the coin lands tails, then she'll wake you up without creating any copies. You wake up in a white, windowless room just like the one you remember. Knowing the setup of the experiment, at what odds should you be willing to bet that the coin landed heads? Should your odds just be 50/50, since the coin was fair? Or should they be biased 1000 : 1 in favor of the coin having landed heads – since if it *did* land heads, then there are a thousand of you confronting the same situation, compared to only one if the coin landed tails?

Many people immediately respond that the odds should be 50/50: they consider it a metaphysical absurdity to adjust the odds based on the number of copies of yourself in existence. (Are we to imagine a 'warehouse full of souls,' with the odds of any particular soul being taken out of the warehouse proportional to the number of suitable bodies for it?) However, those who consider 50/50 the obvious answer should consider a slight variant of the puzzle. Suppose that, if the coin lands tails then, as before, the experimenter leaves a single copy of you in a white room. If the coin lands heads, then the experimenter creates a thousand copies of you and places them in a thousand windowless rooms. Now, though, 999 of the rooms are painted blue; only one of the rooms is white like you remember.

You wake up from the anesthesia and find yourself in a white room. *Now* what posterior probability should you assign to the coin having landed heads? *If* you answered 50/50 to the first puzzle, then a simple application of Bayes' rule implies that, in the *second* puzzle, you should consider it overwhelmingly likely that the coin landed tails. For if the coin landed heads, then presumably you had a 99.9% probability of being one of the 999 copies who woke up in a blue room. So the fact that you woke up in a white room furnishes powerful evidence about the coin. Not surprisingly, many people find *this* result just as metaphysically unacceptable as the 1000 : 1 answer to the first puzzle! Yet as Bostrom points out, it seems mathematically inconsistent to insist on 50/50 as the answer to *both* puzzles.

Probably the most famous 'paradox of brain-copying' was invented by Simon Newcomb, then popularized by Robert Nozick (1969) and Martin Gardner (1974). In *Newcomb's paradox*, a superintelligent 'Predictor' presents you with two closed boxes and offers you a choice between opening the first box only or opening both boxes. Either way, you get to keep whatever you find in the box or boxes that you open. The contents of the first box can vary – sometimes it contains $1,000,000, sometimes nothing – but the second box always contains $1,000.

Just from what was already said, it seems that it must be preferable to open both boxes. For whatever you would get by opening the first box only, you can get $1,000 more by opening the second box as well. But here's the catch: using a detailed brain model, the Predictor has already foreseen your choice. *If* it predicted that you would open both boxes then the Predictor left the first box empty; while if it predicted that you would open the first box only then the Predictor put $1,000,000 in the first box. Furthermore, the Predictor has played this game hundreds of times before, both with you and with other people, and its predictions have been right every time. Everyone who opened the first box ended up with $1,000,000, while everyone who opened both boxes ended up with only $1,000. Knowing all of this, what do you do?

Some people dismiss the problem as contradictory – arguing that, if the assumed Predictor exists, then you have no free will, so there's no use fretting over how many boxes to open since your choice is already predetermined anyway. Among those willing to play along, opinion has been split for decades between 'one-boxers' and 'two-boxers.' Lately, though, the one-boxers seem to have been gaining the upper hand – and reasonably so in my opinion, since by the assumptions of the thought experiment, the one-boxers *do* always walk away richer!

As I see it, the real problem is to *explain* how one-boxing could possibly be rational, given that, at the time you're contemplating your decision, the million dollars are either in the first box or not. Can a last-minute decision to open both boxes somehow 'reach backwards in time,' causing the million dollars that 'would have been' in the first box to disappear? Do we need to distinguish between your 'actual' choices and your 'dispositions,' and say that, while one-boxing is admittedly irrational, *making yourself into the sort of person* who one-boxes is rational?

While I consider myself a one-boxer, the only justification for one-boxing that makes sense to me goes as follows.[9] In principle, you could base your decision of whether to one-box or two-box on anything you like: for example, on whether the name of some obscure childhood friend had an even or odd number of letters. However, this suggests that the problem of predicting whether you will one-box or two-box is 'you-complete.'[10] In other words, if the Predictor can solve this problem reliably then it seems to me that it must possess a simulation of you so detailed as to constitute another *copy* of you (as discussed previously).

But in that case, to whatever extent we want to think about Newcomb's paradox in terms of a freely willed decision at all, we need to imagine *two* entities separated in space and time – the 'flesh-and-blood you' and the simulated version being run by the Predictor – that are nevertheless 'tethered together' and share common interests. *If* we think this way, then we can easily explain why one-boxing can be rational, even without backwards-in-time causation. Namely, as you contemplate whether to open one box or two, who's to say that you're not 'actually' the simulation? If you are, then of course your decision can affect what the Predictor does in an ordinary, causal way.

For me, the takeaway is this. *If* any of these technologies – brain-uploading, teleportation, the Newcomb predictor, etc. – were actually realized then all sorts of 'woolly metaphysical questions' about personal identity and free will would start to have *practical consequences*. Should you fax yourself to Mars or not? Sitting in the hospital room, should you bet that the coin landed heads or tails? Should you expect to 'wake up' as one of your backup copies, or as a simulation being run by the Newcomb Predictor? These questions all seem 'empirical,' yet one can't answer them without taking an implicit stance on questions that many people would prefer to regard as outside the scope of science.

Thus, the idea that we can 'escape all that philosophical crazy-talk' by declaring that the human mind is a computer program running on the hardware of the brain, and that's all there is to it, strikes me as ironically backwards. Yes, we can say that, and we might even be right. But far from bypassing all philosophical perplexities, such a move lands us in a *swamp* of them! For now we need to give some account of how a rational agent ought to make decisions and scientific predictions, in situ-

[9] I came up with this justification around 2002 and set it out in a blog post in 2006: see www.scottaaronson.com/blog/?p=30 . Later, I learned that Radford Neal (2006) had independently proposed similar ideas.

[10] In theoretical computer science, a problem belonging to a class C is called *C-complete* if solving the problem would suffice to solve any other problem in C.

ations where it knows it's only one of several exact copies of itself inhabiting the same universe.

Many will try to escape the problem, by saying that such an agent, being (by assumption) 'just a computer program,' simply does whatever its code determines it does given the relevant initial conditions. For example, if a piece of code says to bet heads in a certain game, then all agents running that code will bet heads; if the code says to bet tails, then the agents will bet tails. Either way, an *outside* observer who knew the code could easily calculate the probability that the agents will win or lose their bet. So what's the philosophical problem?

For me, the problem with this response is simply that it gives up on science as *something agents can use to predict their future experiences.* The agents wanted science to tell them, 'given such-and-such physical conditions, here's what you *should* expect to see, and why.' Instead they're getting the worthless tautology, 'if your internal code causes you to expect to see X, then you expect to see X, while if your internal code causes you to expect to see Y, then you expect to see Y.' But the same could be said about *anything*, with no scientific understanding needed! To paraphrase Democritus,[11] it seems like the ultimate victory of the mechanistic worldview is also its defeat.

As far as I can see, the only hope for avoiding these difficulties is if – because of chaos, the limits of quantum measurement, or whatever other obstruction – minds *can't* be copied perfectly from one physical substrate to another, as can programs on standard digital computers. So that's a possibility that this essay explores at some length. To clarify, we can't use any philosophical difficulties that would arise if minds were copyable as evidence for the empirical claim that they're *not* copyable. The universe has never shown any particular tendency to cater to human philosophical prejudices! But I'd say the difficulties provide more than enough reason to *care* about the copyability question.

12.2.6 Determinism versus predictability

I'm a determinist: I believe, not only that humans lack free will, but that everything that happens is completely determined by prior causes. So why should an analysis of 'mere unpredictability' change my thinking at all? After all, I readily admit that, despite being metaphysically determined, many future events are unpredictable in practice. But for me, the fact that we can't predict something is *our* problem, not Nature's!

There's an observation that doesn't get made often enough in free-will debates,

[11] In Democritus's famous dialogue between intellect and the senses, intellect declares: "By convention there is sweetness, by convention bitterness, by convention color, in reality only atoms and the void." To which the senses reply: "Foolish intellect! Do you seek to overthrow us, while it is from us that you take your evidence? Your victory is your defeat."

but that seems extremely pertinent here. Namely: *if you stretch the notion of 'determination' far enough then events become 'determined' so trivially that the notion itself becomes vacuous.*

For example, a religious person might maintain that all events are predetermined by God's Encyclopedia, which of course only God can read. Another, secular person might maintain that *by definition*, 'the present state of the universe' contains all the data needed to determine future events, even if those future events (such as quantum measurement outcomes) aren't *actually* predictable via present-day measurements. In other words: if, in a given conception of physics, the present state does *not* fully determine all the future states then such a person will simply add 'hidden variables' to the present state until it does so.

Now, if our hidden-variable theorist isn't careful, and piles on additional requirements like spatial locality, then she'll quickly find herself contradicting one or more of quantum mechanics' no-hidden-variable theorems (such as the Bell (1987), Kochen and Specker (1967), or Pusey et al. (2012) theorems). But the bare assertion that 'everything is determined by the current state' is no more disprovable than the belief in God's Encyclopedia.

To me, this immunity from any possible empirical discovery shows just how *feeble* a concept 'determinism' really is, unless it's supplemented by further concepts like locality, simplicity, or (best of all) actual predictability. A form of 'determinism' that applies not merely to our universe, but to any *logically possible* universe, is not a determinism that has 'fangs' or that could credibly threaten any notion of free will worth talking about.

12.2.7 Quantum mechanics and hidden variables

Forget about free will or Knightian uncertainty: I deny even that *probability* plays any fundamental role in physics. For me, like for Einstein, the much-touted 'randomness' of quantum mechanics merely shows that we humans haven't yet discovered the underlying deterministic rules. Can you prove that I'm wrong?

With minimal use of Occam's razor, yes, I can! In 1926, when Einstein wrote his famous aphorism about God and dice, the question of whether quantum events were 'truly' random or merely pseudorandom could still be considered metaphysical. After all, common sense suggests we can never say with confidence that *anything* is random: the most we can ever say is that *we* failed to find a pattern in it.

But common sense is flawed here. A large body of work, starting with that of Bell in the 1960s (see Bell, 1987), has furnished evidence that quantum measurement outcomes *can't* be governed by any hidden pattern but must be random in just the way quantum mechanics says they are. Crucially, this evidence doesn't circularly assume that quantum mechanics is the final theory of nature. Instead, it assumes just a few general principles (such as spatial locality and 'no cosmic conspiracies'), together with the results of specific experiments that have already been done. Since these points are often misunderstood, it might be worthwhile to spell them out in more detail.

Consider the *Bell inequality*, whose violation by entangled particles (in accord with quantum mechanics) has been experimentally demonstrated more and more firmly since the 1980s (Aspect et al., 1982). From a modern perspective, Bell simply showed that certain games, played by two cooperating but non-communicating players Alice and Bob, can be won with greater probability if Alice and Bob share entangled particles than if they merely share correlated *classical* information.[12] Bell's theorem is usually presented as ruling out a class of theories called *local hidden-variable theories*. Those theories sought to explain Alice's and Bob's measurement results in terms of ordinary statistical correlations between two random variables X and Y, which are somehow associated with Alice's and Bob's particles respectively and which have the properties that nothing Alice does can affect Y and that nothing Bob does can affect X. (One can imagine the particles flipping a coin at the moment of their creation, whereupon one of them declares, "OK, if anyone asks, I'll be spinning up and you'll be spinning down!")

In popular treatments, Bell's theorem is usually presented as demonstrating the reality of what Einstein called "spooky action at a distance."[13] However, as many people have pointed out over the years – see, for example, my 2002 critique (Aaronson, 2002) of Stephen Wolfram's *A New Kind of Science* (Wolfram, 2002) – one can also see Bell's theorem in a different way: as using the *assumption* of no instantaneous communication to address the even more basic issue of *determinism*. From this perspective, Bell's theorem says the following:

> *Unless* Alice's and Bob's particles communicate faster than light, the results of all possible measurements that Alice and Bob could make on those particles *cannot*

[12] The standard example is the *CHSH game* (Clauser et al., 1969). Here Alice and Bob are given bits x and y respectively, which are independent and uniformly random. Their goal is for Alice to output a bit a, and Bob a bit b, such that $a+b \pmod 2 = xy$. Alice and Bob can agree on a strategy in advance, but can't communicate after receiving x and y. Classically, it's easy to see that the best they can do is always to output $a = b = 0$, in which case they win the game with probability 3/4. By contrast, if Alice and Bob own one qubit each of the entangled state $\frac{1}{\sqrt{2}}(|00\rangle + |11\rangle)$, then there exists a strategy by which they can win with probability $\cos^2(\pi/8) \approx 0.85$. That strategy has the following form: Alice measures her qubit in a way that depends on x and outputs the result as a, while Bob measures his qubit in a way that depends on y and outputs the result as b.

[13] The reason why many people (including me) cringe at that sort of talk is the *no-communication theorem*, which explains why, despite Bell's theorem, entanglement *can't* be used to send actual messages faster than light. (Indeed, if it could, then quantum mechanics would flat-out contradict special relativity.)
The situation is this: *if* one wanted to violate the Bell inequality using classical physics *then* one would need faster-than-light communication. But that doesn't imply that quantum mechanics' violation of the same inequality should *also* be understood in terms of faster-than-light communication! We're really dealing with an intermediate case here – 'more than classical locality, but less than classical nonlocality' – which I don't think anyone even recognized as a logical possibility until quantum mechanics forced it on them.

have been determined prior to measurement– not even by some bizarre, as-yet-undiscovered, uncomputable law – assuming the statistics of all the possible measurements agree with the quantum predictions. Instead, the results *must* be 'generated randomly on-the-fly' in response to whichever measurement is made, just as quantum mechanics says they are.

The above observation was popularized in 2006 by John Conway and Simon Kochen, who called it the 'free will theorem' (Conway and Kochen, 2009). Conway and Kochen put the point as follows: if there's no faster-than-light communication, *and* Alice and Bob have the 'free will' to choose how to measure their respective particles, then the particles must have their own 'free will' to choose how to respond to the measurements.

Alas, Conway and Kochen's use of the term 'free will' has generated confusion. For the record, *what Conway and Kochen mean by 'free will' has only the most tenuous connection to what most people (including me, in this essay) mean by it!* Their result might more accurately be called the 'freshly generated randomness theorem.'[14] For the indeterminism that's relevant here is 'only' probabilistic: indeed, Alice and Bob could be replaced by simple dice-throwing or quantum-state-measuring automata without affecting the theorem at all.[15]

Another recent development has made the conflict between quantum mechanics and determinism particularly vivid. It's now known how to exploit Bell's theorem to generate so-called 'Einstein-certified random bits' for use in cryptographic applications (Pironio et al., 2010; Vazirani and Vidick, 2012), starting from a much smaller number of 'seed' bits that are known to be random.[16] Here 'Einstein-certified' means that *if* the bits pass certain statistical tests then they *must* be close to uniformly random, unless nature resorted to 'cosmic conspiracies' between separated physical devices to bias the bits.

Thus, if one wants to restore determinism while preserving the empirical success of quantum mechanics then one has to posit a conspiracy in which every elementary particle, measuring device, and human brain potentially colludes. Furthermore, this conspiracy needs to be so diabolical as to leave essentially no trace of its existence! For example, in order to explain why we can't exploit the conspiracy to send faster-than-light signals, we have to imagine that the conspiracy prevents our own brains (or the quantum-mechanical random number generators in our computers, etc.) from making the choices that would cause those signals to be sent. To my

[14] Or the 'free whim theorem,' as Conway likes to suggest when people point out the irrelevance of human free will to the theorem.

[15] This point was recently brought out by Fritz, in his paper 'Bell's theorem without free will' (Fritz, 2012). Fritz replaces the so-called 'free will assumption' of the Bell inequality – that is, the assumption that Alice and Bob get to choose which measurements to perform – by an assumption about the *independence* of separated physical devices.

[16] The protocol of Vazirani and Vidick (2012) needs only $O(\log n)$ seed bits to generate n Einstein-certified output bits.

mind, this is no better than the creationists' God, who planted fossils in the ground to confound the paleontologists.

I should say that at least one prominent physicist, Gerard 't Hooft, actually advocates such a cosmic conspiracy (Hooft, 2007) (under the name 'superdeterminism'); he speculates that a yet-undiscovered replacement for quantum mechanics will reveal its workings.[17] For me, though, the crux is that once we start positing conspiracies between distant regions of spacetime, or between the particles we measure and our own instruments or brains, determinism becomes consistent with *any* possible scientific discovery, and therefore retreats into vacuousness. As the extreme case, as pointed out in the question in Section 12.2.6, someone could always declare that everything that happens was 'determined' by God's unknowable book listing everything that will ever happen! That sort of determinism can never be falsified, but has zero predictive or explanatory power.

In summary, I think it's fair to say that *physical indeterminism is now a settled fact to roughly the same extent as evolution, heliocentrism, or any other discovery in science*. So *if* that fact is considered relevant to the free-will debate, then all sides might as well just accept it and move on! (Of course, we haven't yet touched on the question whether physical indeterminism *is* relevant to the free-will debate.)

12.2.8 The consequence argument

How does your perspective respond to Peter van Inwagen's consequence argument?

Some background for non-philosophers: the consequence argument (Inwagen, 1983) is an attempt to formalize most people's intuition for why free will is incompatible with determinism. The argument consists of the following steps:

(i) If determinism is true then our choices today are determined by whatever the state of the universe was (say) 100 million years ago, when dinosaurs roamed the earth.
(ii) The state of the universe 100 million years ago is clearly outside our ability to alter.
(iii) Therefore, if determinism is true, then our choices today are outside our ability to alter.
(iv) Therefore, if determinism is true, then we don't have free will.

[17] Some people might argue that Bohmian mechanics (Bohm, 1952), the interpretation of quantum mechanics that originally inspired Bell's theorem, is also 'superdeterministic.' But Bohmian mechanics is empirically equivalent to standard quantum mechanics – from which fact it follows immediately that the 'determinism' of Bohm's theory is a formal construct that, whatever else one thinks about it, has no actual consequences for prediction. To put it differently: at least in its standard version, Bohmian mechanics buys its 'determinism' via the mathematical device of pushing all the randomness back to the beginning of time. It then accepts the nonlocality that such a tactic inevitably entails because of Bell's theorem.

(An aside: as discussed in the question in Section 12.2.3, the traditional obsession with 'determinism' here seems unfortunate to me. What people really mean to ask, I think, is whether free will is compatible with *any* mechanistic account of the universe, regardless of whether the account happens to be deterministic or probabilistic. However, one could easily rephrase the consequence argument to allow for this, with the state of the universe 100 million years ago now fully determining the *probabilities* of our choices today, if not the choices themselves. And I don't think that substitution would make any essential difference to what follows.)

One can classify beliefs about free will according to how they respond to the consequence argument. If you accept the argument as well as its starting premise of determinism (or mechanism), and hence also the conclusion of no free will, then you're a *hard determinist (or mechanist)*. If you accept the argument, but reject the conclusion by denying the starting premise of determinism or mechanism, then you're a *metaphysical libertarian*. If you reject the argument by denying that steps (iii) or (iv) follow from the previous steps, then you're a *compatibilist*.

What about the perspective I explore here? It denies step (ii) – or rather, it denies the usual notion of 'the state of the universe 100 million years ago,' insisting on a distinction between 'macrofacts' and 'microfacts' about that state. It agrees that the past *macro*facts – such as whether a dinosaur kicked a particular stone – have an objective reality that is outside our ability to alter. But it denies that we can always speak in straightforward causal terms about the past *micro*facts, such as the quantum state $|\psi\rangle$ of some particular photon impinging on the dinosaur's tail.

As such, my perspective can be seen as an example of a little-known view that Fischer (1995) calls *multiple-pasts compatibilism*. As I'd put it, multiple-pasts compatibilism agrees that the past microfacts about the world determine its future, and it also agrees that the past *macro*facts are outside our ability to alter. However, it maintains that there might be many possible settings of the past microfacts – the polarizations of individual photons, etc. – that all coexist in the same 'past-ensemble.' By definition, such microfacts can't possibly have made it into the history books, and a multiple-pasts compatibilist would deny that they're necessarily 'outside our ability to alter.' Instead, our choices today might play a role in selecting *one* past from a giant ensemble of macroscopically identical but microscopically different pasts.

While I take the simple idea above as a starting point, there are two main ways in which I try to go further than it. First, I insist that whether we *can* make the needed distinction between 'microfacts' and 'macrofacts' is a *scientific* question – one that can only be addressed through detailed investigation of the quantum decoherence process and other aspects of physics. Second, I change the focus, from unanswerable metaphysical conundrums about what 'determines' what, toward empirical questions about the actual *predictability* of our choices (or at least the probabilities of those choices) given the past macrofacts. I argue that, by making progress on the predictability question, we can learn something about whether multiple-pasts compatibilism is a *viable* response to the consequence argument even if we can never know for certain whether it's the *right* response.

12.2.9 Paradoxes of prediction

You say you're worried about the consequences for rational decision-making, Bayesian inference, and so forth if our choices were all mechanically predictable. Why isn't it reassurance enough that, logically, predictions of an agent's behavior can never be known ahead of time to the agent itself? For if they *were* known ahead of time, then in real life – as opposed to Greek tragedies, or stories like Philip K. Dick's *Minority Report* (Dick, 1998) – the agent could simply defy the prediction by doing something else! Likewise, why isn't it enough that, because of the time hierarchy theorem from computational complexity, predicting an agent's choices might require as much computational power as the agent itself expends in making those choices?

The obvious problem with such computational or self-referential arguments is that they can't possibly prevent one agent, say Alice, from predicting the behavior of a *different* agent, say Bob. And in order to do that, Alice doesn't need unlimited computational power: *she only needs a bit more computational power than Bob has.*[18] Furthermore, Bob's free will actually seems *more* threatened by Alice predicting his actions than by Bob predicting his own actions, supposing the latter were possible! This explains why I won't be concerned in this essay with computational obstacles to prediction but only with obstacles that arise from Alice's physical inability to gather the requisite information about Bob.

Admittedly, as MacKay (1960), Lloyd (2012), and others have stressed, if Alice wants to predict Bob's choices then she needs to be careful not to *tell* Bob her predictions before they come true! And that does indeed make predictions of Bob's actions an unusual sort of 'knowledge': knowledge that can be falsified by the very fact of Bob's learning it.

But unlike some authors, I don't make much of these observations. For even if Alice can't tell Bob what he's going to do, it's easy enough for her to demonstrate to him afterwards that *she* knew. For example, Alice could put her prediction into a sealed envelope, and let Bob open the envelope only after the prediction came true. Or she could send Bob a *cryptographic commitment* to her prediction, withholding the decryption key until afterward. If Alice could do these things reliably then

[18] This is analogous to how, in computational complexity theory, there exists a program that that uses (say) $n^{2.0001}$ time steps and that simulates any n^2-step program provided to it as input. The time hierarchy theorem, which is close to tight, rules out only the simulation of n^2-step programs using significantly *less* than n^2 time steps.

it seems likely that Bob's self-conception would change just as radically as if he knew the predictions in advance.

12.2.10 Singulatarianism

How could it possibly make a difference to anyone's life whether his or her neural computations were buffeted around by microscopic events subject to Knightian uncertainty? Suppose that only 'ordinary' quantum randomness and classical chaos turned out to be involved: how on earth could that matter outside the narrow confines of free-will debates? Is the variety of free will that apparently interests you – one based on the physical unpredictability of our choices – really a variety 'worth wanting,' in Daniel Dennett's famous phrase (Dennett, 1984)?

As a first remark, if there's *anything* in this debate that all sides can agree on, hopefully they can agree that the truth (whatever it is) doesn't care what we want, consider 'worth wanting,' or think is necessary or sufficient to make our lives meaningful!

But, to lay my cards on the table, my interest in the issues discussed in this essay was sparked, in part, by considering arguments of the so-called *Singulatarians*. These are people who look at current technological trends – including advances in neuroscience, nanotechnology, and artificial intelligence, as well as the dramatic increases in computing power – and foresee a 'technological singularity' perhaps 50–200 years in our future (not surprisingly, the projected timescales vary). By this, they mean not a mathematical singularity, but a rapid 'phase transition,' perhaps analogous to the appearance of the first life on earth, the appearance of humanity, or the invention of agriculture or of writing. In the Singulatarians' view, the next such change will happen around the time when humans manage to build artificial intelligences (AIs) that are smarter than the smartest humans. It stands to reason, the Singulatarians say, that such AIs will realize they can best further their goals (whatever those might be) by building AIs even smarter than themselves – and then the super-AIs will build yet smarter AIs, and so on, presumably until the fundamental physical limits on intelligence (whatever they are) are reached.

Following the lead of science-fiction movies, of course one might wonder about the role of humans in the resulting world. Will the AIs wipe us out, treating us as callously as humans have treated each other and most animal species? Will the AIs keep us around as pets, or as revered (if rather dimwitted) creators and ancestors? Will humans be invited to upload their brains to the post-singularity computer cloud – with each human, perhaps, living for billions of years in his or her own simulated paradise? Or will the humans simply merge their consciousnesses into the AI's hive-mind, losing their individual identities but becoming part of something unimaginably greater?

Hoping to find out, many Singulatarians have signed up to have their brains cryogenically frozen upon their (biological) deaths, so that some future intelligence (before or after the singularity) might be able to use the information therein to revive them.[19] One leading Singulatarian, Eliezer Yudkowsky, has written at length about the irrationality of people who *don't* sign up for cryonics: how they value social conformity and 'not being perceived as weird' over a non-negligible probability of living for billions of years.[20]

With some notable exceptions, academics in neuroscience and other relevant fields have tended to dismiss Singulatarians as nerds with hyperactive imaginations: people who have no idea how great the difficulties are in modeling the human brain or building a general-purpose AI. Certainly, one could argue that the Singulatarians' *timescales* might be wildly off. And even if one accept their timescales, one could argue that (almost by definition) the unknowns in their scenario are so great as to negate any *practical* consequences for the humans alive now. For example, suppose we conclude – as many Singulatarians have – that the greatest problem facing humanity today is how to ensure that, when superhuman AIs are finally built, those AIs will be 'friendly' to human concerns. The difficulty is: *given our current ignorance about AI, how on earth should we act on that conclusion?* Indeed, how could we have any confidence that whatever steps we *did* take wouldn't backfire, and increase the probability of an *un*friendly AI?

Yet on questions of principle – that is, of what the laws of physics could ultimately allow – I think the uncomfortable truth is that it's the Singulatarians who are the scientific conservatives, while those who reject their vision as fantasy are scientific radicals. For at some level, all the Singulatarians are doing is taking conventional thinking about physics and the brain to its logical conclusion. If the brain is a 'meat computer,' then given the right technology, why *shouldn't* we be able to copy its program from one physical substrate to another? And why couldn't we then run multiple copies of the program in parallel, leading to all the philosophical perplexities discussed in Section 12.2.5?

Maybe the conclusion is that we should all become Singulatarians! But given the stakes, it seems worth exploring the possibility that there are scientific reasons why human minds *can't* be casually treated as copyable computer programs: not just practical difficulties, or the sorts of question-begging appeals to human specialness that are child's-play for Singulatarians to demolish. If one likes, the origin of this essay was my own refusal to accept the lazy cop-out position, which answers the question of whether the Singulatarians' ideas are *true* by repeating that their ideas

[19] The largest cryonics organization is Alcor, www.alcor.org.

[20] See, for example, lesswrong.com/lw/mb/lonely_dissent/.

are crazy and weird. If uploading our minds to digital computers is indeed just a fantasy, then I demand to know what it is about the physical universe that *makes* it a fantasy.

12.2.11 The Libet experiments

Haven't neuroscience experiments already proved that our choices aren't nearly as unpredictable as we imagine? Seconds before a subject is conscious of making a decision, EEG recordings can already detect the neural buildup to that decision. Given that empirical fact, isn't any attempt to ground freedom in unpredictability doomed from the start?

It's important to understand what experiments have and haven't shown, since the details tend to get lost in popularization. The celebrated experiments by Libet (see Libet, 1999) from the 1970s used EEGs to detect a 'readiness potential' building up in a subject's brain up to a second and a half before the subject made the 'freely-willed decision' to flick her finger. The implications of this finding for free will were avidly discussed – especially since the subject might not have been *conscious* of any intention to flick her finger until (say) half a second before the act. So, did the prior appearance of the readiness potential prove that what we perceive as 'conscious intentions' are just window-dressing, which our brains add after the fact?

However, as Libet acknowledged, an important gap in the experiment was that it had inadequate 'control.' That is, how often did the readiness potential form, *without* the subject flicking her finger (which might indicate a decision that was 'vetoed at the last instant')? Because of this gap, it was unclear to what extent the signal Libet found could actually be used for prediction.

More recent experiments – see especially Soon et al. (2008) – have tried to close this gap, by using fMRI scans to predict *which* of two buttons a person would press. Soon et al. reported that they were able to do so four or more seconds in advance, with success probability significantly better than chance (around 60%). The question is, how much should we read into these findings?

My own view is that the quantitative aspects are crucial when discussing these experiments. For, compare a (hypothetical) ability to predict human decisions a full minute in advance with an ability to predict the same decisions 0.1 seconds in advance, in terms of the intuitive 'threat' to free will. The two cases seem utterly different! A minute seems like clearly enough time for a deliberate choice, while 0.1 seconds seems like clearly *not* enough time; on the latter scale, we are only talking about physiological reflexes. (For intermediate scales such as 1 second, intuition – or at least my intuition – is more conflicted.)

Similarly, compare a hypothetical ability to predict human decisions with 99%

accuracy, against an ability to predict them with 51% accuracy. I expect that only the former, and not the latter, would strike anyone as threatening or even uncanny. For it's obvious that human decisions are *somewhat* predictable: if they weren't, there would be nothing for advertisers, salespeople, seducers, or demagogues to exploit! Indeed, with zero predictability, we couldn't even talk about *personality* or *character* as having any stability over time. So *better-than-chance* predictability is just too low a bar to clear for it to have any relevance to the free-will debate. One wants to know: *how much* better than chance? Is the accuracy better than what my grandmother, or an experienced cold-reader, could achieve?

Even within the limited domain of button-pressing, years ago I wrote a program that invited the user to press the 'G' or 'H' keys in any sequence – 'GGGHGHHHHGHG' – and that tried to predict which key the user would press next. The program used only the crudest pattern-matching – "in the past, was the subsequence GGGH more likely to be followed by G or H?" Yet humans are so poor at generating 'random' digits that the program regularly achieved prediction accuracies of around 70% – no fMRI scans needed![21]

In summary, I believe neuroscience might *someday* advance to the point where it completely rewrites the terms of the free-will debate, by showing that the human brain is 'physically predictable by outside observers' in the same sense as a digital computer. But it seems nowhere close to that point today. Brain-imaging experiments have succeeded in demonstrating predictability with better-than-chance accuracy, in limited contexts and over short durations. Such experiments are deeply valuable and worth trying to push as far as possible. However, the mere *fact* of limited predictability is something that humans knew millennia before brain-imaging technologies became available.

12.2.12 Mind and morality

Notwithstanding your protests that you won't address the 'mystery of consciousness,' your entire project seems geared toward the old, disreputable quest to find some sort of dividing line between 'real, conscious, biological' humans and digital computer programs, even supposing that the latter could perfectly emulate the former. Many thinkers have sought such a line before, but most scientifically minded people regard the results as dubious. For Roger Penrose, the dividing line involves neural structures called microtubules

[21] Soon et al. (2008) – see the Supplementary Material, pp. 14–15 – argued that, by introducing a long delay between trials and other means, they were able to rule out the possibility that their prediction accuracy was due purely to 'carryover' between successive trials. They also found that the probability of a run of N successive presses of the same button decreased exponentially with N, as would be expected if the choices were independently random. However, it would be interesting for future research to compare fMRI-based prediction 'head-to-head' against prediction using carefully designed machine learning algorithms that see only the sequence of previous button presses.

harnessing exotic quantum-gravitational effects (see Section 12.6). For the philosopher John Searle (1992), the line involves the brain's unique 'biological causal powers': powers whose existence Searle loudly asserts but never actually explains. For you, the line seems to involve hypothesized limits on predictability imposed by the no-cloning theorem. But, regardless of where the line gets drawn, let's discuss where the rubber meets the road. Suppose that in the far future, there are trillions of emulated brains (or 'ems') running on digital computers. In such a world, would you consider it acceptable to 'murder' an em (say, by deleting its file), simply because it lacked the 'Knightian unpredictability' that biological brains might or might not derive from amplified quantum fluctuations? If so, then isn't that a cruel, arbitrary, 'meatist' double standard – one that violates the most basic principles of your supposed hero, Alan Turing?

For me, this *moral* objection to my project is possibly the most pressing objection of all. Will *I* be the one to shoot a humanoid robot pleading for its life, simply because the robot lacks the supposedly crucial 'freebits,' or 'Knightian unpredictability,' or whatever else the magical stuff is supposed to be that separates humans or machines?

Thus, perhaps my most important reason to take the freebit picture seriously is that it *does* suggest a reply to the objection: one that strikes me as both intellectually consistent and moral. I simply need to adopt the following ethical stance: *I'm against any irreversible destruction of knowledge, thoughts, perspectives, adaptations, or ideas, except possibly by their owner.* Such destruction is worse the more valuable the thing destroyed, the longer it took to create, and the harder it is to replace. From this basic revulsion to *irreplaceable loss*, a hatred of murder, of genocide, of the hunting of endangered species to extinction, and even (say) of the burning of the Library of Alexandria can all be derived as consequences.

Now, what about the case of 'deleting' an emulated human brain from a computer memory? The same revulsion applies in full force – *if the copy deleted is the last copy in existence.* If, however, there are other extant copies, then the deleted copy can always be 'restored from backup,' so deleting it seems at worst like property damage. For biological brains, by contrast, whether such backup copies *can* be physically created is of course exactly what's at issue, and the freebit picture conjectures a negative answer.

These considerations suggest that the moral status of ems really *could* be different than that of organic humans, but for straightforward practical reasons that have nothing to do with 'meat chauvinism' or with question-begging philosophical appeals to human specialness. The crucial point is that even a program that passed the Turing test would revert to looking 'crude and automaton-like' *if you could read, trace, and copy its code.* And whether the code *could* be read and copied might depend strongly on the machine's physical substrate. Destroying something that's both as complex as a human being *and* one-of-a-kind could be regarded as an especially heinous crime.

I see it as the great advantage of this reply that it makes no direct reference to the first-person experience of *anyone*, neither biological humans nor ems. On this account, we don't *need* to answer probably unanswerable questions about whether or not ems would be conscious, in order to constructed a workable moral code that applies to them. Deleting the last copy of an em in existence should be prosecuted as murder, *not* because doing so snuffs out some inner light of consciousness (who else is to know?), but rather because it deprives the rest of society of a unique, irreplaceable, store of knowledge and experiences, precisely as murdering a human would.

12.3 Knightian uncertainty and physics

Having spent almost half the essay answering *a priori* objections to the investigation I want to undertake, I'm finally ready to start the investigation itself! In this section, I'll set out and defend two propositions, both of which are central to what follows.

The first proposition is that *probabilistic uncertainty (like that of quantum measurement outcomes) can't possibly, by itself, provide the sort of 'indeterminism' that could be relevant to the free-will debate.* In other words, *if* we see a conflict between free will and the deterministic predictability of human choices then we should see the same conflict between free will and *probabilistic* predictability, assuming the probabilistic predictions are as accurate as those of quantum mechanics. Conversely, if we hold free will to be compatible with 'quantum-like predictability,' then we might as well hold free will to be compatible with *deterministic* predictability also. In my perspective, for a form of uncertainty to be relevant to free will, a necessary condition is that it be what the economists call *Knightian uncertainty*. Knightian uncertainty simply refers to uncertainty that we lack a clear, agreed-upon way to quantify – like our uncertainty about the existence of extraterrestrial life, as opposed to our uncertainty about the outcome of a coin toss.

The second proposition is that, in current physics, there appears to be only one source of Knightian uncertainty that could possibly be both fundamental and relevant to human choices. That source is *uncertainty about the microscopic, quantum-mechanical, details of the universe's initial conditions (or the initial conditions of our local region of the universe).* In classical physics, there's no known fundamental principle that prevents a predictor from learning the relevant initial conditions to whatever precision it likes without disturbing the system to be predicted. But in quantum mechanics there *is* such a principle, namely the uncertainty principle (or from a more 'modern' standpoint, the no-cloning theorem). It's crucial to understand that this source of uncertainty is separate from the randomness of quantum measurement *outcomes*: the latter is much more often invoked in free-will speculations, but in my opinion it shouldn't be. If we know a system's quantum state ρ, then quantum mechanics lets us calculate the probability of any outcome of any measurement that might later be made on the system. But if we *don't* know the state ρ, then ρ itself can be thought of as subject to Knightian uncertainty.

In the next two subsections, I'll expand on and justify the above claims.

12.3.1 Knightian uncertainty

A well-known argument maintains that the very concept of free will is logically incoherent. The argument goes like this:

> Any event is either determined by earlier events (like the return of Halley's comet), or else not determined by earlier events (like the decay of a radioactive atom). If the event is determined, then clearly it isn't 'free.' But if the event is *un*determined, it isn't 'free' either: it's merely arbitrary, capricious, and random. Therefore no event can be 'free.'

I'm far from the first to point out that the above argument has a gap, contained in the vague phrase 'arbitrary, capricious, and random.' An event can be 'arbitrary,' in the sense of being undetermined by previous events, without being random in the narrower technical sense of being generated by some known or knowable probabilistic process. The distinction between arbitrary and random is not just word-splitting: it plays a huge practical role in computer science, economics, and other fields. To illustrate, consider the following two events:

E_1 = Three weeks from today, the least significant digit of the Dow Jones average will be even;

E_2 = Humans will make contact with an extraterrestrial civilization within the next 500 years.

For both events, we are ignorant about whether they will occur, but we are ignorant in completely different ways. For E_1, we can quantify our ignorance by a probability distribution, in such a way that almost any reasonable person would agree with our quantification. For E_2, we can't.

For another example, consider a computer program which has a bug that only appears when a call to the random number generator returns the result 3456. That's not necessarily a big deal – since, with high probability, the program would need to be run thousands of times before the bug reared its head. Indeed, many problems today are solved using *randomized algorithms* (such as Monte Carlo simulation), which *do* have a small but nonzero probability of failure.[22] However, if the program has a bug that only occurs when the user *inputs* 3456, that's a much more serious problem. For how can the programmer know, in advance, whether 3456 is an input (maybe even the *only* input) that the user cares about? So a programmer *must* treat the two types of uncertainty differently: she can't just toss them both into a bin labeled 'arbitrary, capricious, and random.' And indeed, the difference between the two types of uncertainty shows up constantly in theoretical computer science and information theory.[23]

In economics, the 'second type' of uncertainty – the type that can't be objectively quantified using probabilities – is called *Knightian uncertainty*, after Frank Knight, who wrote about it extensively in the 1920s (Knight, 2010). Knightian uncertainty has been invoked to explain phenomena from risk-aversion in behavioral economics to the 2008 financial crisis (and was popularized by Taleb (2010) under the name 'black swans'). An agent in a state of Knightian uncertainty might describe its beliefs using a convex set of probability distributions rather than a single distribution.[24] For example, it might say that a homeowner will default on a mortgage with some probability between 0.1 and 0.3 but, within that interval, be unwilling to quantify its uncertainty further. The notion of probability intervals leads naturally to various generalizations of probability theory, of which the best known is the *Dempster–Shafer theory of belief* (Shafer, 1976).

What does any of this have to do with the free-will debate? As I said in Section 12.1.1, from my personal perspective Knightian uncertainty seems like a *precondition* for free will as I understand the latter. In other words, I *agree* with the free-will-is-incoherent camp when it says that a random event can't be considered 'free' in any meaningful sense. Several writers, such as Fischer et al. (2007), Balaguer (2009), Satinover (2001), and Koch (2012), have speculated that the randomness inherent in quantum-mechanical wavefunction collapse, were it relevant to brain function, could provide all the scope for free will that's needed. But I think those writers are mistaken on this point.

For me, the bottom line is simply that it seems like a sorry and pathetic 'free will' that's ruled by ironclad, externally knowable statistical laws, and that retains only a 'ceremonial' role, analogous to spinning a roulette wheel or shuffling a deck of cards. Should we say that a radioactive nucleus has 'free will,' just because (according to quantum mechanics) we can't predict exactly when it will decay, but can

[22] The probability of failure can be made arbitrarily small by the simple expedient of running the algorithm over and over and taking a majority vote.

[23] For example, it shows up in the clear distinctions made between random and adversarial noise, between probabilistic and nondeterministic Turing machines, and between average-case and worst-case analyses of algorithms.

[24] In Appendix 12C, I briefly explain why I think convex sets provide the 'right' representation of Knightian uncertainty, though this point doesn't much matter for the rest of the essay.

only calculate a precise probability distribution over the decay times? That seems perverse – especially since given *many* nuclei, we can predict almost perfectly what *fraction* will have decayed by such-and-such a time. Or imagine an artificially intelligent robot that used nuclear decays as a source of random numbers. Does anyone seriously maintain that, if we swapped out the actual decaying nuclei for a 'mere' pseudorandom computer simulation of such nuclei (leaving all other components unchanged), the robot would suddenly be robbed of its free will? While I'm leery of 'arguments from obviousness' in this subject, it really *does* seem to me that if we say the robot has free will in the first case then we should also say so in the second.

And thus, I think that the free-will-is-incoherent camp would be right, *if* all uncertainty were probabilistic. But I consider it far from obvious that all uncertainty *is* (usefully regarded as) probabilistic. Some uncertainty strikes me as Knightian, in the sense that rational people might never even reach agreement about how to assign the probabilities. And while Knightian uncertainty might or might not be relevant to predicting human choices, I definitely (for reasons I'll discuss later) don't think that current knowledge of physics or neuroscience lets us exclude the possibility.

At this point, though, we'd better hear from those who reject the entire concept of Knightian uncertainty. Some thinkers – I'll call them *Bayesian fundamentalists* – hold that Bayesian probability theory provides the only sensible way to represent uncertainty. On that view, 'Knightian uncertainty' is just a fancy name for someone's failure to carry a probability analysis far enough.

In support of their position, Bayesian fundamentalists often invoke the so-called *Dutch book arguments* (see for example Savage (1954)), which say that any rational agent satisfying a few axioms must behave *as if* its beliefs were organized using probability theory. Intuitively, even if you claim not to have any opinion whatsoever about (say) the probability of life being discovered on Mars, I can still 'elicit' a probabilistic prediction from you, by observing which bets about the question you will or won't accept.

However, a central assumption on which the Dutch book arguments rely – basically, that a rational agent shouldn't mind taking at least one side of any bet – has struck many commentators as dubious. And if we drop that assumption, then the path is open to Knightian uncertainty (involving, for example, convex *sets* of probability distributions).

Even if we accept the standard derivations of probability theory, the bigger problem is that Bayesian agents can have different 'priors.' If one strips away the philosophy, Bayes' rule is just an elementary mathematical fact about how one should update one's prior beliefs in light of new evidence. So one can't use Bayesianism to justify a belief in the existence of *objective* probabilities underlying all events unless one is also prepared to defend the existence of an 'objective prior.' In economics, the idea that all rational agents can be assumed to start with the same prior is called the *Common Prior Assumption*, or CPA. Assuming the CPA leads to some wildly counterintuitive consequences, most famously *Aumann's agreement theorem* (Aumann, 1976). That theorem says that two rational agents with common priors (but differing information) can never 'agree to disagree': as soon as their opinions on any subject become common knowledge, their opinions must be the same.

The CPA has long been controversial; see Morris (1995) for a summary of arguments for and against. To my mind, though, the real question is: *what could possibly have led anyone to take the CPA seriously in the first place?*

Setting aside methodological arguments in economics[25] (which don't concern us here), I'm aware of two substantive arguments in favor of the CPA. The first argument is that, if two rational agents (call them Alice and Bob) have different priors, then Alice will realize that *if she had been born Bob*, she would have had Bob's prior, and Bob will realize that if he had been born Alice, he would have had Alice's. But if Alice and Bob are indeed rational then why should they assign any weight to personal accidents of their birth, which are clearly irrelevant to the objective state of the universe? (See Cowen and Hanson (2012) for a detailed discussion of this argument.)

The simplest reply is that, even if Alice and Bob accepted this reasoning, they would *still* generally end up with different priors, unless they furthermore shared the same *reference class*; that is, the set of all agents who they imagine they 'could have been' if they weren't themselves. For example, if Alice includes all humans in her reference class, while Bob includes only those humans capable of understanding Bayesian reasoning such as he and Alice are engaging in now, then their beliefs will differ. But requiring agreement on the reference class makes the argument circular – presupposing, as it does, a 'God's-eye perspective' transcending the individual agents, the very thing whose existence or relevance was in question. Section 12.8 will go into more detail about 'indexical' puzzles (that is, puzzles concerning the probability of your own existence, the likelihood of having been born at one time rather than another, etc). But I hope this discussion already makes clear how much debatable metaphysics lurks in the assumption that a single Bayesian prior governs (or *should* govern) every probability judgment of every rational being.

The second argument for the CPA is more ambitious: it seeks to tell us *what* the true prior is, not merely that it exists. According to this argument, any sufficiently intelligent being ought to use what's called the *universal prior* from algorithmic

[25] For example, that even if the CPA is wrong we should assume it because economic theorizing would be too unconstrained without it, or because many interesting theorems need it as a hypothesis.

information theory. This is basically a distribution that assigns a probability proportional to 2^{-n} to every possible universe describable by an n-bit computer program. In Appendix 12B, I'll examine this notion further, explain why some people (Hutter, 2007; Schmidhuber, 1997) have advanced it as a candidate for the 'true' prior, but also explain why, despite its mathematical interest, I don't think it can fulfill that role. (Briefly, a predictor using the universal prior can be thought of as a superintelligent entity that figures out the right probabilities almost as fast as is information-theoretically possible. But that's conceptually very different from an entity that *already knows* the probabilities.)

12.3.2 Quantum mechanics and the no-cloning theorem

While defending the meaningfulness of Knightian uncertainty, the previous subsection left an obvious question unanswered: where, in a law-governed universe, could we possibly *find* Knightian uncertainty?

Granted, in almost any part of science, it's easy to find systems that are 'effectively' subject to Knightian uncertainty, in that we don't yet have models for the systems that capture all the important components and their probabilistic interactions. The Earth's climate, a country's economy, a forest ecosystem, the early universe, a high-temperature superconductor, or even a flatworm or a cell are examples. Even if our probabilistic models of many of these systems are improving over time, none of them comes anywhere close to (say) the quantum-mechanical model of the hydrogen atom – a model that answers essentially *everything* one could think to ask within its domain, modulo an unavoidable (but precisely quantifiable) random element.

However, in all these cases (the Earth's climate, the flatworm, etc.), the question arises: what grounds could we ever have to think that Knightian uncertainty was *inherent* to the system, rather than an artifact of our own ignorance? Of course, one could have asked the same question about probabilistic uncertainty, before the discovery of quantum mechanics and its no-hidden-variable theorems (see Section 12.2.7). But the fact remains that, today, we don't have any physical theory that demands Knightian uncertainty in anything like the way that quantum mechanics demands probabilistic uncertainty. And, as I said in Section 12.3.1, I insist that the 'merely probabilistic' aspect of quantum mechanics can't do the work that many free-will advocates have wanted it to do for nearly a century.

On the other hand, no matter how much we've learned about the dynamics of the physical world, there remains one enormous source of Knightian uncertainty that's been 'hiding in plain sight,' and that receives surprisingly little attention in the free-will debate. This is our ignorance of the relevant *initial conditions*. By this I mean both the initial conditions of the entire universe or multiverse (say, at the

Big Bang) and the 'indexical conditions' that characterize the part of the universe or multiverse in which *we* happen to reside. To make a prediction, of course one needs initial conditions as well as dynamical laws: indeed, beyond idealized toy problems, the initial conditions are typically the 'larger' part of the input. Yet leaving aside recent cosmological speculations (and 'genericity' assumptions, like those of thermodynamics), the specification of initial conditions is normally not even considered the *task* of physics. So, if there are no laws that fix the initial conditions, or even a distribution over possible initial conditions – if there aren't even especially promising *candidates* for such laws – then why isn't this just what free-will advocates have been looking for?

It will be answered immediately that there's an excellent reason why not: namely, whatever else we do or don't know about the universe's initial state (e.g., at the Big Bang), clearly nothing about it was determined by any of *our* choices! (This is the assumption made explicit in step (ii) of van Inwagen's 'consequence argument' from Section 12.2.8.)

The above answer might strike the reader as conclusive. Yet *if* our interest is in actual, physical predictability – rather than in the metaphysical concept of 'determination' – then notice that it's no longer conclusive at all. For we still need to ask: how much can we *learn* about the initial state by making measurements? This, of course, is where quantum mechanics might become relevant.

It's actually easiest for our purposes to forget the famous *uncertainty principle*, and talk instead about the *no-cloning theorem*. The latter is simply the statement that there's no physical procedure, consistent with quantum mechanics, that takes as input a system in an arbitrary quantum state $|\psi\rangle$,[26] and outputs two systems *both* in the state $|\psi\rangle$.[27] Intuitively, it's not hard to see why: a measurement of (say) a qubit $|\psi\rangle = \alpha|0\rangle + \beta|1\rangle$ reveals only a single, probabilistic bit of information about the continuous parameters α and β; the rest of the information vanishes forever. The proof of the no-cloning theorem (in its simplest version) is as easy as observing that the 'cloning map,'

$$\begin{aligned}(\alpha|0\rangle + \beta|1\rangle)|0\rangle &\longrightarrow (\alpha|0\rangle + \beta|1\rangle)(\alpha|0\rangle + \beta|1\rangle)\\ &= \alpha^2|0\rangle|0\rangle + \alpha\beta|0\rangle|1\rangle + \alpha\beta|1\rangle|0\rangle + \beta^2|1\rangle|1\rangle,\end{aligned}$$

acts nonlinearly on the amplitudes; but, in quantum mechanics, unitary evolution must be linear.

[26] Here we assume for simplicity that we're talking about pure states, not mixed states.

[27] The word 'arbitrary' is needed because, if we knew how $|\psi\rangle$ was prepared, then of course we could simply run the preparation procedure a second time. The no-cloning theorem implies that, if we *don't* already know a preparation procedure for $|\psi\rangle$, then we can't learn one just by measuring $|\psi\rangle$. (And conversely, the inability to learn a preparation procedure implies the no-cloning theorem. If we could copy $|\psi\rangle$ perfectly, we could keep repeating to make as many copies as we wanted, and then use quantum state tomography on the copies to learn the amplitudes to arbitrary accuracy.)

Yet despite its mathematical triviality, the no-cloning theorem has the deep consequence that quantum states have a certain 'privacy': unlike classical states, they can't be copied promiscuously around the universe. One way to gain intuition for the no-cloning theorem is to consider some striking cryptographic protocols that rely on it. In *quantum key distribution* (Bennett and Brassard, 1984) – something that's already (to a small extent) been commercialized and deployed – a sender, Alice, transmits a secret key to a receiver, Bob, by encoding the key in a collection of qubits. The crucial point is that if an eavesdropper, Eve, tries to learn the key by measuring the qubits then the very fact that she measured the qubits will be detectable by Alice and Bob – so Alice and Bob can simply keep trying until the channel is safe. *Quantum money*, proposed decades ago by Wiesner (1983) and developed further in recent years (Aaronson, 2009; Aaronson and Christiano, 2012; Farhi et al., 2012), would exploit the no-cloning theorem even more directly, to create cash that can't be counterfeited according to the laws of physics, but that can nevertheless be verified as genuine.[28] A closely related proposal, *quantum software copy-protection* (see Aaronson, 2009; Aaronson and Christiano, 2012), would exploit the no-cloning theorem in a still more dramatic way: to create quantum states $|\psi_f\rangle$ that can be used to evaluate some function f but that can't feasibly be used to create more states with which f can be evaluated. Research on quantum copy-protection has shown that, at least in a few special cases (and maybe more broadly), it's possible to create a physical object that

(a) interacts with the outside world in an interesting and nontrivial way, yet
(b) effectively hides from the outside world the information needed to predict how the object will behave in future interactions.

When put that way, the *possible* relevance of the no-cloning theorem to free-will discussions seems obvious! And indeed, a few bloggers and others[29] have previously speculated about a connection. Interestingly, their motivation for doing so has usually been to *defend compatibilism* (see Section 12.2.3). In other words, they've invoked the no-cloning theorem to explain why, despite the mechanistic nature of physical law, human decisions will nevertheless remain unpredictable *in practice*. In a discussion of this issue, one commenter[30] opined that, while the no-cloning theorem *does* put limits on physical predictability, human brains will also remain unpredictable for countless more prosaic reasons that have nothing to do with quantum mechanics. Thus, he said, invoking the no-cloning theorem in free-will discussions is "like hiring the world's most high-powered lawyer to get you out of a parking ticket."

[28] Unfortunately, unlike quantum key distribution, quantum money is not yet practical, because of the difficulty of protecting quantum states against decoherence for long periods of time.

[29] See, for example, www.daylightatheism.org/2006/04/on-free-will-iii.html.

[30] Sadly, I no longer have the reference.

Personally, though, I think the world's most high-powered lawyer might ultimately be needed here! For the example of the Singulatarians (see Section 12.2.10) shows why, in these discussions, it doesn't suffice to offer 'merely practical' reasons why copying a brain state is hard. Every practical problem can easily be countered by a speculative technological answer – one that assumes a future filled with brain-scanning nanorobots and the like. If we want a proposed obstacle to copying to survive unlimited technological imagination, then the obstacle had better be grounded in the laws of physics.

So as I see it, the real question is: once we disallow quantum mechanics, does there remain any *classical* source of 'fundamental unclonability' in physical law? Some might suggest, for example, the impossibility of keeping a system perfectly isolated from its environment during the cloning process or the impossibility of measuring continuous particle positions to infinite precision. But either of these would require very nontrivial arguments (and, if one wanted to invoke continuity, one would also have to address the apparent breakdown of continuity at the Planck scale). There are also formal analogues of the no-cloning theorem in various classical settings, but none of them seem able to do the work required here.[31] As far as current physics can say, *if* copying a bounded physical system is fundamentally impossible, then the reason for the impossibility seems ultimately traceable to quantum mechanics.

Let me end this subsection by discussing *quantum teleportation*, since nothing more suggestively illustrates the 'philosophical work' that the no-cloning theorem can potentially do. Recall, from Section 12.2.5, the 'paradox' raised by teleportation machines. Namely, after a perfect copy of you has been reconstituted on Mars from the information in radio signals, what should be done with the 'original' copy of you back on Earth? Should it be euthanized?

Like other such paradoxes, this one need not trouble us if (because of the no-cloning theorem, or whatever other reason) we drop the assumption that such copying is possible – at least with great enough fidelity for the copy on Mars to be a 'second instantiation of you.' However, what makes the situation more interesting is that there *is* a famous protocol, discovered by Bennett et al. (1993), for 'teleporting' an arbitrary quantum state $|\psi\rangle$ by sending classical information only. (This protocol also requires quantum entanglement, in the form of *Bell pairs*, $\frac{|00\rangle+|11\rangle}{\sqrt{2}}$,

[31] For example, no-cloning holds for classical probability distributions: there's no procedure that takes an input bit b that equals 1 with unknown probability p, and produces two output bits that are both 1 with probability p independently. But this observation lacks the import of the quantum no-cloning theorem, because regardless of what one wanted to do with the bit b, one might as well have *measured* it immediately, thereby 'collapsing' it to a deterministic bit – which *can* of course be copied.
Also, seeking to clarify foundational issues in quantum mechanics, Spekkens (2007) constructed an 'epistemic toy theory' that's purely classical, but where an analogue of the no-cloning theorem holds. However, the toy theory involves 'magic boxes' that we have no reason to think can be physically realized.

shared between the sender and receiver in advance, and one Bell pair gets 'used up' for every qubit teleported.)

Now, a crucial feature of the teleportation protocol is that, in order to determine which classical bits to send, the sender needs to perform a measurement on her quantum state $|\psi\rangle$ (together with her half of the Bell pair) that *destroys* $|\psi\rangle$. In other words, in quantum teleportation the destruction of the original copy is not an extra decision that one needs to make; rather, it happens as an inevitable byproduct of the protocol itself! Indeed this must be so, since otherwise quantum teleportation could be used to violate the no-cloning theorem.

12.3.3 The freebit picture

At this point one might interject: theoretical arguments about the no-cloning theorem are well and good but, even if accepted, they still don't provide any concrete picture of how Knightian uncertainty could be relevant to human decision making.

So let me sketch a possible picture (the only one I can think of, consistent with current physics), which I call the 'freebit picture.' At the Big Bang, the universe had some particular quantum state $|\Psi\rangle$. If known, $|\Psi\rangle$ would of course determine the universe's future history, modulo the probabilities arising from quantum measurement. However, *because* $|\Psi\rangle$ is the state of the whole universe (including us), we might refuse to take a 'God's-eye view' of $|\Psi\rangle$, and insist on considering $|\Psi\rangle$ different from an ordinary state that we prepare for ourselves in the lab. In particular, we might regard at least some (not all) of the qubits that constitute $|\Psi\rangle$ as what I'll call *freebits*. A freebit is simply a qubit for which the most complete physical description possible involves Knightian uncertainty. While the details aren't so important, I give a brief mathematical account of freebits in Appendix 12C. For now, suffice it to say that a *freestate* is a convex set of quantum mixed states, and a freebit is a 2-level quantum system in a freestate.

Thus, by the *freebit picture*, I mean the picture of the world according to which

(i) owing to Knightian uncertainty about the universe's initial quantum state $|\Psi\rangle$, at least some of the qubits found in nature are regarded as freebits, and
(ii) the presence of these freebits makes predicting certain future events – possibly including some human decisions – physically impossible, even probabilistically and even with arbitrarily advanced future technology.

I will say more about the 'biological' aspects of the freebit picture in Section 12.3.4; that is, the actual chains of causation that could in principle connect freebits to (say) a neuron firing or not firing. In the rest of this section, I'll discuss some physical and conceptual questions about freebits themselves.

Firstly, why is it important that freebits be qubits, rather than *classical* bits subject to Knightian uncertainty? The answer is that only for qubits can we appeal to the no-cloning theorem. Even if the value of a classical bit b can't be determined by measurements on the entire rest of the universe, a superintelligent predictor could always learn b by non-invasively measuring *b itself*. But the same is not true for a qubit.

Secondly, isn't Knightian uncertainty in the eye of the beholder? That is, why couldn't one observer regard a given qubit as a 'freebit' while a different observer, with more information, described the same qubit by an ordinary quantum mixed state ρ? The answer is that our criterion for what counts a freebit is extremely stringent. Given a 2-level quantum system S, *if a superintelligent demon could reliably learn the reduced density matrix ρ of S, via arbitrary measurements on anything in the universe (including S itself), then S is not a freebit.* Thus, to qualify as a freebit, S must be a 'freely moving part' of the universe's quantum state $|\Psi\rangle$: it must not be possible (even in principle) to trace S's causal history back to any physical process that generated S according to a known probabilistic ensemble. Instead, our Knightian uncertainty about S must (so to speak) go 'all the way back,' and be traceable to uncertainty about the initial state of the universe.

To illustrate the point: suppose we detect a beam of photons with varying polarizations. For the most part, the polarizations look uniformly random (i.e., like qubits in the maximally mixed state). But there is a slight bias toward the vertical axis, and the bias is slowly changing over time, in a way not fully accounted for by our model of the photons. So far, we can't rule out the possibility that freebits might be involved. However, suppose we later learn that the photons are coming from a laser in another room, and that the polarization bias is due to drift in the laser that can be characterized and mostly corrected. Then the scope for freebits is correspondingly reduced.

Someone might interject: "but *why* was there drift in the laser? Couldn't freebits have been responsible for the drift itself?" The difficulty is that, *even if so*, we still couldn't use those freebits to argue for Knightian uncertainty in the laser's *output*. For between the output photons and whatever freebits might have caused the laser to be configured as it was, there stands a classical, macroscopic, intermediary: the laser itself. If a demon had wanted to predict the polarization drift in the output photons, the demon could simply have traced the photons back to the laser then *non-invasively measured the laser's classical degrees of freedom* – cutting off the causal chain there and ignoring any further antecedents. In general, given some quantum measurement outcome Q that we're trying to predict, if there exists a classical observable C that could have been non-invasively measured long before Q and that, if measured, would have let the demon probabilistically predict Q to arbitrary accuracy (in the same sense that radioactive decay is probabilistically predictable) then I'll call C a *past macroscopic determinant (PMD)* for Q.

In the freebit picture, we're exclusively interested in the quantum states – if there are any! – that *can't* be grounded in PMDs but can only be traced all the way back to the early universe, with no macroscopic intermediary along the way that 'screen off' the early universe's causal effects. The reason is simple: such states, if they exist, are the only ones that our superintelligent demon, able to measure all the macroscopic observables in the universe, would *still* have Knightian uncertainty about. In other words, such states are the only possible freebits.

Of course this immediately raises a question:

($*$) **In the actual universe, *are* there any quantum states that can't be grounded in PMDs?**

A central contention of this essay is that pure thought doesn't suffice to answer question ($*$): here we've reached the limits of where conceptual analysis can take us. There are possible universes consistent with the rules of quantum mechanics where the requisite states exist, and other such universes where they don't exist, and deciding which kind of universe *we* inhabit seems to require scientific knowledge that we don't have.

Some people, while agreeing that logic and quantum mechanics don't suffice to settle question ($*$), would nevertheless say we can settle it using simple facts about astronomy. At least near the surface of the earth, they'd ask, what quantum states could there possibly be that *didn't* originate in PMDs? Most of the photons impinging on the Earth come from the Sun, whose physics is exceedingly well understood. Of the subatomic particles that could conceivably 'tip the scales' of (say) a human neural event, causing it to turn out one way rather than another, others might have causal pathways that lead back to other astronomical bodies (such as supernovae), the Earth's core, etc. But it seems hard to imagine how any of the latter possibilities *wouldn't* serve as PMDs: that is, how they wouldn't effectively 'screen off' any Knightian uncertainty from the early universe.

To show that the above argument is inconclusive, one need only mention the cosmic microwave background (CMB) radiation: CMB photons pervade the universe, famously accounting for a few percent of the static in old-fashioned TV sets. Furthermore, many CMB photons are believed to reach the Earth having maintained quantum coherence ever since being emitted at the so-called *time of last scattering*, roughly 380,000 years after the Big Bang. Finally, unlike (say) neutrinos or dark matter, CMB photons readily interact with matter. In short, we're continually bathed with at least one type of radiation that seems to satisfy most of the freebit picture's requirements!

However, no sooner is the CMB suggested for this role than we encounter two serious objections. The first is that the time of last scattering, when the CMB photons were emitted, is separated from the Big Bang itself by 380,000 years. So if we wanted to postulate CMB photons as freebit carriers then we'd also need a story about why the hot early universe should *not* be considered a PMD and about how a qubit might have 'made it intact' from the Big Bang – or at any rate, from as far back as current physical theory can take us – to the time of last scattering. The second objection asks us to imagine someone *shielded* from the CMB: for example, someone in a deep underground mine. Would such a person be 'bereft of Knightian freedom,' at least while he or she remained in the mine?

Because of these objections, I find that, while the CMB might be one *piece* of a causal chain conveying a qubit to us from the early universe (without getting screened off by a PMD), it can't possibly provide the full story. It seems to me that convincingly answering question ($*$) would require something like a census of the possible causal chains from the early universe to ourselves that are allowed by particle physics and cosmology. I don't know whether the requisite census is beyond present-day science, but it's certainly beyond *me*! Note that, if it could be shown that *all* qubits today can be traced back to PMDs, and that the answer to ($*$) is negative, then the freebit picture would be falsified.

12.3.4 Amplification and the brain

We haven't yet touched on an obvious question: *once freebits have made their way into (say) a brain, by whatever means, how could they then tip the scales of a decision?* But it's not hard to suggest plausible answers to this question, without having to assume anything particularly exotic about either physics or neurobiology. Instead, one can appeal to the well-known phenomenon of chaos (i.e., sensitive dependence on initial conditions) in dynamical systems.

The idea that chaos in brain activity might somehow underlie free will is an old one. However, that idea has traditionally been rejected, on the sensible ground that a classical chaotic system is still perfectly deterministic! Our inability to measure the initial conditions to unlimited precision, and our consequent inability to predict very far into the future, seem at best like practical limitations.

Thus, a revised idea has held that the role of chaos for free will might be to take quantum fluctuations – which, as we know, are *not* deterministic (see Section 12.2.7) – and amplify those fluctuations to macroscopic scale. However, this revised idea has also been rejected, on the (again sensible) ground that, even if true, it would only make the brain a *probabilistic* dynamical system, which still seems 'mechanistic' in any meaningful sense.

The freebit picture makes a further revision: namely, it postulates that chaotic dynamics in the brain can have the effect of amplifying *freebits* (i.e., microscopic

Knightian uncertainty) to macroscopic scale. If nothing else, this overcomes the elementary objections above. Yes, the resulting picture might still be wrong, but not for some simple *a priori* reason – and to me, that represents progress!

It's long been recognized that neural processes relevant to cognition can be sensitive to microscopic fluctuations. An important example is the opening and closing of a neuron's sodium-ion channels, which normally determine whether and for how long a neuron fires. This process is modeled probabilistically (in particular, as a Markov chain) by the standard *Hodgkin–Huxley equations* (Hodgkin and Huxley, 1952) of neuroscience. Of course, one then has to ask: is the apparent randomness in the behavior of the sodium-ion channels ultimately traceable back to quantum mechanics (and if so, by what causal routes)? Or does the 'randomness' merely reflect our ignorance of relevant classical details?

Balaguer (2009) put the above question to various neuroscientists, and was told either that the answer is unknown or that it's outside neuroscience's scope. For example, he quotes Sebastian Seung as saying: "The question of whether [synaptic transmission and spike firing] are 'truly random' processes in the brain isn't really a neuroscience question. It's more of a physics question, having to do with statistical mechanics and quantum mechanics." He also quotes Christof Koch as saying: "At this point, we do not know to what extent the random, i.e. stochastic, neuronal processes we observe are due to quantum fluctuations (à la Heisenberg) that are magnified by chemical and biological mechanisms or to what extent they just depend on classical physics (i.e. thermodynamics) and statistical fluctuations in the underlying molecules."

In his paper 'A scientific perspective on human choice' (Sompolinsky, 2005), the neuroscientist Haim Sompolinsky offers a detailed review of what's known about the brain's sensitivity to microscopic fluctuations. Though skeptical about any role for such fluctuations in cognition, he writes:

> In sum, given the present state of our understanding of brain processes and given the standard interpretation of quantum mechanics, we cannot rule out the possibility that the brain is truly an indeterministic system; that because of quantum indeterminism, there are certain circumstances where choices produced by brain processes are not fully determined by the antecedent brain process and the forces acting on it. If this is so, then the first prerequisite [i.e., indeterminism] of 'metaphysical free will' ... may be consistent with the scientific understanding of the brain.[32]

To make the issue concrete: suppose that, with godlike knowledge of the quantum state $|\Psi\rangle$ of the entire universe, you wanted to change a particular decision

[32] However, Sompolinsky then goes on to reject 'metaphysical free will' as incompatible with a scientific worldview: if, he says, there were laws relevant to brain function beyond the known laws of physics and chemistry then those laws would themselves be incorporated into science, leaving us back where we started. I would agree if it weren't for the logical possibility of 'Knightian laws,' which explains why we couldn't even predict the probability distributions for certain events.

made by a particular human being – and do so *without* changing anything else, except insofar as the other changes flowed from the changed decision itself. Then the question that interests us is: *what sorts of changes to* $|\Psi\rangle$ *would or wouldn't suffice to achieve your aim?* For example, would it suffice to change the energy of a single photon impinging on the subject's brain? Such a photon might get absorbed by an electron, thereby slightly altering the trajectories of a few molecules near one sodium-ion channel in one neuron, thereby initiating a chain of events that ultimately causes the ion channel to open, which causes the neuron to fire, which causes other neurons to fire, etc. If that sort of causal chain is plausible – which, of course, is an empirical question – then, at least as far as neuroscience is concerned, the freebit picture would seem to have the raw material that it needs.

Some people might shrug at this, and regard our story of 'the photon that broke the camel's back' as so self-evidently plausible that the question isn't even scientifically interesting! So it's important to understand that there are two details of the story that matter enormously, if we want the freebit picture to be viable. The first detail concerns the *amount of time* needed for a microscopic change in a quantum state to produce a macroscopic change in brain activity. Are we talking seconds? hours? days?[33] The second detail concerns *localization*. We'd like our change to the state of a single photon to be 'surgical' in its effects: it should change a person's neural firing patterns enough to alter that person's actions, but any other macroscopic effects of the changed photon state should be mediated through the change to the brain state. The reason for this requirement is simply that, if it fails, then a superintelligent predictor might 'non-invasively' learn about the photon by measuring its *other* macroscopic effects, and ignoring its effects on the brain state.

To summarize our questions:

> What are the causal pathways by which a microscopic fluctuation can get chaotically amplified, in human or other animal brains? What are the characteristic timescales for those pathways? What 'side effects' do the pathways produce, separate from their effects on cognition?[34]

In Section 12.9, I'll use these questions – and the freebit picture's dependence on their answers – to argue that the picture makes falsifiable predictions. For now, I'll simply say that these questions strike me as wide open to investigation, and not only in principle. That is, I can easily imagine that in (say) 50 years, neuroscience, molecular biology, and physics will be able to say more about these questions than they can today. And, crucially, the questions strike me as scientifically interesting regardless of one's philosophical predilections. Indeed, one could reject the freebit

[33] Or one could ask: if we model the brain as a chaotic dynamical system, what's the Lyapunov exponent?

[34] Another obvious question is whether brains *differ* in any interesting way from other complicated dynamical systems such as lava lamps or the Earth's atmosphere in terms of their response to microscopic fluctuations. This question will be taken up in Sections 12.5.2 and 12.5.3.

picture completely, and still see progress on these questions as a 'rising tide that lifts all boats' in the scientific understanding of free will. The freebit picture seems unusual only in *forcing* us to address these questions.

12.3.5 Against homunculi

A final clarification is in order about the freebit picture. One might worry that freebits are playing the role of a homunculus: the 'secret core' of a person; a smaller person inside who directs the brain like a manager, puppeteer, or pilot; Ryle's ghost in the machine. But in philosophy and cognitive science, the notion of a homunculus has been rightly laughed off the stage. Like the theory that a clock can only work if there's a smaller clock inside, the homunculus theory blithely offers a black box where a *mechanism* is needed, and it leads to an obvious infinite regress: who's in charge of the homunculus?

Furthermore, if this were really the claim at issue, one would want to know: why do humans (and other animals) *have* such complicated brains in the first place? Why shouldn't our skulls be empty, save for tiny 'freebit antennae' for picking up signals from the Big Bang?

But whatever other problems the freebit picture has, I think it's innocent of the charge of homunculism. On the freebit picture – as, I'd argue, on *any* sane understanding of the world – the physical activity of the brain retains its starring role in cognition. To whatever extent your 'true self' has *any* definite location in spacetime, that location is in your brain. To whatever extent your behavior is predictable, that predictability ultimately derives from the predictability of your brain. And to whatever extent your choices have an author, *you* are their author, not some homunculus secretly calling the shots.

But if this is so, one might ask, then *what role could possibly be left for freebits* besides the banal role of an unwanted noise source, randomly jostling neural firing patterns this way and that? Perhaps the freebit picture's central counterintuitive claim is that freebits *can* play a more robust role than that of glorified random-number generator, without usurping the brain's causal supremacy. Or more generally: *an organized, complex system can include a source of 'pure Knightian unpredictability,' which foils probabilistic forecasts made by outside observers yet need not play any role in explaining the system's organization or complexity.* While I confess that this claim is strange, I fail to see any *logical* difficulty with it, nor do I see any way to escape it if the freebit picture is accepted.

To summarize, on the freebit picture, freebits are simply part of the explanation for how a brain can reach decisions that are not probabilistically predictable by outside observers, and that are therefore 'free' in the sense that interests us. As such, on this picture freebits are just one ingredient among many in the physical substrate of the mind. I'd no more consider them the 'true essence' of a person than the myelin coating that speeds transmission between neurons.

12.4 Freedom from the inside out

The discussion in Section 12.3.5 might remind us about the importance of stepping back. Setting aside any other concerns, *isn't it anti-scientific insanity to imagine that our choices today could correlate nontrivially with the universe's microstate at the Big Bang? Why shouldn't this idea just be thrown out immediately?*

In this section, I'll discuss an unusual perspective on time, causation, and boundary conditions: one that, *if* adopted, makes the idea of such a correlation seem not particularly crazy at all. The interesting point is that, despite its strangeness, this perspective seems perfectly compatible with everything we know about the laws of physics. The perspective is not new; it was previously suggested by Carl Hoefer (2002) and independently by Cristi Stoica (2008; 2012). (As Hoefer discusses, centuries before either of them, Kant appears to have made related observations in his *Critique of Practical Reason*, while trying to reconcile moral responsibility with free will! However, Kant's way of putting things strikes me as obscure and open to other interpretations.) Adopting Hoefer's terminology, I'll call the perspective 'Freedom from the Inside Out,' or FIO for short.

The FIO perspective starts from the familiar fact that the known equations of physics are *time-reversible*: any valid solution to the equations is still a valid solution if we replace t by $-t$.[35] Hence, there seems to be no particular reason to imagine time as 'flowing' from past to future. Instead, we might as well adopt what philosophers call the 'block universe' picture, where the whole 4-dimensional spacetime manifold is laid out at once as a frozen block.[36] Time is simply one more coordinate parameterizing that manifold, along with the spatial coordinates x, y, z. The equations *do* treat the t coordinate differently from the other three, but not in any way that seems to justify the intuitive notion of t 'flowing' in a particular direction, any more than the x, y, z coordinates 'flow' in a particular direction. Of course, with the discovery of special relativity, we learned that the choice of t coordinate is no more unique than the choice of x, y, z coordinates; indeed, an event in a faraway galaxy that you judge as years in the future, might well be judged as years in the past by someone walking past you on the street. To many philosophers, this seems to make the argument for a block-universe picture even stronger than it had been in Newtonian physics.[37]

[35] One also needs to interchange left with right and particles with antiparticles, but that doesn't affect the substance of the argument.

[36] Of course, if the physical laws were probabilistic, then we'd have a probability distribution over possible blocks. This doesn't change anything in the ensuing discussion.

[37] As Einstein himself famously wrote to Michele Besso's family in 1955: "Now Besso has departed from this

The block-universe picture is sometimes described as 'leaving no room for free will' – but that misses the much more important point, that the picture also leaves no room for *causation*! If we adopt this mentality consistently, then *it's every bit as meaningless to say that "Jack and Jill went up the hill because of the prior microstate of the universe," as it is to say that "Jack and Jill went up the hill because they wanted to."* Indeed, we might as well say that Jack and Jill went up the hill because of the *future* microstate of the universe! Or rather, the concept of 'because' plays no role in this picture: there are the differential equations of physics, and the actual history of spacetime is one particular solution to those equations, and that's that.

Now, the idea of 'freedom from the inside out' is simply to embrace the block-universe picture of physics, then turn it on its head. We say: if this frozen spacetime block has no 'intrinsic' causal arrows anyway, then *why not annotate it with causal arrows ourselves*, in whichever ways seem most compelling and consistent to us? And, at least *a priori*, who's to say that some of the causal arrows we draw can't point 'backwards' in time – for example, from human mental states and decisions to 'earlier' microstates consistent with them? Thus, we might decide that yesterday your brain absorbed photons in certain quantum states *because* today you were going to eat tuna casserole, and running the differential equations backward from the tuna casserole produced the photons. In strict block-universe terms, this seems *absolutely* no worse than saying that you ate tuna casserole today because of the state of the universe yesterday.

I'll let Hoefer explain further:

> The idea of freedom from the inside out is this: we are perfectly justified in viewing our own actions *not* as determined by the past, *nor* as determined by the future, but rather as simply determined (to the extent that this word sensibly applies) *by ourselves, by our own wills*. In other words, they need not be viewed as *caused* or *explained* by the physical states of other, vast regions of the block universe. Instead, we can view our own actions, *qua* physical events, as primary explainers, determining – in a very partial way – physical events outside ourselves to the past and future of our actions, in the block. We adopt the perspective that the determination or explanation that matters is from the *inside* (of the block universe, where we live) *outward*, rather than from the *outside* (e.g. the state of things on a time slice 1 billion years ago) *in*. And we adopt the perspective of downward causation, thinking of our choices and intentions as primary explainers of our physical actions, rather than letting microstates of the world usurp this role. We are free to adopt these perspectives because, quite simply, physics – including [a] postulated, perfected deterministic physics – is perfectly compatible with them. (Hoefer, 2002, pp. 207–208, emphases in original)

Some readers will immediately object as follows:

strange world a little ahead of me. That means nothing. People like us, who believe in physics, know that the distinction between past, present and future is only a stubbornly persistent illusion."

> Yes, 'causality' in the block universe might indeed be a subtle, emergent concept. But doesn't the FIO picture take that observation to a ludicrous extreme, by turning causality into a free-for-all? For example, why couldn't an FIO believer declare that the dinosaurs went extinct 65 million years ago *because* if they hadn't, today I might not have decided to go bowling?

The reply to this objection is interesting. To explain it, we first need to ask: if the notion of 'causality' appears nowhere in fundamental physics, then why does it *look* like past events constantly cause future ones, and never the reverse? Since the late nineteenth century, physicists have had a profound answer to that question, or at least a reduction of the question to a different question.

The answer goes something like this: causality is an 'emergent phenomenon,' associated with the Second Law of Thermodynamics. Though no one really knows why, the universe was in a vastly 'improbable,' low-entropy state at the Big Bang – which means, for well-understood statistical reasons, that the further in time we move *away* from the Big Bang, the greater the entropy we'll find. Now, the creation of reliable memories and records is essentially always associated with an increase in entropy (some would argue by definition). And in order for us, as observers, to speak sensibly about 'A causing B,' we must be able to create records of A happening *and then* B happening. But by the above, this will essentially never be possible unless A is closer in time than B to the Big Bang.

We're now ready to see how the FIO picture evades the unacceptable conclusion of a causal free-for-all. It does so by *agreeing* with the usual account of causality based on entropy increase, in all situations where the usual account is relevant. While Hoefer (2002) and Stoica (2008; 2012) are not explicit about this point, I would say that on the FIO picture, it can only make sense to draw causal arrows 'backwards in time,' in those rare situations where entropy is *not* increasing with time.

What are those situations? To a first approximation, they're the situations where physical systems are allowed to evolve reversibly, free from contact with their external environments. In practice, such perfectly isolated systems will almost always be microscopic, and the reversible equation relevant to them will just be the Schrödinger equation. One example of such a system would be a photon of cosmic background radiation, which was emitted in the early universe and has been propagating undisturbed ever since. But these are precisely the sorts of systems that I'd consider candidates for 'freebits'! As far as I can see, if we want the FIO picture not to lead to absurdity, then we can entertain the possibility of backwards-in-time causal arrows *only* for these systems. For this is where the 'normal' way of drawing causal arrows breaks down, and we have nothing else to guide us.

12.4.1 The harmonization problem

There's another potential problem with the FIO perspective – Hoefer (2002) calls it the 'harmonization problem' – that is so glaring that it needs to be dealt with immediately. The problem is this: once we let certain causal arrows point backward in time, say from events today to the microstate at the Big Bang, we've set up what a computer scientist would recognize as a giant *constraint satisfaction problem*. Finding a 'solution' to the differential equations of physics is no longer just a matter of starting from the initial conditions and evolving them forward in time. Instead, we can now also have constraints involving *later* times – for example, that a person makes a particular choice – from which we're supposed to propagate backward. But if this is so, then *why should a globally consistent solution even exist, in general?* In particular, what happens if there are *cycles* in the network of causal arrows? In such a case, we could easily face the classic grandfather paradox of time travel to the past: for example, if an event A causes another event B in its future but B also causes $\text{not}(A)$ in its past. Furthermore, even if globally consistent solutions *do* exist, one is tempted to ask what 'algorithm' Nature uses to build a solution up. Should we imagine that Nature starts at the initial conditions and propagates them forward, but then 'backtracks' if a contradiction is found, adjusting the initial conditions slightly in an attempt to avoid the contradiction? What if it gets into an infinite loop in this way?

With hindsight, we can see the discussion from Section 12.2.5 about brain-uploading, teleportation, Newcomb's paradox, and so on as highlighting the same concerns about the consistency of spacetime – indeed, as using speculative future technologies to make those concerns more vivid. Suppose, for example, that your brain has been scanned and that a complete physical description of it (sufficient to predict all your future actions) has been stored on a computer on Pluto. And suppose you then make a decision. Then from an FIO perspective, the question is: can we take your decision to explain, not only the state of your brain at the moment of your choice, and not only various microstates in your past lightcone[38] (insofar as they need to be compatible with your choice), but *also what's stored in the computer memory billions of kilometers away*? Should we say that you and the computer now make your decisions in synchrony, whatever violence that might seem to do to the locality of spacetime? Or should we say that the very act of copying your brain state removed your freedom where previously you had it, or proved that you were never free in the first place?

Fortunately, it turns out that in the freebit picture *none of these problems arises.* The freebit picture might be rejected for other reasons, but it can't be charged with

[38] Given a spacetime point x, the past lightcone of x is the set of points from which x can receive a signal, and the future lightcone of x is the set of points to which it can send one.

logical consistency or with leading to closed timelike curves. The basic observation is simply that we have to distinguish between what I'll call *macrofacts* and *microfacts*. A macrofact is a fact about a 'decohered, classical, macroscopic' property of a physical system S at a particular time. More precisely, a macrofact is a fact F that could, in principle, be learned by an external measuring device without disturbing the system S in any significant way. Note that, for F to count as a macrofact, it's not necessary that anyone has ever known F or will ever know F: only that F *could have been* known, if a suitable measuring device had been around at the right place and the right time. So for example, there is a macrofact about whether a Stegosaurus kicked a particular rock 150 million years ago, even if no human will ever see the rock, and even if the relevant information can no longer be reconstructed, even in principle, from the quantum state of the entire solar system. There are also macrofacts constantly being created in the interiors of stars and in interstellar space.

By contrast, a microfact is a fact about an undecohered quantum state ρ: a fact that couldn't have been learned, even in principle, without the potential for altering ρ (if the measurement were performed in the 'wrong' basis). For example, the polarization of some particular photon of the cosmic microwave background radiation is a microfact. A microfact might or might not concern a freebit, but the quantum state of a freebit is always a microfact.

Within the freebit picture, the solution to the 'harmonization problem' is now simply to impose the following two rules.

(1) Backwards-in-time causal arrows can point only to microfacts, never to macrofacts.
(2) No microfact can do 'double duty': if it is caused by a fact to its future, then it is not itself the cause of anything, nor is it caused by anything else.

Together, these rules readily imply that no cycles can arise, and more generally, that the 'causality graph' never produces a contradiction. For whenever they hold, the causality graph with be a *directed acyclic graph* (a dag), with all arrows pointing forward in time, except for some 'dangling' arrows pointing backward in time that never lead anywhere else.

Rule (1) is basically imposed by fiat: it just says that, for all the events we actually observe, we must seek their causes only to their past, never to their future. This rule might someday be subject to change (for example, if closed timelike curves were discovered), but for now, it seems like a pretty indispensable part of a scientific worldview. By contrast, rule (2) can be justified by appealing to the no-cloning theorem. If a microfact f is directly caused by a macrofact F to its future then at some point a *measurement* must have occurred (or more generally, some decoherence event that we can *call* a 'measurement'), in order to amplify f to macroscopic scale and correlate it with F. In the freebit picture, we think of the correlation with F as completely fixing f, which explains why f can't also be caused by anything other than F. But why can't f itself cause some macrofact F' (which, by rule (1), would need to be to f's future)? Here there are two cases: $F' = F$ or $F' \neq F$. On the one hand, if $F' = F$ then we've simply created a 'harmless' 2-element cycle, where f and F cause each other. It's purely by convention that we disallow such cycles and declare that F causes f rather than the reverse. On the other hand, if $F' \neq F$ then we have two independent measurements of f to f's future, violating the no-cloning theorem.[39] Note that this argument wouldn't have worked if f had been a macrofact, since macroscopic information *can* be measured many times independently.

Subtle questions arise when we ask about the possibility of microfacts causing other microfacts. Rules (1) and (2) allow that sort of causation, as long as it takes place forward in time – or, if backward in time, as long as it consists of a single 'dangling' arrow only. If we wanted, without causing any harm we could allow long chains (and even dags) of microfacts causing other microfacts backward in time, possibly originating at some macrofact to their future. We would need to be careful, though, that none of those microfacts ever caused any facts to their future, since that would create the possibility of cycles. A simpler option is just to declare the entire concept of causality irrelevant to the microworld. On that view, whenever a microfact f 'causes' another microfact f', unitarity makes it just as legitimate to say that f' causes f, or that neither causes the other. Because of the reversibility of microscopic laws, the temporal order of f and f' is irrelevant: if $U|\psi\rangle = |\varphi\rangle$ then $U^\dagger|\varphi\rangle = |\psi\rangle$. This view would regard causality as inextricably bound up with the thermodynamic arrow of time, and therefore with *ir*reversible processes and therefore with macrofacts.

12.4.2 Microfacts and macrofacts

An obvious objection to the distinction between microfacts and macrofacts is that it's poorly-defined. The position of a rock might be 'obviously' a macrofact, and the polarization of a photon 'obviously' a microfact, but there is a continuous transition between the two. Exactly how decohered and classical does a fact have to be before it counts as a 'macrofact' for our purposes? This, of course, is reminiscent of the traditional objection to Bohr's Copenhagen interpretation of quantum mechanics and, in particular, to its unexplained yet crucial distinction between the

[39] For the same reason, we also have the rule that a microfact cannot cause two macrofacts to its future via disjoint causal pathways. The only reason this rule wasn't mentioned earlier is that it plays no role in eliminating cycles.

'microworld' of quantum mechanics and the 'macroworld' of classical observations.

Here, my response is basically to admit ignorance. The freebit picture is not particularly sensitive to the precise *way* we distinguish between microfacts and macrofacts. But if the picture is to make sense, it does require that there *exist* a consistent way to make this distinction. (Or at least, it requires that there exist a consistent way to *quantify* the macro-ness of a fact f. The degree of macro-ness of f might then correspond to the 'effort of will' needed to affect f retrocausally, with the effort becoming essentially infinite for ordinary macroscopic facts!)

One obvious way to enforce a macro/micro distinction would be via a *dynamical collapse theory*, such as that of Ghirardi, Rimini, and Weber (1986) or of Penrose (1996). In these theories, all quantum states periodically undergo 'spontaneous collapse' to some classical basis, at a rate that grows with their mass or some other parameter, and in such a way that the probability of spontaneous collapse is close to 0 for all quantum systems that have so far been studied but close to 1 for ordinary 'classical' systems. Unfortunately, the known dynamical-collapse theories tend to suffer from technical problems, such as small violations of conservation of energy, and of course there is as yet no experimental evidence for them. More fundamentally, I personally cannot believe that Nature would solve the problem of the 'transition between microfacts and macrofacts' in such a seemingly ad hoc way, a way that does so much violence to the clean rules of linear quantum mechanics.

As I'll discuss further in Section 12.5.5, my own hope is that a principled distinction between microfacts and macrofacts could ultimately emerge from cosmology. In particular, I'm partial to the idea that, in a deSitter cosmology like our own, *a 'macrofact' is simply any fact of which the news is already propagating outward at the speed of light*, so that the information can never, even in principle, be gathered together again in order to 'uncause' the fact. A microfact would then be any fact for which this propagation *hasn't* yet happened. The advantage of this distinction is that it doesn't involve the slightest change to the principles of quantum mechanics. The disadvantage is that the distinction is 'teleological': the line between microfacts and macrofacts is defined by what *could* happen arbitrarily far in the future. In particular, this distinction implies that if, hypothetically, we could surround the solar system by a perfect reflecting boundary, then within the solar system, there would no longer be *any* macrofacts! It also implies that there can be no macrofacts in an *anti*-deSitter (adS) cosmology, which *does* have such a reflecting boundary. Finally, it suggests that there can probably be few if any macrofacts inside the event horizon of a black hole. For even if the information in the black hole interior *eventually* emerges, in coded form, in the Hawking radiation, the Hawking evaporation process is so slow ($\sim 10^{67}$ years for a solar-mass black hole) that there would seem to be plenty of time for an observer outside the hole to gather most of the radiation and thereby prevent the information from spreading further.[40]

Because I can't see a better alternative – and also, because I rather like the idea of cosmology playing a role in the foundations of quantum mechanics! – my current inclination is to bite the bullet, and accept these and other implications of the macro/micro distinction I've suggested. But there's enormous scope here for better ideas (or, of course, new developments in physics) to change my thinking.

12.5 Further objections

In this section, I'll present five objections to the freebit picture together with my responses. Some of these objections are obvious, and are generic to *any* analysis of 'freedom' that puts significant stock in the actual unpredictability of human choices. Others are less obvious and are more specific to the freebit picture.

12.5.1 The advertiser objection

The first objection is simply that human beings are depressingly predictable in practice: if they weren't, then they wouldn't be so easily manipulable! So, does surveying the sorry history of humankind – in which most people, most of the time, did exactly the boring, narrowly self-interested, things one might have expected them to do – furnish any evidence at all for the *existence* of freebits?

Response I already addressed a related objection in Section 12.2.11, but this one seems so important that it's worth returning to it from a different angle.

It's obvious that humans are at least partly predictable – and sometimes *extremely* predictable, depending on what one is trying to predict. For example, it's vanishingly unlikely that tomorrow, the CEO of General Motors will show up to work naked, or that Noam Chomsky will announce his support for American militarism. Nevertheless, that doesn't mean we know how to program a computer to *simulate* these people, anticipating every major or minor decision they make throughout their lives! It seems crucial to maintain the distinction between the partial predictability that even the most vocal free-will advocate would concede, and the 'physics-like' predictability of a comet. To illustrate, imagine a machine that correctly predicted *most* decisions of *most* people: what they'll order for dinner, which movies they'll select, which socks they'll take out of the drawer, and so on. (In a few domains, this goal is already being approximated by recommendation

[40] To be more precise here, one would presumably need to know the detailed *mapping* between the qubits of Hawking radiation and the degrees of freedom inside the hole, which in turn would require a quantum theory of gravity.

systems, such as those of Amazon and Netflix.) But imagine that the machine was regularly blindsided by the most *interesting, consequential, life-altering* decisions. In that case, I suspect most people's intuitions about their own 'freedom' would be shaken slightly, but would basically remain intact. By analogy, for most computer programs that arise in practice, it's easy to decide whether they halt, but that hardly decreases the importance of the *general* unsolvability of the halting problem. Perhaps, as Fischer et al. (2007) speculated, we truly exercise freedom only for a relatively small number of 'self-forming actions' (SFAs) – that is, actions that help to *define* who we are – and the rest of the time are essentially 'running on autopilot.' Perhaps these SFAs are common in childhood and early adulthood, but become rare later in life, as we get set in our ways and correspondingly more predictable to the people around us.

Having said this, I concede that the intuition in favor of humans' predictability is a powerful one. Indeed, even supposing humans *did* have the capacity for Knightian freedom, one could argue that that capacity can't be particularly important, if almost all humans choose to 'run on autopilot' almost all of the time!

However, against the undeniable fact of humans so often being manipulable like lab rats, there's a *second* undeniable fact, which should be placed on the other side of the intuitive ledger. This second fact is the conspicuous failure of investors, pundits, intelligence analysts, and so on *actually to predict*, with any reliability, what individuals or even entire populations will do. Again and again the best forecasters are blindsided (though it must be admitted that *after the fact* forecasters typically excel at explaining the inevitability of whatever people decided to do!).

12.5.2 The weather objection

What's special about the human brain? If we want to describe it as having 'Knightian freedom,' then why not countless *other* complicated dynamical systems, such as the Earth's weather, or a lava lamp?

Response For systems like the weather or a lava lamp, I think a plausible answer can actually be given. These systems are indeed unpredictable, but the unpredictability seems much more probabilistic than Knightian in character. To put it another way, the famous 'butterfly effect' seems likely to be an artifact of *deterministic* models of those systems; one expects it to get washed out as soon as we model the systems' microscopic components probabilistically. To illustrate this, imagine that the positions and momenta of all the molecules in the Earth's atmosphere had been measured to roughly the maximum precision allowed by quantum mechanics; and that, on the basis of those measurements, a supercomputer had predicted a 23% probability of a thunderstorm in Shanghai at a specific date next year. Now suppose we changed the initial conditions by adding a single butterfly. In classical, deterministic physics, that could certainly change whether the storm happens, due to chaotic effects – but that isn't the relevant question. The question is: how much does adding the butterfly change the *probability* of the storm? The answer seems likely to be 'hardly at all.' After all, the original 23% was already obtained by averaging over a huge number of possible histories. So unless the butterfly somehow changes the *general features* of the statistical ensemble, its effects should be washed out by the unmeasured randomness in the millions of *other* butterflies (and whatever else is in the atmosphere).

Yet with brains, the situation seems plausibly different. For brains seem 'balanced on a knife-edge' between order and chaos: were they as orderly as a pendulum, they couldn't support interesting behavior; were they as chaotic as the weather, they couldn't support rationality. More concretely, a brain is composed of neurons, each of which (in the crudest idealization) has a firing rate dependent on whether or not the sum of signals from incoming neurons exceeds some threshold. As such, one expects there to be many molecular-level changes one could make to a brain's state that don't affect the overall firing pattern at all, but a few changes – for example, those that push a critical neuron 'just over the edge' to firing or not firing – that affect the overall firing pattern drastically. So for a brain – unlike for the weather – a single freebit *could* plausibly influence the probability of some macroscopic outcome, even if we model all the system's constituents quantum mechanically.

A closely related difference between brains and the weather is that, while both are presumably chaotic systems able to amplify tiny effects, only in the case of brains are the amplifications likely to have 'irreversible' consequences. Even if a butterfly flapping its wings can cause a thunderstorm halfway around the world, a butterfly almost certainly won't change the average *frequency* of thunderstorms – at least, not without changing something other than the weather as an intermediary. To change the frequency of thunderstorms, one needs to change the trajectory of the earth's *climate*, something less associated with butterflies than with macroscopic 'forcings' (for example, increasing the level of carbon dioxide in the atmosphere, or hitting the earth with an asteroid). With brains, by contrast, it seems perfectly plausible that a single neuron's firing or not firing could affect the rest of a person's life (for example, if it caused the person to make a *very* bad decision).

12.5.3 The gerbil objection

Even if it's accepted that brains are very different from lava lamps or the weather, considered purely as dynamical systems, one could attempt a different *reductio ad absurdum* of the freebit picture, by constructing a physical system that *behaves* almost exactly like a brain yet whose Knightian uncertainty is 'decoupled' from its intelligence in what seems like an absurd way. Thus, consider the following thought experiment: on one side of a room, we have a digital computer, whose internal operations are completely deterministic. The computer is running an AI program that easily passes the Turing test: many humans, let's say, have maintained long Internet correspondences with the AI on diverse subjects and not one ever suspected its consciousness or humanity. On the other side of the room, we have a gerbil in a box. The gerbil scurries in its box, in a way that we can imagine to be subject to at least some Knightian uncertainty. Meanwhile, an array of motion sensors regularly captures information about the gerbil's movements and transmits it across the room to the computer, which uses the information as a source of random bits for the AI. Being extremely sophisticated, of course the AI doesn't make *all* its decisions randomly. But if it can't decide between two roughly equal alternatives, then it sometimes uses the gerbil movements to break a tie, much as an indecisive human might flip a coin in a similar situation.

The problem should now be obvious. By assumption, we have a system that *acts* with human-level intelligence (i.e., it passes the Turing test), and that's *also* subject to Knightian uncertainty, arising from amplified quantum fluctuations in a mammalian nervous system. So if we believe that humans have a 'capacity for freedom' because of those two qualities then we seem *obligated* to believe that the AI/gerbil hybrid has that capacity as well. Yet if we simply disconnect the computer from the gerbil box, then the AI loses its 'capacity for freedom'! For then the AI's responses, though they might still *seem* intelligent, could be 'unmasked as mechanistic' by anyone who possessed a copy of the AI's program. Indeed, even if we replaced the gerbil box by an 'ordinary' quantum-mechanical random number generator, the AI's responses would still be *probabilistically* predictable; they would no longer involve Knightian uncertainty.

Thus, a believer in the freebit picture seems forced to an insane conclusion: that the gerbil, though presumably oblivious to its role, is like a magic amulet that gives the AI a 'capacity for freedom' it wouldn't have had otherwise. Indeed, the gerbil seems uncomfortably close to the soul-giving potions of countless children's stories (stories that always end with the main character realizing that she *already had a soul*, even without the potion!). Yet, if we reject this sort of thinking in the gerbil-box scenario, then why shouldn't we also reject it for the human brain? With the brain, it's true, the 'Knightian-indeterminism-providing gerbil' has been moved physically closer to the locus of thought: now it scurries around in the synaptic junctions and neurons, rather than in a box on the other side of the room. But why should proximity make a difference? *Wherever* we put the gerbil, it just scurries around aimlessly! Maybe the scurrying is probabilistic, maybe it's 'Knightian,' but either way, it clearly plays no more *explanatory* role in intelligent decision making than the writing on the Golem's forehead.

In summary, it seems the only way to rescue the freebit picture is to paper over an immense causal gap – between the brain's 'Knightian noise' and its cognitive information processing – with superstition and magic.

Response Of all the arguments against the freebit picture, this one strikes me as by far the most serious, which is why I tried to present it in a way that would show its intuitive force.

On reflection, however, there *is* at least one potentially important difference between the AI/gerbil system and the brain. In the AI/gerbil system, the 'intelligence' and 'Knightian noise' components were *cleanly separable* from one another. That is, the computer *could* easily be disconnected from the gerbil box, and reattached to a *different* gerbil box, or an ordinary random-number generator, or nothing at all. And after this was done, there's a clear sense in which the AI would still be running 'the exact same program': only the 'indeterminism-generating peripheral' would have been swapped out. For this reason, it seems best to think of the gerbil as simply *yet another part of the AI's external environment* – along (for example) with all the questions sent to the AI over the Internet, which could *also* be used as a source of Knightian indeterminism.

With the brain, by contrast, it's not nearly so obvious that the 'Knightian indeterminism source' *can* be physically swapped out for a different one, without destroying or radically altering the brain's *cognitive* functions as well. Yes, we can imagine futuristic nanorobots swarming through a brain, recording all the 'macroscopically measurable information' about the connections between neurons and the strengths of synapses, then building a new brain that was 'macroscopically identical' to the first, differing only in its patterns of microscopic noise. But how would we know whether the robots had recorded *enough* information about the original brain? What if, in addition to synaptic strengths, there was also cognitively relevant information at a smaller scale? Then we'd need more advanced nanorobots, able to distinguish even smaller features. Ultimately, we could imagine robots able to record *all* 'classical' or even 'quasi-classical' features. But by definition, the robots would then be measuring features that were *somewhat* quantum mechanical, and therefore they would inevitably change those features.

Of course, this is hardly a conclusive argument, since maybe there's a gap of many orders of magnitude between (a) the smallest possible scale of cognitive relevance, and (b) the scale where the no-cloning theorem becomes relevant. In that case, at least in principle, the nanorobots really *could* complete their scan of the brain's 'cognitive layer' without risking the slightest damage to it; only the (easily-replaced?) 'Knightian indeterminism layer' would be altered by the nanorobots' presence. Whether this is possible is an empirical question for neuroscience.

However, the discussion of brain scanning raises a broader point: that, against the gerbil-box scenario, we need to weigh some other, older thought experiments where many people's intuitions go the other way. Suppose the nanorobots *do* eventually complete their scan of all the 'macroscopic, cognitively relevant' information in *your* brain, and suppose they then transfer the information to a digital computer, which proceeds to run a macroscopic-scale simulation of your brain. Would that simulation *be* you? If your 'original' brain were destroyed in this process, or simply anesthetized, would you expect to wake up as the digital version? (Arguably, this is not even a philosophical question, just a straightforward empirical question asking you to predict a future observation!)

Now, suppose you believe there's some conceivable digital doppelgänger that would *not* 'be you,' and you also believe that the difference between you and it resides somewhere in the physical world. Then since (by assumption) the doppelgänger is functionally indistinguishable from you, it would seem to follow that the difference between you and it *must* reside in 'functionally irrelevant' degrees of freedom, such as microscopic ones. Either that or else the boundary between the 'functional' and 'non-functional' degrees of freedom is not even sharp enough for the doppelgängers to be created in the first place.

My conclusion is that *either* you can be uploaded, copied, simulated, backed up, and so forth, leading to all the puzzles of personal identity discussed in Section 12.2.5, *or else* you can't bear the same sort of 'uninteresting' relationship to the 'non-functional' degrees of freedom in your brain that the AI bore to the gerbil box.

To be clear, nothing I've said even hints at any *sufficient condition* on a physical system, for the system to have attributes such as free will or consciousness. That is, even if human brains were subject to Knightian noise at the microscopic scale, and even if the sources of such noise could *not* be cleanly separated from the cognitive functions, human beings might still fail to be 'truly' conscious or free – whatever those things mean! – for other reasons. At most, I'm investigating plausible *necessary* conditions for Knightian freedom as defined in this essay (and hence, in my personal perspective, for 'free will' also).

12.5.4 The initial-state objection

The third objection holds that the notion that 'freebits' from the early universe nontrivially influence present-day events is not merely strange, but inconsistent with known physics. More concretely, Stenger (2009) has claimed that it *follows* from known physics that the initial state at the Big Bang was essentially random, and can't have encoded any 'interesting' information.[41] His argument is basically that the temperature at the Big Bang was enormous; and in quantum field theory (neglecting gravity), such extreme temperatures seem manifestly incompatible with any non-trivial structure.

Response Regardless of whether Stenger's *conclusion* holds, today there are strong indications, from cosmology and quantum gravity, that something has to be wrong with the above *argument* for a thermal initial state.

Firstly, by this argument, the universe's entropy should have been maximal at the Big Bang, but the Second Law tells us that the entropy was minimal! Stenger (2009) noticed the obvious problem, and tried to solve it by arguing that the entropy at the Big Bang really *was* maximal, given the tiny size of the observable universe at that time. On that view, the reason why entropy can increase as we move away from the Big Bang is simply that the universe is expanding, and with it the dimension of the state space. But others, such as Carroll (2010) and Penrose (2007), have pointed out severe problems with that answer. For one thing, if the dimension of the state space can increase, then we give up on *reversibility*, a central feature of quantum mechanics. For another, this answer has the unpalatable implication that the entropy should turn around and decrease in a hypothetical Big Crunch. The alternative, which Carroll and Penrose favor, is to hold that *despite* the enormous temperature at the Big Bang, the universe's state then was every bit as 'special' and low-entropy as the Second Law demands, but its specialness must have resided in *gravitational* degrees of freedom that we don't yet fully understand.

The second indication that the 'thermal Big Bang' picture is incomplete is that *quantum field theory has misled us in a similar way before*. In 1975, Hawking (1974) famously used quantum field theory in curved spacetime to calculate the temperature of black holes and to propose the existence of the Hawking radiation by which black holes eventually lose their mass and disappear. However, Hawking's calculation seemed to imply that the radiation was thermal – so that, in particular, it couldn't encode any non-trivial information about objects that fell into the black hole. This led to the *black-hole information-loss paradox*, since quantum-mechanical reversibility forbids the quantum states of the infalling objects simply to 'disappear' in this way. Today, largely because of the *AdS/CFT correspondence* (see Boer 2003), there's a near consensus among experts that the quantum states of infalling objects *don't* disappear as Hawking's calculation suggested. Instead, at least from the perspective of someone outside the black hole, the infalling information should 'stick around' on or near the event horizon, in not-quite-understood

[41] Here Stenger's concern was not free will or human predictability but, rather, ruling out the possibility (discussed by some theologians) that God could have arranged the Big Bang with foreknowledge about life on Earth.

quantum-gravitational degrees of freedom, before finally emerging in garbled form in the Hawking radiation. And if quantum field theory says otherwise, that's because quantum field theory is only a limiting case of a quantum theory of gravity.

The AdS/CFT correspondence is just one realization of the *holographic principle*, which has emerged over the last two decades as a central feature of quantum theories of gravity. It's now known that many physical theories have both a D-dimensional 'bulk' formulation and a $(D-1)$-dimensional 'boundary' formulation. In general, these two formulations *look* completely different: states that are smooth and regular in one formulation might look random and garbled in the other; questions that are trivial to answer in one formulation might seem hopeless in the other.[42] Nevertheless, there exists an isomorphism between states and observables, by which the boundary formulation 'holographically encodes' everything that happens in the bulk formulation. As a classic example, if Alice jumps into a black hole, then she might perceive herself falling smoothly toward the singularity. Meanwhile Bob, far from the black hole, might describe exactly the same physical situation using a 'dual description' in which Alice never makes it past the event horizon, and instead her quantum information gets 'pancaked' across the horizon in a horrendously complicated way. In other words, it is not absurd to suppose that a 'disorganized mess of entropy' on the boundary of a region could be 'isomorphic' to a richly structured state in the region's interior; there are now detailed examples where that's exactly what happens.[43]

The bottom line is that, when discussing extreme situations like the Big Bang, it's *not okay* to ignore quantum-gravitational degrees of freedom simply because we don't yet know how to model them. And including those degrees of freedom seems to lead straight back to the unsurprising conclusion that *no one knows* what sorts of correlations might have been present in the universe's initial microstate.

12.5.5 The Wigner's-friend objection

A final objection comes from the Many-Worlds Interpretation of quantum mechanics – or, more precisely, from taking seriously the possibility that a conscious observer could be measured as being in a coherent superposition of states.

Let's start with the following thought experiment, called *Wigner's friend* (after Eugene Wigner, who wrote about it in 1962). An intelligent agent A is placed in a

[42] As the simplest example, the boundary formulation makes it obvious that the total entropy in a region should be upper-bounded by its *surface area*, rather than its volume. In the bulk formulation, that property is strange and unexpected.

[43] Admittedly, the known examples involve isomorphisms between two theories with different numbers of spatial dimensions but both with a time dimension. There don't seem to be any non-trivial examples where the boundary theory lives on an initial spacelike or null hypersurface of the bulk theory. (One could, of course, produce a trivial example, by simply defining the 'boundary theory' to consist of the initial conditions of the bulk theory, with no time evolution! By 'non-trivial,' I mean something more interesting than that.)

coherent superposition of two different mental states, like so:

$$|\Phi\rangle = \frac{|1\rangle|A_1\rangle + |2\rangle|A_2\rangle}{\sqrt{2}}.$$

We imagine that these two states correspond to two different questions that the agent could be asked: for example, $|1\rangle|A_1\rangle$ represents A being asked its favorite color, while $|2\rangle|A_2\rangle$ denotes A being asked its favorite ice cream flavor. Then, crucially, a second agent B comes along and measures $|\Phi\rangle$ in the basis

$$\left\{ \frac{|1\rangle|A_1\rangle + |2\rangle|A_2\rangle}{\sqrt{2}}, \frac{|1\rangle|A_1\rangle - |2\rangle|A_2\rangle}{\sqrt{2}} \right\}, \tag{12.1}$$

in order to verify that A really *was* in a superposition of two mental states, not just in one state or the other.

Now, there are many puzzling questions one can ask about this scenario: most obviously, "what is it like for A to be manipulated in this way?" If A perceives itself in a definite state – either A_1 *or* A_2, but not both – then will B's later measurement of A in the basis (12.1) appear to A as a violation of the laws of physics?

However, let's pass over this well-trodden ground, and ask a different question more specific to the freebit picture. According to that picture, A's 'free decision' of how to answer whichever question it was asked should be correlated with one or more freebits w. But if we write out the combined state of the superposed A and the freebits,

$$\frac{|1\rangle|A_1\rangle + |2\rangle|A_2\rangle}{\sqrt{2}} \otimes |w\rangle,$$

then a problem becomes apparent. Namely, the same freebits w need to do 'double duty,' correlating with A's decision in both the $|1\rangle$ branch and the $|2\rangle$ branch! In other words, even supposing microscopic details of the environment could somehow 'explain' what happens in one branch, how could the *same* details explain both branches? As A_1 contemplated its favorite color, would it find itself oddly constrained by A_2's simultaneous contemplations of its favorite ice cream flavor, or vice versa?

One might think we could solve this problem by stipulating that w is split into two collections of freebits, $|w\rangle = |w_1\rangle \otimes |w_2\rangle$, with $|w_1\rangle$ corresponding to A_1's response and $|w_2\rangle$ corresponding to A_2's. But this solution quickly runs into an exponential explosion: if we considered a state like

$$\frac{1}{2^{500}} \sum_{x \in \{0,1\}^{1000}} |x\rangle|A_x\rangle,$$

we would find we needed 2^{1000} freebits to allow each A_x to make a yes/no decision independently of all the other A_x's.

Another 'obvious' way out would be if the freebits were entangled with A:

$$|\Phi'\rangle = \frac{|1\rangle |A_1\rangle |w_1\rangle + |2\rangle |A_2\rangle |w_2\rangle}{\sqrt{2}}.$$

The problem is that there seems to be no way to *produce* such entanglement, without violating quantum mechanics. If w is supposed to represent microscopic details of A's environment, ultimately traceable to the early universe, then it would be extremely mysterious to find A and w entangled. Indeed, in a Wigner's-friend experiment, such entanglement would show up as a *fundamental decoherence source*: something which was *not* traceable to any 'leakage' of quantum information from A yet which somehow prevented B from observing quantum interference between the $|1\rangle |A_1\rangle$ and $|2\rangle |A_2\rangle$ branches, when B measured in the basis (12.1). If, however, B *did* ultimately trace the entanglement between A and w to a 'leakage' of information from A, then w would have been revealed to have never contained freebits at all! For in that case, w would 'merely' be the result of unitary evolution coupling A to its environment – so it could presumably be predicted by B, who could even verify its hypothesis by measuring the state $|\Phi'\rangle$ in the basis

$$\left\{ \frac{|1\rangle |A_1\rangle |w_1\rangle + |2\rangle |A_2\rangle |w_2\rangle}{\sqrt{2}}, \frac{|1\rangle |A_1\rangle |w_1\rangle - |2\rangle |A_2\rangle |w_2\rangle}{\sqrt{2}} \right\}.$$

Response As in Section 12.4.2 – where asking about the definition of 'microfacts' and 'macrofacts' led to a closely related issue – my response to this important objection is to *bite the bullet*. That is, I accept the existence of a deep incompatibility between the freebit picture and the physical feasibility of the Wigner's-friend experiment. To state it differently: *if* the freebit picture is correct, *and* the Wigner's-friend experiment can be carried out, then I think we're forced to conclude that – at least for the duration of the experiment – *the subject no longer has the 'capacity for Knightian freedom,'* and is now a 'mechanistic,' externally characterized physical system similar to a large quantum computer.

I realize that the position above sounds crazy, but it becomes less so once one starts thinking about what would *actually* be involved in performing a Wigner's-friend experiment on a human subject. Because of the immense couplings between a biological brain and its external environment (see Tegmark, 1999, for example), the experiment is likely to be impossible with any entity we would currently recognize as 'human.' Instead, as a first step (!), one would presumably need to solve the problem of *brain-uploading*: that is, transferring a human brain into a digital substrate. Only then could one even *attempt* the second part, of transferring the now-digitized brain onto a quantum computer and preparing and measuring it in a superposition of mental states. I submit that, while the resulting entity *might* be 'freely willed,' 'conscious,' etc., it certainly wouldn't be uncontroversially so, nor can we reach any clear verdict by reasoning by analogy with our own case.

Notice in particular that, if the agent A could be manipulated in superposition, then, as a direct byproduct of those manipulations, A would presumably undergo *the same mental processes over and over, forwards in time as well as backwards in time*. For example, the 'obvious' way for B to measure A in the basis (12.1), would simply be for B to 'uncompute' whatever unitary transformation U had placed A in the superposed state $|\Phi\rangle$ in the first place. Presumably, the process would then be repeated many times, as B accumulated more statistical evidence for the quantum interference pattern. So, during the uncomputation steps, would A 'experience time running backwards'? Would the inverse map $U^\dagger$ feel different from U? Or would all the applications of U and $U^\dagger$ be 'experienced simultaneously,' being functionally indistinguishable from one another? I hope I'm not alone in feeling a sense of vertigo about these questions! To me, it's at least a plausible speculation that A doesn't experience *anything* and that the reasons why it doesn't are related to B's very ability to manipulate A in these ways.

More broadly, the view I've taken here on superposed agents strikes me as almost a *consequence* of the view I took earlier, on agents whose mental states can be perfectly measured, copied, simulated, and predicted by other agents. For there's a close connection between being able to *measure* the exact state $|S\rangle$ of a physical system and being able to detect quantum interference in a superposition of the form

$$|\psi\rangle = \frac{|S_1\rangle + |S_2\rangle}{\sqrt{2}},$$

consisting of two 'slight variants' of $|S\rangle$. If we know $|S\rangle$ then, among other things, we can load a copy of $|S\rangle$ onto a quantum computer, and thereby prepare and measure a superposition like $|\psi\rangle$ – provided, of course, that one counts the quantum computer's *encodings* of $|S_1\rangle$ and $|S_2\rangle$ as 'just as good as the real thing.' Conversely, the ability to detect interference between $|S_1\rangle$ and $|S_2\rangle$ presupposes that we know, and can control, *all* the degrees of freedom that make them different: quantum mechanics tells us that if any degrees of freedom are left unaccounted for, we will simply see a probabilistic mixture of $|S_1\rangle$ and $|S_2\rangle$, not a superposition.

But if this is so, one might ask, then *what makes humans any different?* According to the most literal reading of quantum mechanics' unitary evolution rule – which some call the Many-Worlds Interpretation – don't we *all* exist in superpositions of enormous numbers of branches, and isn't our inability to measure the interference between those branches merely a 'practical' problem, caused by rapid decoherence? Here I reiterate the speculation put forward in Section 12.4.2: that the decoherence of a state $|\psi\rangle$ should be considered 'fundamental' and 'irreversible' precisely when $|\psi\rangle$ becomes entangled with degrees of freedom that are receding toward our deSitter horizon at the speed of light and that can no longer be collected together even in principle. That sort of decoherence could be avoided, at least in principle, by a fault-tolerant quantum computer, as in the Wigner's-friend thought experiment above. But it plausibly *can't* be avoided by any entity that we would currently recognize as 'human.'

One could also ask: *if* the freebit picture were accepted, what would be the implications for the foundations of quantum mechanics? For example, would the Many-Worlds Interpretation then have to be rejected? Interestingly, I think the answer is no. Since I haven't suggested any change to the formal rules of quantum mechanics, *any* interpretations that accept those rules – including Many-Worlds, Bohmian mechanics, and various Bayesian or subjectivist interpretations – would in some sense 'remain on the table' (to whatever extent they were on the table before!). As far as we're concerned in this essay, if one wants to believe that different branches of the wavefunction, in which one's life followed a different course than what one observes, 'really exist' then that's fine; if one wants to deny the reality of those branches, that's fine as well. Indeed, if the freebit picture were correct then *Nature would have conspired so that we had no hope, even in principle, of distinguishing the various interpretations by experiment.*

Admittedly, one might think it's obvious that the interpretations can't be distinguished by experiment, with or without the freebit picture. Isn't that why they're *called* 'interpretations'? But as Deutsch (1985a) pointed out, scenarios like Wigner's friend seriously challenge the idea that the interpretations are empirically equivalent. For example, if one could perform a quantum interference experiment on *one's own mental state*, then couldn't one directly experience what it was like for the different components of the wavefunction describing that state to evolve unitarily?[44] And wouldn't that, essentially, vindicate a Many-Worlds-like perspective while ruling out the subjectivist views that refuse to countenance the reality of 'parallel copies of oneself'? By denying that the subject of a Wigner's-friend experiment has the 'capacity for Knightian freedom,' the freebit picture suggests that maybe there's nothing comparable with *being* such a subject – and hence, that debates about quantum interpretation can freely continue forever, with not even the in-principle possibility of an empirical resolution.

[44] A crucial caveat is that, after the interference experiment was over, one would retain no reliable memories or publishable records about 'what it was like'! For the very fact of such an experiment implies that one's memories are being created and destroyed at will. Without the destruction of memories, we can't get interference. After the experiment is finished, one might have *something* in one's memory, but what it is could have been probabilistically predicted even before the experiment began, and can in no way depend on 'what it was like' in the middle. Still, *at least while the experiment was under way*, maybe one would know which interpretation of quantum mechanics was correct!

12.6 Comparison with Penrose's views

Probably the most original thinker to have speculated about physics and mind in the last half-century has been Roger Penrose. In his books, including *The Emperor's New Mind* (1989) and *Shadows of the Mind* (1996),[45] Penrose has advanced three related ideas, all of them extremely controversial:

(1) Arguments related to Gödel's Incompleteness Theorem imply that the physical action of the human brain cannot be algorithmic (i.e., that it must violate the Church–Turing Thesis).
(2) There must be an 'objective' physical process that collapses quantum states and produces the definite world we experience, and that the best place to look for such a process is in the interface of quantum mechanics, general relativity, and cosmology (the 'specialness' of the universe's initial state providing the only known source of time-asymmetry in physics, not counting quantum measurement).
(3) Objective collapse events, possibly taking place in the cytoskeletons of the brain's neurons (and subsequently amplified by 'conventional' brain activity), provide the best candidate for the source of noncomputability demanded by (1).

An obvious question is how Penrose's views relate to the ones discussed here. Some people might see the freebit picture as 'Penrose lite.' For, on the one hand, it embraces Penrose's core belief in a relationship between the mysteries of mind and those of modern physics and even follows Penrose in focusing on certain *aspects* of physics, such as the 'specialness' of the initial state. On the other hand, the account here rejects almost all Penrose's further speculations: for example, about noncomputable dynamics in quantum gravity and the special role of the cytoskeleton in exploiting those dynamics.[46] Let me now elaborate on six differences between my account and Penrose's.

(1) **I make no attempt to 'explain consciousness.'** Indeed, that very goal seems misguided to me, at least if 'consciousness' is meant in the phenomenal sense rather than the neuroscientists' more restricted senses.[47] For as countless

[45] A later book, *The Road to Reality* (2007), says little directly about mind, but is my favorite. I think it makes Penrose's strongest case for a gap in our understanding of quantum mechanics, thermodynamics, and cosmology, which radical new ideas will be needed to fill.

[46] Along another axis, though, some people might see the freebit picture as *more* radical, in that it suggests the impossibility of *any* non-tautological explanation for certain events and decisions, even an explanation invoking oracles for Turing-uncomputable problems.

[47] Just like 'free will,' the word 'consciousness' has been the victim of ferocious verbal overloading, having been claimed for everything from 'that which disappears under anesthesia,' to 'that which a subject can give verbal reports about,' to 'the brain's executive control system'! Worse, 'consciousness' has the property that, even if one specifies *exactly* what one means by it, readers are nevertheless certain to judge anything one says against their own preferred meanings. For this reason, just as I ultimately decided to talk about 'freedom'

philosophers have pointed out over the years (see McGinn, 2000, for example), *all* scientific explanations seem equally compatible with a 'zombie world,' which fulfills the right causal relations but where no one 'really' experiences anything. More concretely, even if Penrose were right that the human brain has 'super-Turing' computational powers – and I see no reason to think he is right – I've never understood how that would help with what Chalmers (1996) calls the 'hard problem' of consciousness. For example, could a Turing machine equipped with an oracle for the halting problem perceive the 'redness of red' any better than a Turing machine without such an oracle?

Given how much Penrose says about consciousness, I find it strange that he says almost nothing about the related mystery of *free will*. My central claim in this essay is that there exists an 'empirical counterpart' of free will (what I call 'Knightian freedom'), whose investigation really *does* lead naturally to questions about physics and cosmology, in a way that does not, as far as I know, happen for any of the usual empirical counterparts of consciousness.

(2) **I make no appeal to the Platonic perceptual abilities of human mathematicians.** Penrose's arguments rely on human mathematicians' supposed power to 'see' the consistency of (for example) the Peano axioms of arithmetic (rather than simply *assuming* or *asserting* that consistency, as a computer program engaging in formal reasoning might do). As far as I can tell, to whatever extent this 'power' exists at all, it's just a particularly abstruse type of *qualia* or *subjective experience*, as empirically inaccessible as any other type. In other words, instead of talking about the consistency of Peano arithmetic, I believe Penrose might as well have fallen back on the standard arguments about how a robot could never 'really' enjoy fresh strawberries but at most *claim* to enjoy them.

In both cases, the reply seems obvious: *how do you know* that the robot doesn't really enjoy the strawberries that it claims to enjoy, or see the consistency of arithmetic that it claims to see? And how do you know other people *do* enjoy or see those things? In any case, none of the arguments in this essay turn on these sorts of considerations. If any important difference is to be claimed between a digital computer and a human brain, then I insist that the difference correspond to something empirical: for example, that computer memories can be reliably measured and copied without disturbing them, while brain states quite possibly can't. The difference must *not* rely, even implic-

(or 'Knightian freedom') rather than 'free will' in this essay, so I'd much rather use less fraught terms for executive control, verbal reportability, and so on, and restrict the word 'consciousness' to mean 'the otherwise undefinable thing that people have tried to get at for centuries with the word 'consciousness,' supposing that thing exists.'

itly, on a question-begging appeal to the author's or reader's own subjective experience.

(3) **I make no appeal to Gödel's Incompleteness Theorem.** Let me summarize Penrose's (and earlier, Lucas's 1961) 'Gödelian argument for human specialness.' Consider any finitely describable machine M for deciding the truth or falsehood of mathematical statements, which never errs but which might sometimes output 'I don't know.' Then, by using the code of M, it's possible to construct a mathematical statement S_M – one example is a formal encoding of 'M will never affirm this statement' – that we humans, 'looking in from the outside' can clearly see is true yet that M itself can't affirm without getting trapped in a contradiction.

The difficulties with this argument have been explained at length elsewhere (Aaronson, 2013; Dennett, 1995; Russell and Norvig, 2009); some standard replies to it are given in a footnote.[48]

Here, I'll simply say that I think the Penrose–Lucas argument establishes *some* valid conclusion, but the conclusion is much weaker than what Penrose wants, and it can also be established much more straightforwardly, without Gödelian considerations. The valid conclusion is that, *if you know the code of an AI* then, regardless of how intelligent the AI seems to be, you can 'unmask' it as an automaton, blindly following instructions. To do so, however, you don't need to trap the AI in a self-referential paradox: it's enough to verify that the AI's responses are precisely the ones predicted (or probabilistically predicted) by the code that you possess! Both with the Penrose–Lucas argument and with this simpler argument, it seems to me that the real issue is not whether the AI follows a program, but rather, whether it follows a program that's *knowable by other physical agents.* That's why this essay focusses from the outset on the latter issue.

(4) **I don't suggest any barrier to a suitably programmed digital computer passing the Turing test.** Of course, if the freebit picture were correct, then there *would* be a barrier to duplicating the mental state and predicting the responses of a *specific* human. Even here, though, it's possible that, through non-invasive measurements, one could learn enough to create a digital 'mockup' of a given person that would fool that person's closest friends and relatives,

[48] Why is it permissible to assume that M never errs, if no *human* mathematician (or even, arguably, the entire mathematical community) has ever achieved that standard of infallibility?
Even if M never *did* affirm S_M, or never erred more generally, how could we ever *know* that? Indeed, much like with consciousness itself, even if one person had the mysterious Platonic ability to 'see' M's soundness, how could that person ever convince a skeptical third party?
Finally, if we believed that the human brain was itself finitely describable then why couldn't we construct a similar mathematical statement (e.g., "Penrose will never affirm this statement"), which *Penrose* couldn't affirm without contradicting himself even though a different human, or indeed an AI program, could easily affirm it?

possibly for decades. (For this purpose, it might not matter if the mockup's responses eventually diverged badly from the original person's!) And such a mockup would certainly succeed at the 'weaker' task of passing the Turing test – i.e., of fooling interrogators into thinking it was human, at least until its code was revealed. If these sorts of mockups *couldn't* be built, then it would have to be for reasons well beyond anything explored in this essay.

(5) **I don't imagine anything particularly exotic about the biology of the brain.** In *The Emperor's New Mind* (Penrose, 1989), Penrose speculates that the brain might act as what today we would call an adiabatic quantum computer: a device that generates highly entangled quantum states and might be able to solve certain optimization problems faster than any classical algorithm. (In this case, presumably, the entanglement would be *between neurons*.) In *Shadows of the Mind* (Penrose, 1996), he goes further, presenting a proposal due to him and Stuart Hameroff that ascribes a central role to *microtubules*, a component of the neuronal cytoskeleton. In this proposal, the microtubules would basically be 'antennae' sensitive to yet-to-be-discovered quantum-gravity effects. Since Penrose *also* conjectures that a quantum theory of gravity would include Turing-uncomputable processes (see point (7) below), the microtubules would therefore let human beings surpass the capabilities of Turing machines.

Unfortunately, subsequent research hasn't been kind to these ideas. Calculations of decoherence rates leave almost no room for the possibility that quantum coherence is maintained in the hot, wet, environment of the brain for anything like the timescales relevant to cognition, or for long-range entanglement between neurons (see Tegmark, 1999, for example). As for microtubules, they are common structural elements in cells – not in only neurons – and no clear evidence has emerged that they are particularly sensitive to the quantum nature of spacetime. And this is setting aside the question of the evolutionary pathways by which the 'quantum-gravitational antennae' could have arisen.

The freebit perspective requires none of this: at least from a biological perspective, its picture of the human brain is simply that of conventional neuroscience. Namely, the human brain is a massively parallel, highly connected *classical* computing organ, whose design was shaped by millions of years of natural selection. Neurons perform a role vaguely analogous to that of *gates* in an electronic circuit (though neurons are far more complicated in detail), while synaptic strengths serve as a readable and writable memory. If we restrict to issues of principle then perhaps the most salient *difference* between the brain and today's electronic computers is that the brain is a 'digital/analog hybrid.' This means, for example, that we have no practical way to measure the brain's exact 'computational state' at a given time, to copy the state, or to restore the brain to a previous state; and it is not even obvious whether these things can be done in principle. It also means that the brain's detailed activity might be sensitive to microscopic fluctuations (for example, in the sodium-ion channels) that get chaotically amplified; this amplification might even occur over timescales relevant to human decision-making (say, 30 seconds). Of course, if those fluctuations were quantum-mechanical in origin – and at a small enough scale, they *would* be – then they couldn't be measured even in principle without altering them.

From the standpoint of neuroscience, the last parts of the preceding paragraph are certainly not established, but neither does there seem to be any good evidence against them. I regard them as plausible guesses and hope that future work will confirm or falsify them. To the view above, the freebit picture adds only a single further speculation – a speculation that, moreover, I think *does not even encroach on neuroscience's 'turf.'* This is simply that if we consider the quantum states ρ relevant to the microscopic fluctuations then those states are subject to at least some Knightian uncertainty (i.e., they are 'freestates' as defined in Appendix 12C); and furthermore, at least some of the Knightian uncertainty could ultimately be traced, if we wished, back to our ignorance of the detailed microstate of the early universe. This might or might not be true; however it seems to me that it's not a question for neuroscience at all but for physics and cosmology (see Section 12.9). What the freebit picture 'needs' from neuroscience, then, is extremely modest – certainly compared with what Penrose's picture needs!

(6) **I don't propose an 'objective reduction' process that would modify quantum mechanics.** Penrose speculates that, when the components of a quantum state achieve a large enough energy difference that they induce 'appreciably different' configurations of the spacetime metric (roughly, configurations that differ by one Planck length or more), new quantum-gravitational laws beyond of unitary quantum mechanics should come into effect, and cause an 'objective reduction' of the quantum state. This hypothetical process would underlie what we perceive as a measurement or collapse. Penrose has given arguments that his reduction process, if it existed, would have escaped detection by current quantum interference experiments, but *could* conceivably be detected or ruled out by experiments in the foreseeable future (Marshall et al., 2003). Penrose's is far from the only 'objective reduction' model on the table: for example, there's a well-known earlier model due to Ghirardi, Rimini, and Weber (GRW) (1986), but that model was purely 'phenomenological' rather than being tied to gravity or some other known part of physics.

If an objective reduction process were ever discovered then it would provide a ready-made distinction between microfacts and macrofacts (see Section 12.4.2), of exactly the sort the freebit picture needs. Despite this, I'm profoundly skeptical that any of the existing objective reduction models are close to the truth. The reasons for my skepticism are, firstly, that the models seem too ugly and ad hoc (GRW's more so than Penrose's); and secondly, that the AdS/CFT correspondence now provides evidence that quantum mechanics can emerge unscathed even from the combination with gravity. That's why, in Sections 12.4.2 and 12.5.5, I speculated that the distinction between microfacts and macrofacts might ultimately be defined in terms of deSitter space cosmology, with a macrofact being any fact 'already irreversibly headed toward the deSitter horizon.'

(7) **I don't propose that quantum gravity leads to Turing-uncomputable processes.** One of Penrose's most striking claims is that the laws of physics should involve *uncomputability*; that is, transitions between physical states that cannot in principle be simulated by a Turing machine, even given unlimited time. Penrose arrives at this conclusion via his Gödel argument (see point (3)); he then faces the formidable challenge of where to *locate* the necessary uncomputability in anything plausibly related to physics. Note that this is *separate* from the challenge (discussed in point (5)) of how to make the human brain sensitive to uncomputable phenomena, supposing they exist! In *Shadows of the Mind*, Penrose, seems to admit that this is a weak link in his argument. As evidence for uncomputability, the best he can offer is a theorem of Markov that the 4-manifold homeomorphism problem is undecidable (indeed, equivalent to the halting problem) (Markov, 1958), and a speculation of Geroch and Hartle (1986) that maybe that fact has something to do with quantum gravity, since some attempted formulations of quantum gravity involve sums over 4-manifolds.

Personally, I see no theoretical or empirical reason to think that the laws of physics should let us solve Turing-uncomputable problems – either with our brains or with any other physical system. Indeed, I would go further: in Aaronson (2005), I summarized the evidence that the laws of physics seem to 'conspire' to prevent us from solving NP-complete problems (like the Traveling Salesman Problem) in polynomial time. But NP-complete problems, being solvable in 'merely' exponential time, are child's play compared to Turing-uncomputable problems like the halting problem! For this reason, I regard it as a serious drawback of Penrose's proposals that they demand uncomputability in the dynamical laws and as an advantage of the freebit picture that it suggests nothing of the kind. Admittedly, the freebit picture does require that there be no complete, computationally simple, description of the *initial conditions*.[49] But it seems to me that nothing in established physics should have led us to

[49] More precisely, the initial state, when encoded in some natural way as a binary string, must have non-negligibly large 'sophistication': see Appendix 12B.

expect that such a description would exist *anyway*![50] The freebit picture is silent on whether detailed properties of the initial state can be actually be *used* to solve otherwise-intractable computational problems, such as NP-complete problems, in a reliable way. But the picture certainly gives no reason to *think* this is possible, and I see no evidence for its possibility from any other source.

12.7 'Application' to Boltzmann brains

In this section, I'll explain how the freebit perspective, if adopted, seems to resolve the notorious 'Boltzmann brain problem' of cosmology. No doubt some people will feel that the cure is even worse than the disease! But even if one thinks that, the mere fact of a connection between freebits and Boltzmann brains seems worth spelling out.

First, what is the Boltzmann brain problem? Suppose that – as now seems all but certain (Perlmutter and 31 others; Supernova Cosmology Project, 1999) – our universe will *not* undergo a Big Crunch, but will simply continue to expand forever, its entropy increasing according to the Second Law. Then eventually, after the last black holes have decayed into Hawking radiation, the universe will reach the state known as *thermal equilibrium*: basically an entropy-maximizing soup of low-energy photons, flying around in an otherwise cold and empty vacuum. The difficulty is that, even in thermal equilibrium, there's still a tiny but nonzero probability that *any given (finite) configuration* will arise randomly: for example, via a chance conglomeration of photons, which could give rise to other particles via virtual processes. In general, we expect a configuration with total entropy S to arise at a particular time and place with probability of order $\sim 1/\exp(S)$. But eternity being a long time, even such exponentially unlikely fluctuations should not only occur, but occur *infinitely often*, for the same reason why all Shakespeare's plays presumably appear, in coded form, infinitely often in the decimal expansion of π.[51]

So in particular, we would eventually expect to appear (say) beings physically identical to you, who'd survive just long enough to have whatever mental experiences you're now having, then disappear back into the void. These hypothetical observers are known in the trade as *Boltzmann brains* (see Dyson et al., 2002), after Ludwig Boltzmann, who speculated about related matters in the late nineteenth century. So, how do you know that *you* aren't a Boltzmann brain?

But the problem is worse. Since in an eternal universe you would have infinitely

[50] By contrast, when it comes to dynamical behavior, we have centuries of experience discovering laws that can indeed be simulated on a computer, given the initial conditions as input, and no experience discovering laws that can't be so simulated.

[51] This would follow from the conjecture, as yet unproved, that π is a 'base-10 normal number': that is, that just as for a random sequence, every possible sequence of k consecutive digits appears in π's decimal expansion with asymptotic frequency $1/10^k$.

many Boltzmann-brain doppelgängers, any observer with your memories and experiences seems infinitely *more* likely to be a Boltzmann brain, than to have arisen via the 'normal' processes of Darwinian evolution and so on starting from a Big Bang! Silly as it sounds, this has been a major problem plaguing cosmologists' recent proposals, since they keep wanting to assign enormous probability measure to Boltzmann brains (see Carroll, 2010).

But now suppose you believed the freebit picture, and also believed that possessing Knightian freedom is a necessary condition for counting as an observer. Then I claim that the Boltzmann brain problem would immediately go away. The reason is that, in the freebit picture, Knightian freedom *implies* a certain sort of correlation with the universe's initial state at the Big Bang – so that lack of complete knowledge of the initial state corresponds to lack of complete predictability (even probabilistic predictability) of one's actions by an outside observer. But a Boltzmann brain wouldn't have that sort of correlation with the initial state. By the time thermal equilibrium is reached, the universe will (by definition) have 'forgotten' all details of its initial state, and any freebits will have long ago been 'used up.' In other words, there's no way to make a Boltzmann brain think one thought rather than another by toggling freebits. So, on this account, Boltzmann brains wouldn't be 'free,' even during their brief moments of existence. This, perhaps, invites the further speculation that there's nothing that it's like to be a Boltzmann brain.

12.7.1 What happens when we run out of freebits?

The above discussion of Boltzmann brains leads to a more general observation about the freebit picture. Suppose that

(1) the freebit picture holds,
(2) the observable universe has a finite extent (as it does, assuming a positive cosmological constant), and
(3) the holographic principle (see Section 12.5.4) holds.

Then the number of freebits accessible to any one observer must be finite – simply because the number of bits of *any* kind is then upper bounded by the observable universe's finite holographic entropy. (For details, see Bousso (2000) or Lloyd (2002), both of whom estimate the observable universe's information capacity as roughly $\sim 10^{122}$ bits.)

But the nature of freebits is that they get permanently 'used up' whenever they are amplified to macroscopic scale. So, under the stated assumptions, we conclude that only a finite number of 'free decisions' can possibly be made before the observable universe runs out of freebits! In my view, this should not be too alarming. After all, even *without* the notion of freebits, the Second Law of Thermodynamics (combined with the holographic principle and the positive cosmological constant) already tells us that the observable universe can witness at most $\sim 10^{122}$ 'interesting events,' of any kind, before it settles into thermal equilibrium. For more on the theme of freebits as a finite resource, see Appendix 12B.

12.8 Indexicality and freebits

The Boltzmann brain problem is just one example of what philosophers call an *indexical* puzzle: a puzzle involving the 'first-person facts' of who, what, where, and when *you* are, which seems to persist even after all the 'third-person facts' about the physical world have been specified. Indexical puzzles, and the lack of consensus on how to handle them, underlie many notorious debates in science and philosophy – including the debates surrounding the anthropic principle and the 'fine-tuning' of cosmological parameters, the multiverse and string theory landscape, the Doomsday argument, and the Fermi paradox about where all the extraterrestrial civilizations are. I won't even try to summarize the vast literature on these problems (see Bostrom, 2002, for an engaging introduction). Still, it might be helpful to go through a few examples, just to illustrate what we mean by indexical puzzles.

- When doing Bayesian statistics, it's common to use a *reference class*: roughly speaking, the set of observers from which you consider yourself to have been 'sampled.' For an uncontroversial example, suppose you want to estimate the probability that you have some genetic disease, in order to decide (say) whether it's worth getting tested. In reality, you either have the disease or you don't. Yet it seems perfectly unobjectionable to estimate the *probability* that you have the disease, by imagining that you were 'chosen randomly' from among all people with the same race, sex, and so forth, then looking up the relevant statistics. However, things quickly become puzzling when we ask how *large* a reference class can be invoked. Can you consider yourself to have been 'sampled uniformly at random' from the set of all humans who ever lived or ever will live? If so, then why not also include early hominids, chimpanzees, dolphins, extraterrestrials, or sentient artificial intelligences? Many would simply reply that you're *not* a dolphin or any of those other things, so there's no point worrying about the hypothetical possibility of having been one. The problem is that you're *also* not some other person of the same race and sex – you're *you* – but for medical and actuarial purposes, we clearly *do* reason as if you 'could have been' someone different. So why does your reference class include those other people but not dolphins?
- Suppose you're an astronomer who's trying to use Bayesian statistics to estimate the probability of one cosmological model versus another, conditioned on

the latest data about the cosmic background radiation and so forth. Of course, as discussed in Section 12.3.1, any such calculation requires a specification of *prior probabilities*. The question is: should your prior include the assumption that, all else being equal, we'd be twice as likely to find ourselves in a universe that's twice as large (and thus presumably has twice as many civilizations, in expectation)? If so, then how do we escape the absurd-seeming conclusion that we're *certain* to live in an infinite universe, if such a universe is possible at all – since we expect there to be infinitely many more observers in an infinite universe than in a finite one? Surely we can't deduce the size of the universe without leaving our armchairs, like a medieval scholastic? The trouble is that, as Bostrom (2002) pointed out, *not* adjusting the prior probabilities for the expected number of observers leads to its own paradoxes. As a fanciful example, suppose we're trying to decide between two theories, which on physical grounds are equally likely. Theory A predicts that the universe will contain a single civilization of two-legged creatures, while theory B predicts that the universe will contain a single civilization of two-legged creatures *as well as* a trillion equally advanced civilizations of nine-legged creatures. Observing ourselves to be two-legged, can we conclude that theory A is overwhelmingly more likely – since if theory B were correct, then we would almost certainly have been pondering the question on nine legs? A straightforward application of Bayes' rule seems to imply that the answer should be yes – *unless* we perform the adjustment for the number of civilizations that led to the first paradox!

- Pursuing thoughts like the above quickly leads to the notorious *Doomsday argument*. According to that argument, the likelihood that human civilization will kill itself off in the near future is much larger than one would naïvely think – where 'naïvely' means 'before taking indexical considerations into account.' The logic is simple: suppose human civilization will continue for billions of years longer, colonizing the galaxy and so forth. In that case, our own position near the very beginning of human history would seem absurdly improbable – the more so when one takes into account that such a long-lived, spacefaring, civilization would probably have a much larger population than exists today (just as *we* have a much larger population than existed hundreds of years ago). If we're Bayesians, that presumably means we *must* revise downward the probability that the 'spacefaring' scenario is correct and revise upward the probability of scenarios that give us a more 'average' position in human history. But because of exponential population growth, one expects the latter to be heavily weighted toward scenarios where civilization kills itself off in the very near future. Many commentators have tried to dismiss this argument as flat-out erroneous (see Leslie, 1998, or Bostrom, 2002 for common objections to the argument and responses to them).[52] However, while the 'modern, Bayesian' version of the Doomsday argument might indeed be wrong, it's not wrong because of some trivial statistical oversight. Rather, the argument might be wrong because it embodies an interestingly false way of thinking about indexical uncertainty.

Perplexing though they might be, what do any of these indexical puzzles have to do with *our* subject, free will? After all, presumably no one thinks that we have the 'free will' to choose where and when we're born! Yet, while I've never seen this connection spelled out before, it seems to me that indexical puzzles like those above *do* have some bearing on the free will puzzle. For the indexical puzzles make it apparent that, even if we assume that the laws of physics are completely mechanistic, there remain large aspects of our experience that those laws fail to determine, even probabilistically. Nothing in the laws picks out one particular chunk of suitably organized matter from the immensity of time and space, and says, "here, *this* chunk is you; its experiences are your experiences." Nor does anything in the laws give us even a probability distribution over the possible such chunks. Despite its obvious importance even for empirical questions, our uncertainty about who we are and who we 'could have been' (i.e., our reference class) bears all the marks of Knightian uncertainty. Yet once we've admitted indexical uncertainty into our worldview, it becomes less clear why we should reject the sort of uncertainty that the freebit picture needs! If whether you find yourself born in eighth-century China, twenty-first-century California, or Planet Zorg is a variable that is subject to Knightian uncertainty, then why not what you'll have for dessert tonight?

More concretely, suppose that there are numerous planets nearly identical to Earth, right down to the same books being written, people with the same names and same DNA being born, etc. If the universe is spatially infinite – which cosmologists consider a serious possibility[53] – then there's no need to imagine this scenario: for simple probabilistic reasons, it's almost certainly true! Even if the universe is spatially finite, the probability of such a 'twin Earth' would approach 1 as the

[52] Many people point out that cavemen could have made exactly the same argument, and would have been wrong. This is true but irrelevant: the whole point of the Doomsday argument is that *most* people who make it will be right!
Another common way to escape the argument's conclusion is to postulate the existence of large numbers of extraterrestrial civilizations, which are there regardless of what humans do or don't do. If the extraterrestrials are included in our reference class, they can then 'swamp' the effect of the number of future humans in the Bayesian calculation.

[53] Astronomers can only see as far as light has reached since the Big Bang. If a positive spatial curvature was ever detected on cosmological scales, it would strongly suggest that the universe 'wraps around' – much like hypothetical ancients might have deduced that the earth was round by measuring the curvature of a small patch. So far, though, except for local perturbations, the universe appears perfectly flat to within the limits of measurement, suggesting that it is either infinite or else extends far beyond our cosmological horizon. However, it is logically possible that the universe could be *topologically* closed (and hence finite), despite having zero spatial curvature. Also, assuming a positive cosmological constant, sufficiently far parts of the universe would be forever out of causal contact with us – leading to philosophical debate about whether those parts should figure in scientific explanations or even be considered to 'exist.'

number of stars, galaxies, and so on went to infinity. Naturally, we'd expect any two of these 'twin Earths' to be separated by exponentially large distances – so that, because of the dark energy pushing the galaxies apart, we would *not* expect a given 'Earth-twin' ever to be in communication with any of its counterparts.

Assume for simplicity that there are at most two of these Earth-twins, call them A and B. (That assumption will have no effect on our conclusions.) Let's suppose that, because of (say) a chaotically amplified quantum fluctuation, these two Earths are about to diverge significantly in their histories for the first time. Let's further suppose that *you* – or rather, beings on A and B respectively who look, talk, and act like you – are the proximate cause of this divergence. On Earth A, the quantum fluctuation triggers a chain of neural firing events in 'your' brain that ultimately causes 'you' to take a job in a new city. On Earth B, a different quantum fluctuation triggers a chain of neural firings that causes 'you' to stay where you are.

We now ask: from the perspective of a superintelligence that knows everything above, what's the total probability p that 'you' take the new job? Is it simply $\frac{1}{2}$, since the two actual histories should be weighted equally? What if Earth B had a greater probability than Earth A of having formed in the first place – or did under one cosmological theory but not another? And why do we need to average over the two Earths at all? Maybe 'you$_A$' is the 'real' you, and taking the new job is a defining property of who you are, much as Shakespeare 'wouldn't be Shakespeare' had he not written his plays. So maybe you$_B$ isn't even part of your reference class: it's just a faraway doppelgänger you'll never meet, who looks and acts like you (at least up to a certain point in your life) but *isn't* you. So maybe $p = 1$. Then again, maybe you$_B$ is the 'real' you and $p = 0$. Ultimately, not even a superintelligence could calculate p without knowing something about *what it means to be 'you,'* a topic about which the laws of physics are understandably silent.

Now, someone who accepted the freebit picture would say that the superintelligence's inability to calculate p is no accident. For whatever quantum fluctuation separated Earth A from Earth B could perfectly well have been a freebit. In that case, before you made the decision, the right representation of your physical state would have been a Knightian combination of you$_A$ and you$_B$. (See Appendix 12C for details of how these Knightian combinations fit in with the ordinary density matrix formalism of quantum mechanics.) After you make the decision, the ambiguity is resolved in favor of you$_A$ or you$_B$. Of course, you$_A$ might then turn out to be a Knightian combination of two further entities, you$_{A_1}$ and you$_{A_2}$, and so on.

For me, the appeal of this view is that it 'cancels two philosophical mysteries against each other': free will and indexical uncertainty. As I said in Section 12.1.1, for me free will seems to *require* some source of Knightian uncertainty in the physical world. Meanwhile, indexical uncertainty is a type of Knightian uncertainty that's been considered troublesome and unwanted – though attempts to replace it with probabilistic uncertainty have led to no end of apparent paradoxes, from the Doomsday argument to Bostrom's observer-counting paradoxes to the Boltzmann brain problem. So it seems natural to try to fit the free-will peg into the indexical-uncertainty hole.

12.9 Is the freebit picture falsifiable?

An obvious question about the freebit picture is whether it leads to any new, falsifiable predictions. At one level, I was careful not to 'commit' myself to any such predictions in this essay! My goal was to clarify some conceptual issues about the physical predictability of the brain. Whenever I ran up against an unanswered scientific question – for example, about the role of amplified quantum events in brain activity – I freely confessed my ignorance.

On the other hand, it's natural to ask: *are there empirical conditions that the universe has to satisfy in order for the freebit perspective to have even a* ***chance*** *of being related to 'free will' as most people understand the concept?*

I submit that the answer to the above question is yes. To start with the most straightforward predictions: first, it's necessary that psychology will never become physics. If human beings could be predicted as accurately as comets then the freebit picture would be falsified. For, in such a world, it would have *turned out* that, whatever it is we call 'free will' (or even 'the illusion of free will') in ordinary life, that property is not associated with any fundamental unpredictability of our choices.

It's also necessary that a quantum-gravitational description of the early universe will *not* reveal it to have a simply describable pure or mixed state. Or at least, it's necessary that indexical uncertainty – that is, uncertainty about our own location in the universe or multiverse – will forever prevent us from reducing arbitrary questions about our own future to well-posed mathematical problems. (Such a math problem might ask, for example, for the probability distribution $\mathscr{D}$ that results when the known evolution equations of physics are applied to the known initial state at the Big Bang, marginalized to the vicinity of the earth, and conditioned on some relevant subset of branches of the quantum-mechanical wavefunction in which we happen to find ourselves.)

However, the above 'predictions' have an unsatisfying, 'god-of-the-gaps,' character: they're simply predictions that certain scientific problems will never be completely solved! Can't we do better, and give *positive* predictions? Perhaps surprisingly, I think we can.

The first prediction was already discussed in Section 12.3.4. In order for the freebit picture to work, *it's necessary that quantum uncertainty – for example, in the opening and closing of sodium-ion channels – can not only get chaotically am-*

plified by brain activity but can do so 'surgically' and on 'reasonable' timescales. In other words, the elapsed time between (a) the amplification of a quantum event and (b) the neural firings influenced by that amplification must not be so long that the idea of a connection between the two retreats into absurdity. (Ten seconds would presumably be fine, whereas a year would presumably not be.) Closely related to that requirement, the quantum event must not affect countless *other* classical features of the world, separately from its effects on the brain activity. For if it did, then a prediction machine could in principle measure those other classical features to forecast the brain activity, with no need for potentially invasive measurements on either the original quantum state or the brain itself.

It's tempting to compare the empirical situation for the freebit picture to that for supersymmetry in physics. Both of these frameworks are very hard to falsify – since no matter what energy scale has been explored, or how far into the future neural events have been successfully predicted, a diehard proponent could always hold out for the superparticles or freebits making their effects known at the *next* higher energy or the next longer timescale. Yet, despite this property, supersymmetry and freebits are both 'falsifiable in degrees.' In other words, if the superparticles can be chased up to a sufficiently high energy, then *even if present, they would no longer do most of the work they were originally invented to do*. The same is true of freebits, if the time between amplification and decision is long enough.

Moreover, there's also a *second* empirical prediction of the freebit picture, one that doesn't involve the notion of a 'reasonable timescale.' Recall, from Section 12.3.3, the concept of a past macroscopic determinant (PMD): a set of 'classical' facts (for example, the configuration of a laser) about the causal past of a quantum state ρ that, if known, completely determine ρ. Now consider an omniscient demon, who wants to influence your decision-making process by changing the quantum state of a single photon impinging on your brain. Imagine that there are indeed photons that would serve the demon for this purpose. However, now imagine that *all such photons can be 'grounded' in PMDs*. That is, imagine that the photons' quantum states *cannot* be altered, so maintaining a spacetime history consistent with the laws of physics, without *also* altering classical degrees of freedom in the photons' causal past. In that case, the freebit picture would once again fail. For *if* a prediction machine had simply had the foresight to measure the PMDs then (by assumption) it could also calculate the quantum states ρ and therefore the probability of your reaching one decision rather than another. Indeed, not only could the machine probabilistically predict your actions, it could even provide a complete quantum-mechanical account of where the probabilities came from. Given such a machine, your choices would remain 'unpredictable' only in the sense that a radioactive atom is unpredictable, a sense that doesn't interest us (see Section 12.3). The conclusion is that, for the freebit picture to work, it's necessary that some of

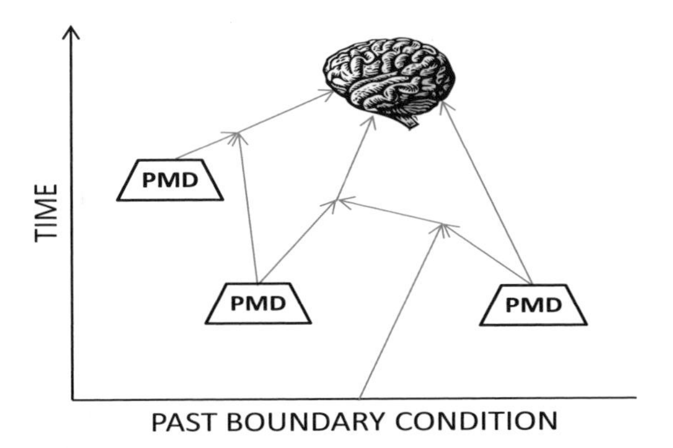

Figure 12.2 Tracing the causal antecedents of a human decision backwards in time, stopping *either* at past macroscopic determinants (PMDs) or at the initial boundary condition

the relevant quantum states *can't* be grounded in PMDs, but only traced back to the early universe (see Figure 12.2).

12.10 Conclusions

At one level, all I have done in this essay is to invoke what David Deutsch, in his 2011 book *The Beginning of Infinity* called the 'momentous dichotomy':

> *Either a given technology is possible, or else there must be some reason (say, of physics or logic) why it isn't possible.*

Granted, the above statement is a near tautology. But as Deutsch points out, the implications of applying the tautology consistently can be enormous. One illustrative application, *slightly* less contentious than free will, involves my own field, quantum computing. The idea there is to apply the principles of quantum mechanics to build a new type of computer, one that could solve certain problems (such as factoring integers) exponentially faster than we know how to solve them with any existing computer. Quantum computing research is being avidly pursued around the world, but there are also a few vocal skeptics of such research. Many of (though not all) the skeptics appear to subscribe to the following three positions.

(a) A useful quantum computer is an almost self-evident absurdity: noise, decoherence, and so forth must conspire to prevent such a machine from working, just as the laws of physics always conspire to prevent perpetual-motion machines from working.
(b) No addition or revision is needed to quantum mechanics: the physical framework that underlies quantum computing and that describes the state of an iso-

lated physical system as a vector in an exponentially large Hilbert space (leading to an apparent exponential *slowdown* when simulating quantum mechanics with a conventional computer) is sufficient.

(c) Reconciling (a) and (b) is not even a particularly interesting problem. At any rate, the burden is on the quantum computing *proponents* to sort these matters out! Skeptics can be satisfied that *something* must prevent quantum computers from working, and leave it at that.

What Deutsch's dichotomy suggests is that such blasé incuriosity, in the face of a glaring conflict between ideas, is *itself* the absurd position. Such a position only pretends to be the 'conservative' one: secretly it is radical, in that it rejects the whole idea that science advances by identifying apparent conflicts and then trying to resolve them.

In this essay, I applied Deutsch's momentous dichotomy to a different question:

> Could there exist a machine, consistent with the laws of physics, that 'non-invasively clones' all the information in a particular human brain that was relevant to behavior – so that the human could emerge from the machine unharmed but would thereafter be fully probabilistically predictable given his or her future sense-inputs, in much the same sense that a radioactive atom is probabilistically predictable?

My central thesis is simply that *there is no 'safe, conservative answer' to this question*. Of course, one can debate what exactly the question means and how we would know whether the supposed cloning machine had succeeded. (See Appendix 12A for my best attempt at formalizing the requirements.) But I contend that philosophical analysis can only take us so far. The question also has an 'empirical core' that could turn out one way or another, depending on details of the brain's physical organization that are not yet known. In particular, does the brain possess what one could call a *clean digital abstraction layer*: that is, a set of macroscopic degrees of freedom that

(1) encode everything relevant to memory and cognition,
(2) can be accurately modeled as performing a classical digital computation, and
(3) 'notice' the microscopic, quantum-mechanical, degrees of freedom at most as pure random-number sources, generating noise according to prescribed probability distributions?

Or, is such a clean separation between the macroscopic and microscopic levels unavailable – so that any attempt to clone a brain would either miss much of the cognitively relevant information or else violate the no-cloning theorem?

In my opinion, *neither* answer to the question should make us wholly comfortable: if it does, then we haven't sufficiently thought through the implications! Suppose, on the one hand, that the brain-cloning device is possible. Then we immediately confront countless 'paradoxes of personal identity' like those discussed in Section 12.2.5. Would you feel comfortable being (painlessly) killed, provided that a perfect clone of your brain's digital abstraction layer remained? Would it matter if the cloned data was moved to a new biological body, or was only 'simulated' electronically? If the latter, what would count as an acceptable simulation? Would it matter if the simulation was run backwards in time, or in heavily-encrypted form, or in only one branch of a quantum computation? Would you literally expect to 'wake up' as your clone? What if two clones were created: would you expect to wake up as each one with 50% probability? When applying Bayesian reasoning, should you, all other things equal, judge yourself twice as likely to wake up in a possible world with twice as many clones of yourself? The point is that, in a world with the cloning device, these would no longer be metaphysical conundrums, but in some sense, just *straightforward empirical questions* about what you should expect to observe! Yet even people who agreed on every possible 'third-person' fact about the physical world and its contents might answer such questions completely differently.

To be clear: it seems legitimate to me, given current knowledge, to conjecture that there is no principled obstruction to perfect brain-cloning and indeed that this essay was misguided even to speculate about such an obstruction. However, *if* one thinks that by taking the 'pro-cloning' route one can sidestep the need for any 'weird metaphysical commitments' of one's own, then I'd say one is mistaken.

So suppose, on the other hand, that the perfect brain-cloning device is *not* possible. Here, exactly like in the case of quantum computing, no truly inquisitive person will ever be satisfied by a bare *assertion* of the device's impossibility, or by a listing of practical difficulties. Instead, such a person will demand to know: what *principle* explains why perfect brain-cloning can't succeed, not even a million years from now? How do we reconcile its impossibility with everything we know about the mechanistic nature of the brain? Indeed, how should we think about the laws of physics *in general*, so that the impossibility of perfect brain-cloning would no longer seem surprising or inexplicable?

As soon as we try to answer these questions, I've argued that we're driven, more or less inevitably, to the view that the brain's detailed evolution would have to be buffeted around by chaotically amplified 'Knightian surprises,' which I called freebits. Before their amplification, these freebits would need to live in quantum-mechanical degrees of freedom, since otherwise a cloning machine could (in principle) non-invasively copy them. Furthermore, our ignorance about the freebits would ultimately need to be traceable back to ignorance about the microstate of the early universe – again because otherwise, cloning would become possible in principle, through vigilant monitoring of the freebits' sources.

Admittedly, all this sounds like a tall order! But, strange though it sounds, I don't see that any of it is ruled out by current scientific understanding – though conceivably it *could* be ruled out in the future. In any case, setting aside one's personal beliefs, it seems worthwhile to understand that *this* is the picture one seems forced to if one starts from the hypothesis that brain-cloning (with all its metaphysical difficulties) should be fundamentally impossible and then tries to make that hypothesis compatible with our knowledge of the laws of physics.

12.10.1 Reason and mysticism

At this point, I imagine some readers might press me: but what do I *really* think? Do I actually take seriously the notion of quantum-mechanical 'freebits' from the early universe playing a role in human decisions? The easiest response is that, in laying out my understanding of the various alternatives – yes, brain states might be perfectly clonable but if we want to avoid the philosophical weirdness that such cloning would entail, then we're led in such-and-such an interesting direction, etc. – I have *already said* what I really think. In pondering these riddles, I don't have any sort of special intuition for which the actual arguments and counterarguments that I can articulate serve as window-dressing. The arguments exhaust my intuition.

I'll observe, however, that even uncontroversial facts can be made to sound incredible when some of their consequences are spelled out; indeed, the spelling out of such consequences has always been a mainstay of popular science writing. To give some common examples: everything you can see in the night sky was once compressed into a space smaller than an atom. The entire story of your life, including the details of your death, is almost certainly encoded (in fact, infinitely many times) somewhere in the decimal expansion of π. If Alois Hitler and Klara Pölzl had moved an inch differently while having intercourse, World War II would probably have been prevented. When you lift a heavy bag of groceries, what you're really feeling is the coupling of gravitons to a stress–energy tensor generated mostly by the rapid movements of gluons. In each of these cases the same point *could* be made more prosaically, and many would prefer that it was. But when we state things vividly, at least we can't be accused of trying to *hide* the full implications of our abstract claims for fear of being laughed at were those implications understood.

Thus, suppose I'd merely argued, in this essay, that it's possible that humans will never become as predictable (even probabilistically predictable) as digital computers, because of chaotic amplification of unknowable microscopic events, our ignorance of which can be traced as far backward in time as one wishes. In that case, some people would likely agree and others would likely disagree, with many averring (as I do) that the question remains open. Hardly anyone, I think, would consider the speculation an absurdity or a gross affront to reason. But if exactly the same idea is phrased in terms of a 'quantum pixie-dust' left over from the Big Bang, which gets into our brains and gives us the capacity for free will – well then, *of course* it sounds crazy! Yet the second phrasing is nothing more than a dramatic rendering of the worldview that the first phrasing implies.

Perhaps some readers will accuse me of mysticism. To this, I can only reply that the view I flirt with in this essay feels like 'mysticism' of an unusually tame sort: one that embraces the mechanistic and universal nature of the laws of physics, as they've played out for 13.7 billion years; that can accept even the 'collapse of the wavefunction' as an effect brought about by ordinary unitary evolution; that's consumed by doubts; that welcomes corrections and improvement; that declines to plumb the cosmos for self-help tips or moral strictures about our sex lives; and that sees science – not introspection, not ancient texts, not 'science' redefined to mean something different, but *just science in the ordinary sense* – as our best hope for making progress on the ancient questions of existence.

To any 'mystical' readers who want human beings to be as free as possible from the mechanistic chains of cause-and-effect, I say: *this picture represents the absolute maximum that I can see how to offer you, if I confine myself to speculations that I can imagine making contact with our current scientific understanding of the world.* Perhaps it's less than you want; on the other hand, it does seem like more than the usual compatibilist account offers! To any 'rationalist' readers, who cheer when consciousness, free will, or similarly woolly notions get steamrolled by the advance of science, I say: you can feel vindicated, if you like, that despite searching (almost literally) to the ends of the universe, I wasn't able to offer the 'mystics' anything more than I was offering! And even what I *do* offer might be ruled out by future discoveries.

Indeed, the freebit picture's falsifiability is perhaps the single most important point about it. Consider the following questions. On what timescales can microscopic fluctuations in biological systems get amplified, and change the probabilities of macroscopic outcomes? What other side effects do those fluctuations have? Is the brain interestingly different in its noise sensitivity from the weather, or other complicated dynamical systems? Can the brain's microscopic fluctuations be fully understood probabilistically or are they subject to Knightian uncertainty? That is, can the fluctuations all be grounded in past macroscopic determinants or are some ultimately cosmological in origin? Can we have a complete theory of cosmological initial conditions? Few things would make me happier than if progress on these questions led to the discovery that the freebit picture was wrong. For then at least we would have learned something.

Acknowledgments. I thank Yakir Aharonov, David Albert, Julian Barbour, Silas Barta, Alex Byrne, Sean Carroll, David Chalmers, Alessandro Chiesa, Andy

Drucker, Owain Evans, Andrew Hodges, Sabine Hossenfelder, Guy Kindler, Seth Lloyd, John Preskill, Huw Price, Haim Sompolinsky, Cristi Stoica, Jaan Tallinn, David Wallace, and others I've no doubt forgotten, for helpful discussions about the subject of this essay; and especially Dana Moshkovitz Aaronson, David Aaronson, Steve Aaronson, Jacopo Tagliabue, and Ronald de Wolf for their comments on drafts. Above all, I thank S. Barry Cooper and Andrew Hodges for 'commissioning' this essay and for their near-infinite patience in humoring my delays with it. It goes without saying that none of the people mentioned necessarily endorse anything I say here (indeed, some of them definitely don't!).

12A Appendix: Defining 'freedom'

In this appendix I'll use the notion of Knightian uncertainty (see Section 12.3) to offer a possible mathematical formalization of 'freedom' for use in free-will discussions. Two caveats are immediately in order. The first is that my formalization only tries to capture what I've called 'Knightian freedom' – a strong sort of in-principle physical unpredictability – and not 'metaphysical free will.' For as discussed in Section 12.1.1, I don't see how *any* definition grounded in the physical universe could possibly capture the latter, to either the believers' or the deniers' satisfaction. Also, as we'll see, formalizing 'Knightian freedom' is *already* a formidable challenge!

The second caveat is that, by necessity, my definition will be in terms of more 'basic' concepts which I'll need to assume as unanalyzed primitives. Foremost among these is the concept of a *physical system*: something that occupies space, exchanges information (and matter, energy, etc.) with other physical systems, and crucially, retains an identity through time even as its internal state changes. Examples of physical systems are black holes, the earth's atmosphere, human bodies, and digital computers. Without some concept like this, it seems to me that we can never specify *whose* freedom we're talking about or even which physical events we're trying to predict.

Yet, as philosophers know well, the concept of a 'physical system' already has plenty of traps for the unwary. As one illustration, should we say that a human body remains 'the same physical system' after its death? If so, then an extremely reliable method to predict a human subject immediately suggests itself: namely, first shoot the subject; then predict that the subject will continue to lie on the ground doing nothing!

Now, it might be objected that this 'prediction method' shouldn't count, since it *changes the subject's state* (to put it mildly), rather than just passively gathering information about the subject. The trouble with that response is that putting the subject in an fMRI machine, interviewing her, or even just having her sign a consent form or walking past her on the street also changes her state! If we don't allow *any* interventions that change the subject's state from what it 'would have been otherwise,' then prediction – at least with the science-fiction accuracy we're imagining – seems hopeless, but in an uninteresting way. So which interventions are allowed and which aren't?

I see no alternative but to take the *set of allowed interventions* as another unanalyzed primitive. When formulating the prediction task (and hence, in this essay, when defining Knightian freedom), we simply declare that certain interventions – such as interviewing the subject, putting her in an fMRI machine, or perhaps even having nanorobots scan her brain state – are allowed; while other interventions, such as killing her, are not allowed.

An important boundary case, much discussed by philosophers, is an intervention that would destroy each neuron of the subject's brain one by one, replacing the neurons by microchips claimed to be functionally equivalent. Is *that* allowed? Note that such an operation could certainly make it easier to predict the subject – since, from that point forward, the predictor would only have to worry about simulating the microchips, not the messy biological details of the original brain. Here I'll just observe that, if we like, we can disallow such drastic interventions without thereby taking any position on the conundrum of what such 'siliconization' would do to the subject's conscious experience. Instead we can simply say that, while the subject might indeed be perfectly predictable after the operation, that fact *doesn't settle the question at hand*, which was about the subject's predictability *before* the operation. For a large part of what we wanted to know was to what extent the messy biological details *do* matter, and we can't answer that question by defining it out of consideration.

But one might object: if the 'messy biological details' need to be left in place when trying to predict a brain, what *doesn't* need to be left in place? After all, brains are not isolated physical systems: they constantly receive inputs from the sense organs, from hormones in the bloodstream, etc. So when modeling a brain, do we also need to model the entire environment in which that brain is immersed – or at least, all aspects of the environment that might conceivably affect behavior? If so, then prediction seems hopeless, but again, not for any 'interesting' reasons: merely for the boring reason that we can't possibly measure *all* relevant aspects of the subject's environment, being embedded in the environment ourselves.

Fortunately, I think there's a way around this difficulty, at the cost of one more unanalyzed primitive. Given a physical system S, denote by $I(S)$ the set of *screenable inputs* to S – by which I mean, the inputs to S with which we judge that any would-be predictor of S should also be provided, in order to ensure a 'fair contest.' For example, if S is a human brain then on the one hand $I(S)$ would probably include (finite-precision digital encodings of) the signals entering the brain through the optic, auditory, and other sensory systems, the levels of various hormones in the blood, and other measurable variables at the interface between the brain and its external environment. On the other hand, $I(S)$ probably *wouldn't* include, for example, the exact quantum state of every photon impinging on the brain. For, arguably, we have no idea how to screen off all those microscopic 'inputs,' short of siliconization or some equally drastic intervention.

Next, call a system S *input-monitorable* if there exists an allowed intervention to S, the result of which is that, after the intervention, all signals in $I(S)$ get 'carbon-copied' to the predictor's computer at the same time as they enter S. For example, using some future technology, a brain might be input-monitored by installing microchips that scan all the electrical impulses in the optic and auditory nerves, the chemical concentrations in the blood–brain barrier, etc., and that faithfully transmit that information to a predictor in the next room via a wireless link. Crucially, and in contrast with siliconization, input-monitoring doesn't strike me as raising any profound issues of consciousness or selfhood. That is, it seems fairly clear that an input-monitored human would still be 'the same human,' just hooked up to some funny devices! Input monitoring also differs from siliconization in that it seems much closer to practical realization.

The definition of freedom that I'll suggest will only make sense for input-monitorable physical systems. If S is not input-monitorable, then I'll simply hold that the problem of 'predicting S's behavior' isn't well enough defined: S is so intertwined with its environment that one can't say where predicting S ends and predicting its environment begins. One consequence is that, in this framework, we can't even *pose the question* of whether humans have Knightian freedom unless we agree (at least provisionally) that humans are input-monitorable. Fortunately, as already suggested, I don't see any major scientific obstacles to supposing that humans *are* input-monitorable, and I even think input-monitoring could plausibly be achieved in 50 or 100 years.

Admittedly, in discussing whether humans are input-monitorable, a lot depends on our choice of screenable inputs $I(S)$. If $I(S)$ is small then input-monitoring S might be easy but predicting S after the monitoring is in place might be hard or impossible, simply because of the predictor's ignorance about crucial features of S's environment. By contrast, if $I(S)$ is large then input-monitoring S might be hard or impossible, but supposing S were input-monitored, predicting S might be easy. Since our main interest is the inherent difficulty of prediction, our preference should always be for the largest $I(S)$ possible.

So, suppose S is input-monitorable and suppose we've arranged things so that all the screenable inputs to S – what S sees, what S hears, etc. – are transmitted in real time to the predictor. We then face the question: what aspects of S are we trying to predict, and what does it mean to predict those aspects?

Our next primitive concept will be that of *S*'s *observable behaviors*. For the earth's atmosphere, observable behaviors might include snow and thunderstorms; for a human brain, they might include the signals sent out by the motor cortex, or even just a high-level description of which words will be spoken and which decisions taken. Fortunately, it seems to me that, for any sufficiently complex system, the prediction problem is *not* terribly sensitive to which observable behaviors we focus on, provided those behaviors belong to a large 'universality class.' By analogy, in computability theory it doesn't matter whether we ask whether a given computer program will ever halt, or whether the program will ever return to its initial state, or whether it will ever print YES to the console, or some other question about the program's future behavior. For, these problems are all *reducible* to each other: if we had a reliable method to predict whether a program would halt, then we could also predict whether the program would print YES to the console, by modifying the program so that it prints YES if and only if it halts. In the same way, if we had a reliable method to predict a subject's hand movements in arbitrary situations, I claim that we could also predict the subject's speech. For 'arbitrary situations' include those where we direct the subject to translate everything she says into sign language! And thus, assuming we've built a 'hand-prediction algorithm' that works in those situations, we must also have built (or had the ability to build) a speech-prediction algorithm as well.

So suppose we fix some set *B* of observable behaviors of *S*. What should count as *predicting* the behaviors? From the outset, we should admit that *S*'s behavior might be inherently probabilistic – as it would, for example, if it depended on amplified quantum events taking place inside *S*. So we should be satisfied if we can predict *S* in 'merely' the same sense that physicists can predict a radioactive atom: namely, by giving a probability distribution over *S*'s possible future behaviors.

Here difficulties arise, which are well known in the fields of finance and weather forecasting. How exactly do we test predictions that take the form of probability distributions, if the predictions apply to events that might not be repeatable? Also, what's to prevent someone from declaring success on the basis of absurdly conservative 'predictions': for example, 50/50 for every yes/no question? Briefly, I'd say that *if* the predictor's forecasts take the form of probability distributions then to whatever extent those forecasts are unimpressive (50/50), *the burden is on the predictor* to convince skeptics that the forecasts nevertheless encoded everything that *could* be predicted about *S* via allowed interventions. That is, the predictor needs to rule out the hypothesis that the probabilities merely reflected ignorance about unmeasured but measurable variables. In my view, this would ultimately require the predictor to give a causal account of *S*'s behavior, which showed explicitly how the observed outcome depended on quantum events – the only sort of events that we know to be probabilistic on physical grounds (see Section 12.2.7).

But it's not enough to let the predictions be probabilistic; we need to scale back our ambitions still further. For even with a system S as simple as a radioactive atom, there's no hope of calculating the *exact* probability that (say) S will decay within a certain time interval – if only because of the error bars in the physical constants that enter into that probability. But, intuitively, this lack of precision doesn't make the atom any less 'mechanistic.' Instead, it seems to me that we should call a probabilistic system S mechanistic *if – and only if – the differences between our predicted probabilities for S's behavior and the 'true' probabilities can be made as small as desired by repeated experiments.*

Yet we are still not done. For what does the predictor P *already know* about S before P's data-gathering process even starts? If P were initialized with a 'magical copy' of S, then of course predicting S would be trivial.[54] However, it also seems unreasonable not to tell P *anything* about S: for example, if P could accurately predict S, but only when given the hint that S is a human being, that would still be rather impressive and intuitively incompatible with S's 'freedom.' So, as our final primitive concept, we assume a *reference class* C of possible physical systems; P is then told only that $S \in C$ and needs to succeed under that assumption. For example, C might be 'the class of all members of the species *Homo sapiens*,' or even the class of all systems macroscopically identifiable as some *particular Homo sapien*.[55]

This, finally, leads to my attempted definition of freedom. Before offering it, let me stress that *nothing in the essay depends much on the details of the definition – and indeed, I'm more than willing to tinker with those details.* So then what's the point of *giving* a definition? One reason is to convince skeptics that the concept of 'Knightian freedom' *can* be made precise, once one has a suitable framework within which to discuss these issues. A second reason is to illustrate just how much little-examined complexity lurks in the commonsense notion of a physical system's being 'predictable' – and to show how non-obvious the questions of freedom and predictability actually become, once we start to unravel that complexity.

Let S be any input-monitorable physical system drawn from the reference class C. Suppose that, as the result of allowed interventions, S is input-monitored by another physical system $P = P(C)$ (the 'predictor'), starting at some time 0.[56] Given times $0 \leq t < u \leq \infty$, let $I_{t,u}$ encode all the information in the screenable inputs that S receives between times t and u, with $I_{t,\infty}$ denoting the information received from time t onwards. Likewise, let $B_{t,u}$ encode all the information in S's observable behaviors

[54] I thank Ronald de Wolf for this observation.

[55] We could also formulate a stronger notion of a 'universal predictor,' which has to work for *any* physical system S (or equivalently, whose reference class C is the set of all physical systems). My own guess is that, *if* there exists a predictor for 'sufficiently complex' systems such as human brains, then there also exists a universal predictor. But I won't attempt to argue for that here.

[56] Here we don't presuppose that time is absolute or continuous. Indeed, all we need is that S passes through a discrete series of 'instants,' which can be ordered by increasing values of t.

between times t and u. (While this is not essential, we can assume that $I_{t,u}$ and $B_{t,u}$ both consist of finite sequences of bits whenever $u < \infty$.)

Let $\mathscr{D}(B_{t,u}|I_{t,u})$ be the 'true' probability distribution[57] over $B_{t,u}$ conditional on the inputs $I_{t,u}$, where 'true' means the distribution that would be predicted by a godlike intelligence who knew the exact physical state of S and its external environment at time t. We assume that $\mathscr{D}(B_{t,u}|I_{t,u})$ satisfies the 'causal property': that $\mathscr{D}(B_{t,v}|I_{t,u}) = \mathscr{D}(B_{t,v}|I_{t,v})$ depends only on $I_{t,v}$ for all $v < u$.

Suppose that, from time 0 to t, the predictor P has been monitoring the screenable inputs $I_{0,t}$ and observable behaviors $B_{0,t}$ and, more generally, interacting with S however it wants via allowed interventions (for example, submitting questions to S by manipulating $I_{0,t}$, and observing the responses in $B_{0,t}$). Then, at time t, we ask P to output a description of a function f which maps the future inputs $I_{t,\infty}$ to a distribution $\mathscr{E}(B_{t,\infty}|I_{t,\infty})$ satisfying the causal property. Here $\mathscr{E}(B_{t,\infty}|I_{t,\infty})$ represents P's best estimate for the distribution $\mathscr{D}(B_{t,\infty}|I_{t,\infty})$. Note that the description of f might be difficult to 'unpack' computationally – for example, it might consist of a complicated algorithm that outputs a description of $\mathscr{E}(B_{t,u}|I_{t,u})$ given as input $u \in (t,\infty)$ and $I_{t,u}$. All we require is that the description be *information-theoretically* complete, in the sense that one *could* extract $\mathscr{E}(B_{t,u}|I_{t,u})$ from it given enough computation time.

Given $\varepsilon, \delta > 0$, we call P a *(t,ε,δ)-predictor* for the reference class C if the following holds. For all $S \in C$, with probability at least $1-\delta$ over any 'random' inputs in $I_{t,u}$ (controlled neither by S nor by P), we have

$$\|\mathscr{E}(B_{t,\infty}|I_{t,\infty}) - \mathscr{D}(B_{t,\infty}|I_{t,\infty})\| < \varepsilon$$

for the actual future inputs $I_{t,\infty}$ (not necessarily for every *possible* $I_{t,\infty}$). Here $\|\cdot\|$ denotes the variation distance.[58]

We call C *mechanistic* if for all $\varepsilon, \delta > 0$, there exists a $t = t_{\varepsilon,\delta}$ and a $P = P_{\varepsilon,\delta}$ such that P is a (t,ε,δ)-predictor for C. We call C *free* if C is not mechanistic.

Two important sanity checks are the following:

(a) According to the above definition, classes C of physical systems such as thermostats, digital computers, and radioactive nuclei are indeed mechanistic (given reasonable sets of screenable inputs, allowed interventions, and observable behaviors). For example, suppose that C is the set of all possible configurations of a particular digital computer; the allowed interventions include reading the entire contents of the disk drives and memory and 'eavesdropping' on all the input ports (all of which is known to be technologically doable without destroying the computer); and the observable behaviors include everything sent to the output ports. In that case, even with no further interaction the predictor can clearly emulate the computer arbitrarily far into the future. Indeed,

[57] Or probability measure over infinite sequences in the case $u = \infty$.

[58] Variation distance is a standard measure of the distance between two probability distributions and is defined by $\|\{p_x\} - \{q_x\}\| := \frac{1}{2}\sum_x |p_x - q_x|$.

even if the computer $S \in C$ has an internal quantum-mechanical random number generator, the probability distribution $\mathscr{D}$ over *possible* future behaviors can still be approximated extremely well.

(b) On the other hand, at least mathematically, one can construct classes of systems C that are free. Indeed, this is trivial to arrange, simply by restricting the screenable inputs so that S's future behavior is determined by some input stream to which P does not have access.

Like many involved definitions in theoretical computer science, cryptography, economics, and other areas, my definition of freedom is 'merely' an attempt to approximate an informal concept that one had prior to formalization. And, indeed, there are many changes to the definition one could contemplate. To give just a few examples, instead of requiring that $\mathscr{E}(B_{t,\infty}|I_{t,\infty})$ approximate $\mathscr{D}(B_{t,\infty}|I_{t,\infty})$ only for the *actual* future inputs $I_{t,\infty}$, one could demand that it do so for all *possible* $I_{t,\infty}$. Or one could assume a distribution over the future $I_{t,\infty}$, and require success on *most* of them. Or one could require success only for most $S \in C$, again assuming a distribution over S's. Or one could switch around the quantifiers, e.g., requiring a single predictor P that achieves greater and greater prediction accuracy $\varepsilon > 0$ the longer it continues. Or one could drop the requirement that P forecast all of $B_{t,\infty}$ requiring only that it forecast $B_{t,u}$ for some large but finite u. It would be extremely interesting to develop the mathematical theory of these different sorts of prediction – something I reluctantly leave to future work. Wouldn't it be priceless if, after millennia of debate, the resolution of the question "are humans free?" turned out to be "yes if you define 'free' with the ε, δ quantifiers inside, but no if you put the quantifiers outside?"

A central limitation of the definition, as it stands, is that it's qualitative rather than quantitative, closer in spirit to computability theory than complexity theory. More concretely, the definition of 'mechanistic' requires only that there *exist* a finite time t after which the predictor succeeds; it puts no limit on the amount of time. But this raises a problem: what if the predictor could succeed in learning to emulate a human subject, but only after observing the subject's behavior for (say) 10^{100} years? Does making the subject immortal, in order to give the predictor enough time, belong to the set of allowed interventions? Likewise, suppose that, after observing the subject's behavior for 20 years, the predictor becomes able to predict the subject's future behavior probabilistically, but *only for the next* 20 *years*, not indefinitely? The definition doesn't consider this sort of 'time-limited' prediction, even though intuitively, it seems almost as hard to reconcile with free will as the unlimited kind. But the actual numbers do matter: a predictor that needed 20 years of data-gathering, in order to learn enough to predict the subject's behavior for the 5 seconds immediately afterward, would seem intuitively compatible with freedom. In any case, in this essay I have mostly ignored quantitative timing issues (except for brief discussions in Sections 12.2.11 and 12.9), and have imagined for simplicity that we have a predictor that after some finite time learns to predict the subject's responses arbitrarily far into the future.

12B Appendix: Prediction and Kolmogorov complexity

As mentioned in Section 12.3.1, some readers will take issue with the entire concept of Knightian uncertainty – that is, with uncertainty that can't even be properly quantified using probabilities. Among those readers, some might be content to assert that there exists a 'true, objective' prior probability distribution $\mathscr{D}$ over all events in the physical world – and while we might not know any prescription to *calculate* $\mathscr{D}$ that different agents can agree on, we can be sure that agents are irrational to whatever extent their own priors deviate from $\mathscr{D}$. However, more sophisticated readers might try to *derive* the existence of a roughly universal prior, using ideas from *algorithmic information theory* (see Li and Vitányi (2008) for an excellent introduction). In this appendix, I'd like to sketch how the latter argument would go and offer a response to it.

Consider an infinite sequence of bits $b_1, b_2, b_3, \ldots$, which might be generated randomly, or by some hidden computable pattern, or by some process with elements of both. (For example, maybe the bits are uniformly random, except that every hundredth bit is the 'majority vote' of the previous 99 bits.) We can imagine, if we like, that these bits represent a sequence of yes-or-no decisions made by a human being. For each $n \geq 1$, a superintelligent predictor is given $b_1, \ldots, b_{n-1}$, and asked to predict b_n. Then the idea of algorithmic statistics is to give a *single* rule, which can be proved to predict b_n 'almost as well' as any other computable rule, in the limit $n \to \infty$.

Here's how it works. Choose any Turing-universal programming language L, which satisfies the technical condition of being *prefix-free*: that is, adding characters to the end of a valid program never yields another valid program. Let P be a program written in L, which runs for an infinite time and has access to an unlimited supply of random bits and which generates an infinite sequence $B = (b_1, b_2, \ldots)$ according to some probability distribution $\mathscr{D}_P$. Let $|P|$ be the number of bits in P. The for its initial guess as to the behavior of B, our superintelligent predictor will use the so-called *universal prior* $\mathscr{U}$, in which each distribution $\mathscr{D}_P$ appears with probability $2^{-|P|}/C$ for some normalizing constant $C = \sum_P 2^{-|P|} \leq 1$. (The reason for the prefix-free condition was to ensure that the sum $\sum_P 2^{-|P|}$ converges.) Then, as the bits $b_1, b_2, \ldots$ start appearing, the predictor repeatedly updates $\mathscr{U}$ using Bayes' rule, so that its estimate for $\Pr[b_n = 1]$ is always

$$\frac{\Pr_{\mathscr{U}}[b_1 \ldots b_{n-1} 1]}{\Pr_{\mathscr{U}}[b_1 \ldots b_{n-1}]}.$$

Now suppose that the 'true' distribution over B is $\mathscr{D}_Q$, for some particular program Q. Then I claim that, in the limit $n \to \infty$, a predictor that starts with $\mathscr{U}$ as its prior will do just as well as if it had started with $\mathscr{D}_Q$. The proof is simple: by definition, $\mathscr{U}$ places a 'constant fraction' of its probability mass on $\mathscr{D}_Q$ from the beginning (where the 'constant,' $2^{-|Q|}/C$, admittedly depends on $|Q|$). So, for all n and $b_1 \ldots b_n$,

$$\frac{\Pr_{\mathscr{U}}[b_1 \ldots b_n]}{\Pr_{\mathscr{D}_Q}[b_1 \ldots b_n]} = \frac{\Pr_{\mathscr{U}}[b_1] \Pr_{\mathscr{U}}[b_2|b_1] \cdots \Pr_{\mathscr{U}}[b_n|b_1 \ldots b_{n-1}]}{\Pr_{\mathscr{D}_Q}[b_1] \Pr_{\mathscr{D}_Q}[b_2|b_1] \cdots \Pr_{\mathscr{D}_Q}[b_n|b_1 \ldots b_{n-1}]} \geq 2^{-|Q|}.$$

Hence

$$\prod_{n=1}^{\infty} \frac{\Pr_{\mathscr{U}}[b_n|b_1 \ldots b_{n-1}]}{\Pr_{\mathscr{D}_Q}[b_n|b_1 \ldots b_{n-1}]} \geq 2^{-|Q|}$$

as well. But for all $\varepsilon > 0$, this means that $\mathscr{U}$ can assign a probability to the correct value of b_n less than $1 - \varepsilon$ times the probability assigned by $\mathscr{D}_Q$, only for $O(|Q|/\varepsilon)$ values of n or fewer.

Thus, an 'algorithmic Bayesian' might argue that there are only two possibilities: either a physical system is predictable by the universal prior $\mathscr{U}$, or else – to whatever extent it isn't – in any meaningful sense the system behaves randomly. There's no third possibility that we could identify with Knightian uncertainty or freebits.

One response to this argument – perhaps the response Penrose would prefer – would be that we can easily defeat the so-called 'universal predictor' $\mathscr{U}$, using a sequence of bits $b_1, b_2, \ldots$ that's deterministic but noncomputable. One way to construct such a sequence is to 'diagonalize against $\mathscr{U}$,' defining

$$b_n := \begin{cases} 0 & \text{if } \Pr_{\mathscr{U}}[b_1 \ldots b_{n-1} 1] > \Pr_{\mathscr{U}}[b_1 \ldots b_{n-1} 0], \\ 1 & \text{otherwise} \end{cases}$$

for all $n \geq 1$. Alternatively, we could let b_n be the nth binary digit of Chaitin's constant Ω (Chaitin, 1975) (basically, the probability that a randomly generated computer program halts).[59] In either case, $\mathscr{U}$ will falsely judge the b_n's to be

[59] For completeness, let me prove that the universal predictor $\mathscr{U}$ fails to predict the digits of $\Omega = 0.b_1 b_2 b_3 \ldots$ Recall that Ω is *algorithmically random*, in the sense that for all n, the shortest program to generate $b_1 \ldots b_n$ has length $n - O(1)$. Now, suppose by contradiction that $\Pr_{\mathscr{U}}[b_1 \ldots b_n] \geq L/2^n$, where $L \gg n$. Let A_n be a program that dovetails over all programs Q, in order to generate better and better lower bounds on $\Pr_{\mathscr{U}}[x]$ for all n-bit strings x (converging to the correct probabilities in the infinite limit). Then we can specify $b_1 \ldots b_n$ by saying: "when A_n is run, $b_1 \ldots b_n$ is the jth string $x \in \{0,1\}^n$ such that A_n's lower bound on $\Pr_{\mathscr{U}}[x]$ exceeds $L/2^n$." Since there are at most $2^n/L$ such strings x, this description requires at most $n - \log_2 L + \log_2 n + O(1)$ bits. Furthermore, it clearly gives us a procedure to generate $b_1 \ldots b_n$. But if $L \gg n$, then this contradicts the

random even in the limit $n \to \infty$. Note that an even more powerful predictor $\mathscr{U}'$, equipped with a suitable oracle, could predict either sequence perfectly. But then we could construct new sequences $b'_1, b'_2, \ldots$ that are unpredictable even by $\mathscr{U}'$, and so on.

The response above is closely related to a notion called *sophistication* from algorithmic information theory (see Antunes and Fortnow, 2009; Gács et al., 2001; Li and Vitányi, 2008). Given a binary string x, recall that the *Kolmogorov complexity* $K(x)$ is the length of the shortest program, in some Turing-universal programming language, whose output (given a blank input) is x. To illustrate, the Kolmogorov complexity of the first n bits of $\pi \approx 11.00100100\ldots_2$ is small ($\log_2 n + O(\log\log n)$), since one only has to provide n (which takes $\log_2 n$ bits), together with a program for computing π to a given accuracy (which takes some small fixed number of bits, independent of n). By contrast, if x is an n-bit string chosen uniformly at random then $K(x) \approx n$ with overwhelming probability, simply by a counting argument. Now, based on those two examples, it's tempting to conjecture that *every string is either highly patterned or random*: that is, either

(i) $K(x)$ is small, or else
(ii) $K(x)$ is large, but only because of 'boring, random, patternless entropy' in x.

Yet the above conjecture, when suitably formalized, turns out to be false. Given a set of strings $S \subseteq \{0,1\}^n$, let $K(S)$ be the length of the shortest program that lists the elements of S. Then given an n-bit string x and a small parameter c, one can define the *c-sophistication* of x, or $\mathrm{Soph}_c(x)$, to be the minimum of $K(S)$, over all sets $S \subseteq \{0,1\}^n$ such that $x \in S$ and

$$K(S) + \log_2 |S| \leq K(x) + c.$$

Intuitively, the sophistication of x is telling us, in a near-minimal program for x, how many bits of the program need to be 'interesting code' rather than algorithmically random data.' Certainly $\mathrm{Soph}_c(x)$ is well-defined and at most $K(x)$, since we can always just take S to be the singleton set $\{x\}$. Because of this, highly patterned strings are unsophisticated. However, random strings are *also* unsophisticated since, for them, we can take S to be the entire set $\{0,1\}^n$. Nevertheless, it's possible to prove (Antunes and Fortnow, 2009; Gács et al., 2001) that there exist highly sophisticated strings: indeed, strings x such that $\mathrm{Soph}_c(x) \geq n - O(\log n)$. These strings could thus be said to inhabit a third category between 'patterned' and

fact that $b_1 \ldots b_n$ has description length $n - O(1)$. Therefore

$$\Pr_{\mathscr{U}}[b_1 \ldots b_n] = \prod_{i=1}^{n} \Pr_{\mathscr{U}}[b_i | b_1 \ldots b_{i-1}] = O\left(\frac{n}{2^n}\right),$$

and $\mathscr{U}$ hardly does better than mere chance.

'random.' Not surprisingly, the construction of sophisticated strings makes essential use of uncomputable processes.

However, for reasons explained in Section 12.6, I'm exceedingly reluctant to postulate uncomputable powers in the laws of physics (such as an ability to generate the digits of Ω). Instead, I would say that, if there's scope for freedom then it lies in the fact that, even when a sequence of bits $b_1, b_2, \ldots$ is computable, the universal predictor is guaranteed to work only in the limit $n \to \infty$. Intuitively, once the predictor has figured out the program Q generating the b_n's, it can *then* predict future b_n's as far as such prediction is possible. However, the number of serious mistakes that the predictor makes before converging on the correct Q could in general be as large as Q's bit-length. Worse yet, there's no finite time after which the predictor can *know* that it's converged on the correct Q. Rather, in principle the predictor can always be surprised by a bit b_n that diverges from the predictions of whatever hypothesis Q it favored in the past, whereupon it needs to find a new hypothesis Q', and so on.

Some readers might object that, in the real world, it's reasonable to assume an upper bound on the number of bits needed to describe a given physical process (for example, a human brain). In that case, the predictor would indeed have an absolute upper bound on $|Q|$, and hence on the number of times it would need to revise its hypothesis substantially.

I agree that such bounds on $|Q|$ almost certainly exist – indeed, they must exist if we accept the holographic principle from quantum gravity (see Section 12.5.4). For me, the issue is simply that the relevant bounds seem too large to be of any practical interest. Suppose, for example, that we believed 10^{14} bits – or roughly one bit per synapse – sufficed to encode everything of interest about a particular human brain. While that strikes me as an underestimate, it still works out to roughly $40,000$ bits per second, assuming an 80-year lifespan. In other words, it seems that a person of normal longevity would have more than enough bits to keep the universal predictor $\mathscr{U}$ on its toes!

The above estimate leads to an amusing thought: *if* one lived forever, then perhaps one's 'store of freedom' would eventually get depleted, much like an n-bit computer program can surprise $\mathscr{U}$ at most $O(n)$ times. (Arguably, this depletion happens to some extent over our actual lifetimes, as we age and become increasingly predictable and set in our ways.) From this perspective, freedom could be considered merely a 'finite-n effect' – but this would be one case where the value of n matters!

12C Appendix: Knightian quantum states

In Section 12.3.2, I introduced the somewhat whimsically-named *freestate*: a representation of knowledge that combines probabilistic, quantum-mechanical, and Knightian uncertainty, thereby generalizing density matrices, which combine probabilistic and quantum uncertainty. (The 'freebits' referred to throughout the essay are then just 2-level freestates.) While there might be other ways to formalize the concept of freestates, in this appendix I'll give a particular formalization that I prefer.

A good starting point is to combine probabilistic and Knightian uncertainty, leaving aside quantum mechanics. For simplicity, consider a bit $b \in \{0,1\}$. In the probabilistic case, we can specify our knowledge of b with a single real number, $p = \Pr[b=1] \in [0,1]$. In the Knightian case, however, we might have a set of *possible* probabilities: for example,

$$p \in \{0.1\} \cup [0.2, 0.3] \cup (0,4.0.5)\,. \qquad (*)$$

This seems rather complicated! Fortunately, we can make several simplifications. Firstly, since we don't care about infinite precision, we might as well take all the probability intervals to be closed. More importantly, I believe we should assume *convexity*: that is, if $p < q$ are both possible probabilities for some event E, then so is every intermediate probability $r \in [p,q]$. My argument is simply that Knightian uncertainty includes probabilistic uncertainty as a special case: if, for example, we have no idea whether the bit b was generated by process P or process Q then, for all we know, b might *also* have been generated by choosing between P and Q with some arbitrary probabilities.

Under the two rules above, the disjunction $(*)$ can be replaced by $p \in [0.1, 0.5]$. More generally, it's easy to see that our states will always be nonempty, convex, regions of the probability simplex: that is, nonempty sets S of probability distributions that satisfy $\alpha\mathscr{D}_1 + (1-\alpha)\mathscr{D}_2 \in S$ for all $\mathscr{D}_1, \mathscr{D}_2 \in S$ and all $\alpha \in [0,1]$. Such a set S can be used to calculate upper and lower bounds on the probability $\Pr[E]$ for any event E. Furthermore, there's no redundancy in this description: if $S_1 \neq S_2$ then it's easy to see that there exists an event E for which S_1 allows a value of $\Pr[E]$ not allowed by S_2 or vice versa.

One might worry about the 'converse' case: probabilistic uncertainty over different states of Knightian uncertainty. However, I believe this case can be 'expanded out' into Knightian uncertainty about probabilistic uncertainty, like so:

$$\frac{(A \text{ OR } B) + (C \text{ OR } D)}{2} = \left(\frac{A+C}{2}\right) \text{ OR } \left(\frac{A+D}{2}\right)$$
$$\text{OR } \left(\frac{B+C}{2}\right) \text{ OR } \left(\frac{B+D}{2}\right).$$

By induction, any hierarchy of probabilistic uncertainty about Knightian uncertainty about probabilistic uncertainty about... etc. can likewise be 'collapsed,' by such a procedure, into simply a convex set of probability distributions.

The quantum case, I think, follows exactly the same lines except that now, instead of a convex set of probability distributions, we need to talk about a convex set of density matrices. Formally, an *n-dimensional freestate* is a nonempty set S of $n \times n$ density matrices such that $\alpha\rho + (1-\alpha)\sigma \in S$ for all $\rho, \sigma \in S$ and all $\alpha \in [0,1]$. Once again, there is no redundancy involved in specifying our knowledge about a quantum system in this way. The argument is simply the following: for all nonempty convex sets $S_1 \neq S_2$, there either exists a state $\rho \in S_1 \setminus S_2$ or a state $\rho \in S_2 \setminus S_1$. Suppose the former without loss of generality. Then by the convexity of S_2, it is easy to find a pure state $|\psi\rangle$ such that $\langle\psi|\rho|\psi\rangle \notin \{\langle\psi|\sigma|\psi\rangle : \sigma \in S_2\}$.

References

Aaronson, S. 2002. Book review of *A New Kind of Science*. *Quantum Inf. Comput.*, **2**(5), 410–423. quant-ph/0206089.

Aaronson, S. 2005. NP-complete problems and physical reality. *SIGACT News*, March. quant-ph/0502072.

Aaronson, S. 2009. Quantum copy-protection and quantum money. pp. 229–242 of: *Proc. IEEE Conf. on Computational Complexity*.

Aaronson, S. 2013. *Quantum Computing Since Democritus*. Cambridge University Press.

Aaronson, S., and Christiano, P. 2012. Quantum money from hidden subspaces. pp. 41–60 of: *Proc. ACM STOC*. arXiv:1203.4740.

Antunes, L., and Fortnow, L. 2009. Sophistication revisited. *Theory Comput. Syst.*, **45**(1), 150–161.

Aspect, A., Grangier, P., and Roger, G. 1982. Experimental realization of Einstein–Podolsky–Rosen–Bohm gedankenexperiment: a new violation of Bell's inequalities. *Phys. Rev. Lett.*, **49**, 91–94.

Aumann, R. J. 1976. Agreeing to disagree. *Ann. Stat.*, **4**(6), 1236–1239.

Balaguer, M. 2009. *Free Will as an Open Scientific Problem*. Bradford Books.

Bell, J. S. 1987. *Speakable and Unspeakable in Quantum Mechanics*. Cambridge University Press.

Bennett, C. H., and Brassard, G. 1984. Quantum cryptography: public key distribution and coin tossing. pp. 175–179 of: *Proc. IEEE Int. Conf. on Computers Systems and Signal Processing*.

Bennett, C. H., Brassard, G., Crépeau, C., Jozsa, R., Peres, A., and Wootters, W. 1993. Teleporting an unknown quantum state by dual classical and EPR channels. *Phys. Rev. Lett.*, **70**, 1895–1898.

Bernstein, E., and Vazirani, U. 1997. Quantum complexity theory. *SIAM J. Comput.*, **26**(5), 1411–1473. First appeared in ACM STOC, 1993.

Bierce, A. 2009. *The Collected Works of Ambrose Bierce: 1909–1912*. Cornell University Library.

Bohm, D. 1952. A suggested interpretation of the quantum theory in terms of 'hidden' variables. *Phys. Rev.*, **85**, 166–193.

Bohr, N. 2010. *Atomic Physics and Human Knowledge*. Dover. First published in 1961.

Bostrom, N. 2002. *Anthropic Bias: Observation Selection Effects in Science and Philosophy*. Routledge.

Bousso, R. 2000. Positive vacuum energy and the N-bound. *J. High Energy Phys.*, **0011**(038). hep-th/0010252.

Carroll, S. 2010. *From Eternity to Here*. Dutton.

Chaitin, G. J. 1975. A theory of program size formally identical to information theory. *J. ACM*, 329–340.

Chalmers, D. J. 1996. *The Conscious Mind: In Search of a Fundamental Theory*. Oxford.

Clauser, J. F., Horne, M. A., Shimony, A., and Holt, R. A. 1969. Proposed experiment to test local hidden-variable theories. *Phys. Rev. Lett.*, **23**, 880–884.

Compton, A. H. 1957. Science and Man's freedom. *The Atlantic*, **200**(4), 71–74.

Conway, J. H., and Kochen, S. 2009. The strong free will theorem. *Notices AMS*, **56**(2), 226–232. arXiv:0807.3286.

Cowen, T., and Hanson, R. 2012. Are disagreements honest? *J. Econ. Methodology*. At hanson.gmu.edu/deceive.pdf.

de Boer, J. 2003. *Introduction to the AdS/CFT correspondence*. University of Amsterdam Institute for Theoretical Physics (ITFA) Technical Report 03-02.

Dennett, D. C. 1984. *Elbow Room: The Varieties of Free Will Worth Wanting*. MIT Press.

Dennett, D. C. 1995. *Darwin's Dangerous Idea: Evolution and the Meanings of Life*. Simon & Schuster.

Deutsch, D. 1985a. Quantum theory as a universal physical theory. *Int. J. Theor. Phys.*, **24**, 1–41.

Deutsch, D. 1985b. Quantum theory, the Church–Turing principle and the universal quantum computer. *Proc. Roy. Soc. London*, **A400**, 97–117.

Deutsch, D. 2011. *The Beginning of Infinity: Explanations that Transform the World*. Allen Lane.

Dick, P. K. 1998. *The Collected Stories of Philip K. Dick. Volume 4: The Minority Report*. Citadel Twilight.

Dyson, L., Kleban, M., and Susskind, L. 2002. Disturbing implications of a cosmological constant. *J. High Energy Phys.*. hep-th/0208013.

Farhi, E., Gosset, D., Hassidim, A., Lutomirski, A., and Shor, P. 2012. Quantum money from knots. pp. 276–289 of: *Proc. Innovations in Theoretical Computer Science (ITCS)*. arXiv:1004.5127.

Fischer, J. M. 1995. *The Metaphysics of Free Will: An Essay on Control*. Wiley–Blackwell.

Fischer, J. M., Kane, R., Pereboom, D., and Vargas, M. 2007. *Four Views on Free Will*. Wiley–Blackwell.

Fritz, T. 2012. Bell's theorem without free will. arXiv:1206.5115.

Gács, P., Tromp, J., and Vitányi, P. M. B. 2001. Algorithmic statistics. *IEEE Trans. Inform. Theory*, **47**(6), 2443–2463.

Gardner, M. 1974. Mathematical Games. *Scientific American*, March, 102. Reprinted with addendum in *The Colossal Book of Mathematics*.

Geroch, R., and Hartle, J. B. 1986. Computability and physical theories. *Found. Phys.*, **16**(6), 533–550.

Ghirardi, G. C., Rimini, A., and Weber, T. 1986. Unified dynamics for microscopic and macroscopic systems. *Phys. Rev. D*, **34**, 470–491.

Hawking, S. W. 1974. Black hole explosions? *Nature*, **248**(5443), 30.

Hodges, A. 2012. *Alan Turing: The Enigma*. Princeton University Press. Centenary edition. First published 1983.

Hodgkin, A., and Huxley, A. 1952. A quantitative description of membrane current and its application to conduction and excitation in nerves. *J. Physiol.*, **117**, 500–544.

Hoefer, C. 2002. Freedom from the inside out. pp. 201–222 of: Callender, C. (ed.), *Time, Reality, and Experience*. Cambridge University Press.

Hooft, G. 't. 2007. The free-will postulate in quantum mechanics. quant-ph/0701097.

Hutter, M. 2007. On universal prediction and Bayesian confirmation. *Theor. Comput. Sci.*, **384**(1), 33–48. arXiv:0709.1516.

Inwagen, P. van. 1983. *An Essay on Free Will*. Oxford University Press.

Knight, F. H. 2010. *Risk, Uncertainty, and Profit*. Nabu Press. First published 1921.

Koch, C. 2012. *Consciousness: Confessions of a Romantic Reductionist*. MIT Press.

Kochen, S., and Specker, S. 1967. The problem of hidden variables in quantum mechanics. *J. of Math. and Mechanics*, **17**, 59–87.

Leslie, J. 1998. *The End of the World: The Science and Ethics of Human Extinction*. Routledge.

Levy, S. 1992. *Artificial Life: A Report from the Frontier Where Computers Meet Biology*. Vintage.

Li, M., and Vitányi, P. M. B. 2008. *An Introduction to Kolmogorov Complexity and Its Applications (3rd ed.)*. First edition published in 1993. Springer.

Libet, B. W. 1999. Do we have free will? *J. Consciousness Studies*, 47–57.

Lloyd, S. 2002. Computational capacity of the universe. *Phys. Rev. Lett.*, **88**. quant-ph/0110141.

Lloyd, S. 2012. A Turing test for free will. *Phil. Trans. Roy. Soc. London A*, **370**, 3597–3610.

Lucas, J. R. 1961. Minds, machines, and Gödel. *Philosophy*, **36**, 112–127.

MacKay, D. M. 1960. On the logical indeterminacy of a free choice. *Mind*, **LXIX**(273), 31–40.

Markov, A. 1958. Unsolvability of the problem of homeomorphy. Pages 300–306 of: *Proc. Int. Cong. Math., Edinburgh*.

Marshall, W., Simon, C., Penrose, R., and Bouwmeester, D. 2003. Towards quantum superpositions of a mirror. *Phys. Rev. Lett.*, **91**(130401). quant-ph/0210001.

McGinn, C. 2000. *The Mysterious Flame: Conscious Minds In A Material World*. Basic Books.

Morris, S. 1995. The Common Prior Assumption in economic theory. *Econ. and Phil.*, **11**, 227–253.

Neal, R. M. 2006. Puzzles of anthropic reasoning resolved using full non-indexical conditioning. Tech. rept. 0607. Dept. of Statistics, University of Toronto. www.cs.toronto.edu/~radford/anth.abstract.html.

Nozick, R. 1969. Newcomb's problem and two principles of choice. pp. 114–115 of: Rescher, Nicholas (ed.), *Essays in Honor of Carl G. Hempel*. Synthese Library.

Penrose, R. 1989. *The Emperor's New Mind*. Oxford.

Penrose, R. 1996. *Shadows of the Mind: A Search for the Missing Science of Consciousness*. Oxford.

Penrose, R. 2007. *The Road to Reality: A Complete Guide to the Laws of the Universe*. Vintage.

Perlmutter, S. and 31 others (Supernova Cosmology Project). 1999. Measurements of Ω and Λ from 42 high-redshift supernovae. *Astrophys. J.*, **517**(2), 565–586. astro-ph/9812133.

Pironio, S., Acín, A., Massar, S., de la Giroday, A. Boyer, Matsukevich, D. N., Maunz, P., Olmschenk, S., Hayes, D., Luo, L., Manning, T. A., and Monroe, C. 2010. Random numbers certified by Bell's theorem. *Nature*, **464**, 1021–1024. arXiv:0911.3427.

Pusey, M. F., Barrett, J., and Rudolph, T. 2012. On the reality of the quantum state. *Nature Physics*, **8**, 475–478. arXiv:1111.3328.

Russell, S., and Norvig, P. 2009. *Artificial Intelligence: A Modern Approach*. 3rd edn. Prentice Hall. First published 1995.

Ryle, G. 2008. *The Concept of Mind*. Kessinger Publishing. First published 1949.

Satinover, J. 2001. *The Quantum Brain*. Wiley.

Savage, L. J. 1954. *The Foundations of Statistics*. Wiley.

Schmidhuber, J. 1997. A computer scientist's view of life, the universe, and everything. pp. 201–208 of: *Foundations of Computer Science: Potential – Theory – Cognition*. Springer.

Searle, J. 1992. *The Rediscovery of the Mind*. MIT Press.

Shafer, G. 1976. *A Mathematical Theory of Evidence*. Princeton University Press.

Shieber, S. M. (ed.). 2004. *The Turing Test: Verbal Behavior as the Hallmark of Intelligence*. Bradford Books.

Sompolinsky, H. 2005. A scientific perspective on human choice. In: Shatz, D., and Berger, Y. (eds.), *Judaism, Science, and Moral Responsibility*. Rowman and Littlefield.

Soon, C. S., Brass, M., Heinze, H.-J., and Haynes, J.-D. 2008. Unconscious determinants of free decisions in the human brain. *Nature Neurosci.*, **11**, 543–545.

Spekkens, R. W. 2007. In defense of the epistemic view of quantum states: a toy theory. *Phys. Rev. A*, **75**(032110). quant-ph/0401052.

Stenger, V. J. 2009. *Quantum Gods: Creation, Chaos, and the Search for Cosmic Consciousness*. Prometheus Books.

Stoica, C. 2008. Flowing with a frozen river. Runner-up in FQXi essay contest on The Nature of Time. fqxi.org/data/essay-contest-files/Stoica_flowzen_time_2.pdf.

Stoica, C. 2012. Modern physics, determinism and free-will. *Noema*, **XI**. www.noema.crifst.ro/doc/2012_5_01.pdf.

Taleb, N. N. 2010. *The Black Swan: The Impact of the Highly Improbable*. 2nd edn. Random House. First published 2007.

Tegmark, M. 1999. The importance of quantum decoherence in brain processes. *Phys. Rev. E*, **61**, 4194–4206.

Turing, A. M. 1950. Computing machinery and intelligence. *Mind*, **59**, 433–460.

Vazirani, U., and Vidick, T. 2012. Certifiable quantum dice – or, true random number generation secure against quantum adversaries. pp. 61–76 of: *Proc. ACM STOC*. arXiv:1111.6054.

von Neumann, J. von. 1932. *Mathematische Grundlagen der Quantenmechanik*. Springer. English translation (1955) *Mathematical Foundations of Quantum Mechanics*. Princeton University Press.

Wiesner, S. 1983. Conjugate coding. *SIGACT News*, **15**(1), 78–88. Original manuscript written circa 1970.

Wigner, E. P. 1962. Remarks on the mind–body problem. In: Good, I. J. (ed.), *The Scientist Speculates*. Heinemann, London.

Wolfram, S. 2002. *A New Kind of Science*. Wolfram Media.

Part Five: Oracles, Infinitary Computation, and the Physics of the Mind

The arc of the book has run from the logical to the physical, and now it returns to the logical. Both Sol Feferman and Philip Welch address the precise modern understanding of the incomputable. It is a remarkable fact that in 1936, when Alan Turing described computability and how to 'invent' a universal machine, he did so in the context of seeing the limitations of the computable. Later, in his machine-intelligence plans, he emphasised the power and scope of the computable. Yet he always knew that the computable to be a countable drop in the incountable ocean, and his 1939 Princeton work, perhaps his deepest and most difficult, is a founding paper for the analysis of the incomputable. As we have seen, Turing's vision takes us to the limits of the computable and beyond, via various routes, mathematical and practical.

This final part of our book reconnects us with the clarifying role of logic, with the physics and thought processes and, at most, the proverbial 'elephant in the room'. The coherence of Turing's by now familiar dialectic, computation versus description, embodied versus ideal, takes us to the core of the organic mystery of information in action. The closed system has to be an idealisation in the real world, with the ideal playing mathematical catch-up via the added ingredient of Turing's oracles. **Sol Feferman**'s article traces the history of the oracle machine through its recursion theoretic evolution, through some very beautiful mathematics, taking in the emergence of its relevance to a range of areas, in logic and beyond. There have to be questions concerning, as Feferman describes it, "the most rarefied and recondite parts of the subject such as degree theory", while his concluding discussion points to the future. Already there are signs that the pathology and mathematical ugliness of the mature mathematics of relative computation – the 'Turing universe' – may be precisely what is needed to support models of embodied computational complexity and their semantics. The 'and Back' of Feferman's title for his chapter may well take on a new and 'Turingesque' significance. Just as the semantics of the universal Turing machine took us up a type to an incomputable real number, it is the semantics of the Turing universe that takes us up the type-theoretical ladder to type-2 information. Mathematically, we model such natural derivations via various 'generalised' computational models able to use appropriate infinitary data, models generally inspired by mathematical analogues of standard classical ones rather than real contexts.

The infinitary computation discussed by **Philip Welch** is rather different and intriguingly related to the physics and, in particular, the mathematics of relativity and the singularities it predicts. Philip's introduction and description of the context is superb, and his chapter is one of the best introductions to the topic of *infinite-time Turing machines* one will encounter. He concludes that

> the above examples illustrate the richness of Turing's original conception; one that goes far beyond his original purposes. We may think in these "mechanistic" terms in ways that he did not intend, but which his model, and his example, has inspired.

Roger Penrose's article starts in 1951. As mentioned in the introduction, Turing was attracted to ideas from Dirac in that period, ideas in mathematical physics which Penrose developed enormously in the 1960s. Here, Penrose returns the compliment and reveals that he was introduced by Max Newman to Turing's ideas in logic. Although Penrose touches tantalisingly on modern developments indicating that essential features of quantum mechanics are involved in biological structure, in this article he remains purely within a logical framework to develop his theme that the 'mathematical objection' to machine intelligence is real, and that the incomputable is of physical significance, especially to brains.

The discussion is very much in the spirit of Turing's 1939 article, with a parallel balance between technical and intuitive content. There is much added to Penrose's previous thoughts on this topic, and the 1939 Turing influence, together with the wide-ranging grasp of ideas, makes for a fascinating revisit to the subject. As Penrose readily points out, a comprehensive understanding of brain functionality is very likely to be tied to a better understanding of the physics. Today, there is growing support for his view that the classical computational modelling of thought processes is of limited value.

We shall probably never know how Turing, in his last years, was himself putting together his picture of computing and his picture of quantum mechanical physics. But this arena of thought, still so open to speculation, makes a natural point at which to conclude our survey of Alan Turing's future.

13

Turing's 'Oracle': From Absolute to Relative Computability and Back[a],[b]

Solomon Feferman

Dedicated to S.C. Kleene for his many fundamental contributions to recursion theory.

13.1 Introduction

We offer here some historical notes on the conceptual routes taken in the development of recursion theory over the last 60 years, and their possible significance for computational practice. These illustrate, incidentally, the vagaries to which mathematical ideas may be susceptible on the one hand, and – once keyed into a research program – their endless exploitation on the other.

At the hands primarily of mathematical logicians, the subject of *effective computability*, or *recursion theory* as it has come to be called (for historical reasons to be explained in the next section), has developed along several interrelated but conceptually distinctive lines. While this began with what were offered as analyses of the *absolute* limits of effective computability, the immediate primary aim was to establish *negative* results of the effective unsolvability of various problems in logic and mathematics. From this the subject turned to refined classifications of unsolvability for which a myriad of techniques were developed. The germinal step, conceptually, was provided by Turing's notion of computability relative to an 'oracle'. At the hands of Post, this provided the beginning of the subject of *degrees of unsolvability*, which became a massive research program of great technical difficulty and combinatorial complexity. Less directly provided by Turing's notion, but

[a] This is a lightly revised and expanded version of Feferman (1992); it is reprinted with the kind permission of Walter de Gruyter Publishing Company. See the postscript for several references to relevant subsequent literature. I wish to thank T. Fernando, M. Lerman, P. Odifreddi and W. Sieg for their useful comments on a draft of the original of this paper.

implicit in it, were notions of *uniform relative computability*, which led to various important *theories of recursive functionals*. Finally the idea of computability has been relativized by extension, in various ways, to more or less *arbitrary structures*, leading to what has come to be called *generalized recursion theory*. Marching in under the banner of degree theory, these strands were to some extent woven together by the recursion theorists, but the trend has been to pull the subject of effective computability even farther away from questions of actual computation. The rise in recent years of *computation theory* as a subject with *that* as its primary concern forces a reconsideration of notions of computability theory both in theory and practice. Following the historical sections, I shall make the case for the primary significance for practice of the various notions of relative (rather than absolute) computability, but *not* of most methods or results obtained thereto in recursion theory.

While a great deal of attention has been paid in the literature to the early history of recursion theory in the 1930s, and to the grounds for the so-called Church–Turing Thesis as to absolute effective computability, hardly any has been devoted to notions of relative computability. The historical sketch here is neither definitive nor comprehensive; rather, the intention is to mark out the principal conceptual routes of development with the end purpose of assessing their significance for computational practice. Nor is any claim made as to the 'right' generalization, if any, of computability to arbitrary structures. However, the time is ripe for a detailed historical study of relative computability in all its guises and for an assessment of proposed generalizations of the Church–Turing thesis.

13.2 'Absolute' effective computability

13.2.1 Machines and recursive functions

The history of the development of what appeared to be the most general concept of effective computability and of the Church–Turing Thesis thereto, is now generally well known; see Kleene (1981); Davis (1982); Gandy (1988) and, especially for Turing's role, Hodges (1983).

By 1936, several very different looking proposals had been made for the explication of this notion: λ-definability (Church), general recursiveness (Herbrand–Gödel), and computability by machine (Turing and, independently, Post). These were proved in short order to lead to co-extensive classes of functions (of natural numbers). In later years, still further notions leading to the same class of functions were introduced; of these we mention (for purposes below) only computability on register machines, introduced by Shepherdson and Sturgis (1963).

The definition generally regarded as being the most persuasive for Church's Thesis is that of computability by Turing machines. This is not described here because of its familiarity and because of its remove from computational practice. The register machine model of Shepherdson–Sturgis is closer to the actual design of computers and even leads to a 'baby' imperative-style programming language. As with all the definitions mentioned above, it makes idealized assumptions to the effect that work-space and computation-time are unlimited. Each register machine is provided with a finite number of memory locations R_i, called registers, each of which has an unlimited capacity in the sense that arbitrarily long sequences of symbols and numbers can be stored in them. Here we restrict attention to computation over the set $N = \{0, 1, 2, 3, \ldots\}$ of natural numbers, to which more general symbolic computation can be reduced. Certain registers $R_1, \ldots, R_n$ are reserved for inputs (when computing a function $f : N^n \to N$), and one, say R_0, is reserved for the output; the other registers provide memory locations for computations in progress. A program is given by a sequence of instructions $I_0, \ldots, I_m$ of which I_0 is the initial instruction and I_m is the final or HALT instruction. The active instructions I_j $(0 \leq j < m)$ are of one of the following forms: (i) increment the contents r_i of R_i by 1, (ii) decrement by 1 (if different from 0), (iii) set the contents of R_i to 0, and (iv) test to see if r_i is 0 and then branch to one or another instruction depending on the answer. These are symbolized respectively by

(i) $r_i := r_i + 1$,
(ii) $r_i := r_i - 1$,
(iii) $r_i := 0$,
(iv) if $r_i = 0$ go to I_k, else go to I_l.

A function $f : N^n \to N$ is computable by such a machine if when we load any natural numbers $x_1, \ldots, x_n$ as input in $R_1, \ldots, R_n$, respectively, and begin with instruction I_0, the output $f(x_0, \ldots, x_n)$ will eventually appear in R_0 at the HALT stage I_m. As mentioned above, it has been shown that the class of register computable functions is the same as that of Turing computable functions.

Returning to the situation in the latter part of the 1930s, the results establishing the (extensional) equivalence of the various proposed definitions of effective computability bolstered Church's Thesis. Church himself had announced this in terms of the Herbrand–Gödel notion of *general recursiveness*. The fact that many effectively computable functions were given in practice by recursive defining equations led logicians to treat effective computability and recursiveness as interchangeable notions. Thus the subject has come to be called *recursive function theory*, or simply *recursion theory*.

As a result especially of Turing's analysis, Church's Thesis has gained almost universal acceptance; the case for it was assembled in Kleene (1952) (especially in

§§62, 63, and 70). For a more recent analysis, see Gandy (1980)[1], and for a comprehensive survey of the literature on this subject see (Odifreddi, 1989, §I,8). Many would agree with Gödel that the importance of general recursiveness (or Turing computability) is that "...with this concept one has for the first time succeeded in giving an *absolute* definition of an interesting epistemological notion, i.e., one not depending on the formalism chosen" (italics mine: quotation from the 1946 article appearing in Gödel (1990)).

13.2.2 Partial recursive functions

In his 1938 paper Kleene made an essential conceptual step forward by introducing the notion of *partial recursive function*. This can be explained in terms of any of the models of effective computability mentioned above, in particular by means of the Turing or register machine approaches. Each instruction sequence or program $I = I_0, \ldots, I_m$ may be coded by a natural number i, and then M_i is used to denote the corresponding machine. For simplicity, consider functions of one argument only; given an arbitrary such x, M_i need not terminate at that input. If it does we write $M_i(x)$ for the output value and say that $M_i(x)$ is defined. This then determines a function f with domain $\mathrm{dom}(f) = \{x \mid M_i(x) \text{ is defined}\}$, whose value for each x in $\mathrm{dom}(f)$ is given by $M_i(x)$. A function is said to be partial recursive just in the case it is one of these fs.

Kleene established a *Normal Form Theorem* for partial recursive functions as follows (adapted to the machine model). Let $C(i,x,y)$ mean that y codes a terminating computation on M_i at input x; it may be seen that C is an effectively decidable relation (in fact, in the subclass of primitive recursive relations). Moreover the function $U(y)$ which extracts the output of y when y represents a terminating computation, and is otherwise 0, is also (primitive) recursive. Hence, for partial recursive f determined by M_i as above, we have

(1) (i) $\mathrm{dom}(f) = \{x \mid (\exists y)C(i,x,y)\}$, and
(ii) $f(x) = U\big((\text{least } y)C(i,x,y)\big)$ for each x in $\mathrm{dom}(f)$.

Moreover every function defined in this way is partial recursive. One may further observe from this result that if we define $g(i,x)$ for all i,x by

(2) $g(i,x) = U\big((\text{least } y)C(i,x,y)\big)$ whenever $M_i(x)$ is defined,

then g is a partial recursive function of two arguments which, for $i = 0,1,2,\ldots$, ranges (as a function of x) over all partial recursive functions of one argument. This is what is called the *Enumeration Theorem* for partial recursive functions.

[1] And the still more recent critical discussion provided in the dissertation Tamburrini (1987); see the postscript for more recent relevant literature.

Now, since g itself is computable by a machine M, we have the consequence – already recognized by Turing in 1937 – that there is a *universal computing machine* M, which can simulate the behavior of any other M_i by acting at input (i,x) as M_i acts at input x. This in turn may be considered to provide the conceptual basis for *general purpose computers* which can store 'software' programs I in the form of data 'i' for any particular application.

13.2.3 Effectively unsolvable problems and the method of reduction

A number of questions had been raised in the period 1900–1930 concerning the uniform effective solvability of certain classes of mathematical problems. Outstanding among these were:

(1) *Diophantine equations.* To decide, if possible, whether a polynomial equation with integer coefficients has integer solutions ('Hilbert's tenth problem').
(2) *Word problem for groups.* To decide, if possible, whether two words in a finitely presented group represent the same element.
(3) *Entscheidungsproblem.* To decide, if possible, whether a given formula in the first-order predicate calculus (1st-order PC) of logic is satisfiable.

While partial progress was made on each of these problems for specific cases, the general problems resisted positive solution. In particular, for (3), initial optimism was tempered by the famous incompleteness results of Gödel in 1931 (see the collection Gödel, 1986) in which it was shown, among other things, that for any sufficiently strong and correct formal system S one can produce a formula A_S of the 1st-order PC such that A_S is satisfiable but that fact is not provable in S. Hence, if there were a decision method D for satisfiability in the 1st-order PC, no such S could verify that D works.

But, in order to obtain negative results concerning these and similar problems, one would need a precise and completely general definition of effective method. This would be analagous to supplying a general definition of ruler-and-compass construction in order to show the non-constructibility of the classical geometric problems (angle trisection, duplication of the cube, etc.), or of solvability by radicals in order to demonstrate the nonsolvability in such terms of various algebraic equations (of 5th degree and beyond). In the case of effective computability and effective decidability, that is just what was supplied by the definitions described in §13.2.1, according to the Church–Turing Thesis. And, indeed, Church and Turing used this to establish the effective unsolvability of the *Entscheidungsproblem.* Their thesis is implicitly taken for granted in the following, where we analyze one aspect of their proofs.

Given a set A of natural numbers, the *membership problem* for A is the question whether one has an effective method to decide, given a natural number x, whether or not x belongs to A. The characteristic function c_A of A is the function defined by $c_A(x) = 1$ if x is in A, and 0 otherwise. The membership problem for A is effectively solvable just in the case where c_A is effectively computable. If this holds we say that A itself is computable or recursive, while if A is not recursive then its membership problem is said to be effectively unsolvable. The very first example of such a problem provided by Turing was that of the *halting problem* (HP) for Turing machines, which is readily adapted to register machines. This is to decide, given i and x, whether or not $M_i(x)$ is defined. The *diagonal HP* is the question whether M_x terminates at input x, and is represented by the set $K = \{x \mid M_x(x) \text{ is defined}\}$. Now one easily shows by a diagonal argument that K cannot be recursive. For if it were, its characteristic function c_K would be recursive. But then so would be the function

$$d(x) = \begin{cases} M_x(x)+1 & \text{if } x \text{ is in } K, \\ 0 & \text{otherwise.} \end{cases}$$

Now, since d is recursive, it is computed by some specific machine M_i, i.e. $d(x) = M_i(x)$ for all x. Then, in particular, $d(i) = M_i(i)$, contradicting $d(i) = M_i(i) + 1$.

The general halting problem is represented by the set

$$H = \{(x,y) \mid M_x(y) \text{ is defined}\}.$$

Clearly x is in K just in the case where (x,x) is in H. If H were recursive then K would be recursive, contrary to what has just been shown. The general situation here is given by the notion of *many–one reduction*, $A \leq_{\mathrm{m}} B$. This is defined to hold just in the case where there is a recursive function f such that, for all x

(1) x is in A if and only if $f(x)$ is in B.

(It is called 'many–one' because the function f might give the same value for many arguments). We have the trivial result:

Theorem *If $A \leq_{\mathrm{m}} B$ and B is recursive then A is recursive. Hence if A is not recursive, B is not recursive.*

In essence Turing (1936) established the negative result for the *Entscheidungsproblem* by taking

$$S = \{x \mid x \text{ is the number of a formula in the 1st-order PC which is satisfiable in some model}\}$$

and showing $K \leq_{\mathrm{m}} S$ where K is the diagonal halting problem. (The argument in Church (1937) instead makes use of reduction from an effectively unsolvable problem in the λ-calculus.)

The relation $\leq_{\mathrm{m}}$ of many–one reduction is one of the most widely applied in practice for effective unsolvability results. Eventually Hilbert's tenth problem and the word problem for groups were shown to be effectively unsolvable[2] by reducing the problem K to them (through a long chain of arguments). However, it is not in principle the most general relation $\leq$ between the sets A and B which will allow one to conclude:

(2) if $A \leq B$ and B is recursive then A is recursive.

For example, we could take $\leq$ to mean that there is a *truth-table reduction* of A to B, i.e. that membership of an x in A is determined by a propositional combination of statements of the form $f_i(x)$ in B, for f_i recursive; in such a case we write $A \leq_{\mathrm{tt}} B$. The most general concept of effective reduction of the membership problem for one set to another is wider still. This was provided by Turing's notion of computation relative to an 'oracle', to which we now turn.

13.3 Relative effective computability over the natural numbers

13.3.1 Turing's 'oracle' and Turing reducibility

The germinal idea of computability relative to an 'oracle' was introduced almost as an aside in Turing (1939), a paper based on Turing's PhD thesis at Princeton University under the direction of Church. The story of how Turing came to do graduate work in Princeton and of the outcome of his studies is told in (Hodges, 1983, pp. 90–146), and again in Feferman (1988), where the contents of the thesis publication are analyzed in some detail. Turing's dissertation work concerned the concepts of ordinal logics, which were introduced in an attempt to overcome Gödel's incompleteness results by the iterated (finite and transfinite) adjunction to each formal system (or logic) S accepted in the process, of such statements as Con_S, expressing the consistency of S, shown by Gödel to be unprovable in S though informally recognized to be true. Turing's aim was to obtain completeness for two-quantifier (Q_2) statements of arithmetic, i.e., those of the form $(\forall x)(\exists y)R(x,y)$ with R recursive, and in this he was only partially successful. Now the section of the published dissertation (Turing, 1939) in which Turing introduces the 'oracle' notion is a brief one, §4, whose main aim is to produce a mathematical problem which is not in Q_2 form. The existence of non-Q_2 definable sets is of course trivial by a cardinality or diagonal argument, but presumably Turing wanted to produce something with more concrete mathematical content. He begins by saying: "Let us suppose that we are supplied with some specified means of solving number-theoretic [Q_2] problems; a kind of oracle as it were... With the help of the oracle we could form a

[2] Through the work, respectively of Davis, Putnam, J. Robinson and Matiyasevich on the former problem, and (with successive improvements) Novikoff, Boone, Britton and Higman on the latter.

new kind of machine (call them O-machines), having as one of its processes that of solving a given number-theoretical problem." Turing then shows (*loc. cit.*) more precisely how to define computability by an O-machine and, by a direct extension of his argument in Turing (1936, 1937), that (in effect) the halting problem for O-machines is not decidable by an O-machine and hence is outside the class of Q_2 problems assumed to be decided by O.

Turing did nothing further with this idea and it was not until Post (1944) that it began to be taken as the basis for systematic investigation. To begin with, the idea of an O-machine is directly generalized to that of a B-machine for any set B. In the register machine model, one simply adds, to the basic instructions, ones of the form

(1) $r_j := 1$ if r_k is in B, else $r_j := 0$.

Given a list I of instructions for computation relative to a set in the sense just explained, and given a specific set B, if the computation by the machine M determined by I terminates at any given x, we write $M^B(x)$ for the output. If, for each x, $M^B(x) = 0$ or 1 then M^B is the characteristic function of a set A with

(2) x in A if and only if $M^B(x) = 1$.

In this case we say that A is *Turing computable from B* or *Turing reducible to B* and write $A \leq_{\mathrm{T}} B$. It is not hard to see that

Theorem *(i)* $A \leq_{\mathrm{m}} B \Rightarrow A \leq_{\mathrm{tt}} B$; *(ii)* $A \leq_{\mathrm{tt}} B \Rightarrow A \leq_{\mathrm{T}} B$.

Moreover, the arguments of the Church–Turing Thesis lead one strongly to accept a relativized version:

(C–T)$^{(\mathrm{r})}$ A is effectively computable from B if (and only if) $A \leq_{\mathrm{T}} B$.

Thus Turing reducibility gives the most general concept of relative effective computability.

The relation of computability of one function from another is even more simply defined by an extension of the definition of register computability. Given a function $g : N \to N$ we add, to the four previous register instructions of §13.2.1.(i), instructions of the form

(3) $r_j := g(r_k)$,

whose meaning is to set the content of register R_j to $g(r_k)$ where r_k is in R_k. Then $f \leq_{\mathrm{T}} g$ if and only if f is register computable from g in this expanded sense. Note that for sets A and B we have

(4) $A \leq_{\mathrm{T}} B$ if and only if $c_A \leq_{\mathrm{T}} c_B$,

where c_A, c_B are the characteristic functions of A and B, respectively. Thus there is really only one basic notion of relative computability involved, namely that for functions relative to functions. However, Post (1944) concentrated on the relation $A \leq_T B$ since the classical effective (un)solvability problems concerned the membership problem for sets and, in particular for a special class of sets, the recursively enumerable ones, to which we turn next.

13.3.2 Recursively enumerable sets, degrees of unsolvability, and Post's problem

A set A of natural numbers is said to be *recursively enumerable (r.e.)* if it is the range of a recursive function, i.e.,

(1) $A = \{f(0), f(1), \ldots, f(x), \ldots\}$

for some recursive f, *or* if A is empty. The empty set is included as a limiting case, so that each r.e. set may also be written in the form

(2) x in A if and only if $(\exists y)R(x,y)$

with R a recursive relation, and conversely. Clearly each set A of the form (1), as well as the empty set, is of the form (2). For the converse, one uses a recursive pairing function p with inverses p_0, p_1 so that $p(p_0(x), p_1(x)) = x$ for each x. Then if A is non-empty, say x_0 in A, and (2) holds, let $f(x) = p_0(x)$ if $R(p_0(x), p_1(x))$ and $f(x) = x_0$ otherwise, so that (1) holds. Note that there may be repetitions in (1), so a non-empty r.e. set could be finite.

Every recursive set A is recursively enumerable, as we see from (2) by taking $R(x,y)$ to be: x in A & $y = 0$. However, the converse is not true: the diagonal halting set K is recursively enumerable but not recursive. The latter was argued above; to see that K is r.e. one simply uses the fact that $K = \{x \mid (\exists y)C(x,x,y)\}$, in the symbolism of §13.2.2 above.

It is not hard to see that all the classical problems of effective (un)solvability mentioned in §13.2.2 concern r.e. sets. For example, Diophantine sets are those in the form

(3) $D_{p,q} = \{x \mid (\exists y_1)\ldots(\exists y_n)(p(x,y_1,\ldots,y_n) = q(x,y_1,\ldots,y_n))\}$,

where p, q are polynomials with coefficients in N, and these reduce to the form (2) by combining the prefix $(\exists y_1)\ldots(\exists y_n)$ into a single $(\exists y)$ using the pairing functions. The *Entscheidungsproblem* is a special case of the general *decision problem* for formal systems S, to decide whether a given formula A is provable from S. Using Gödel-numbering of formulas and derivations this reduces to the question whether the set Prov_S given by

(4) $\text{Prov}_S = \{x \mid (\exists y)\,\text{Proof}_S(x,y)\}$

is recursive. But Prov_S is r.e. whenever S is recursive, since then Proof_S is recursive. The *word problem* for groups is the question whether an equation between words in a finitely presented group is provable from the defining equations of the presentation and the group axioms by the rules of equality, and this leads again to sets of the form (2); similarly for other algebraic systems. Not all natural effectiveness problems are recursively enumerable. The first (*prima facie*) more complicated problem is whether any given unary partial recursive function determined by M_i is total, i.e., whether i belongs to the set

(5) $\text{Tot} = \{z \mid (\forall x)(\exists y)\,C(z,x,y)\}$.

One easily sees that every r.e. set A is reducible to the halting problem, as follows. For A r.e. in the form (2), consider the function

(6) $f(x) = (\text{least } y)R(x,y)$.

Then $f(x)$ is defined just in the case $(\exists y)R(x,y)$, so $\text{dom}(f) = A$. Moreover, f is partial recursive, so $f(x) = M_i(x)$ for some i and all x in $\text{dom}(f)$. Hence we have x in A if and only if $M_i(x)$ is defined, i.e. if and only if (i,x) is in H. In other words

(7) $A \leq_{\text{m}} H$.

A little more detailed argument is required to prove the following:

Lemma *For each r.e. set A, we have* $A \leq_{\text{m}} K$.

(For a proof, see Kleene (1952), p. 343.) Thus K is what Post called *complete* for the class of r.e. sets, i.e., it is r.e. and every r.e. set is reducible to it.

Now Post defined a set A to have *equal* or *lower degree of unsolvability* than B if $A \leq_{\text{T}} B$, and A to have the *same degree of unsolvability* as B if both $A \leq_{\text{T}} B$ and $B \leq_{\text{T}} A$. The latter is an equivalence relation between sets of natural numbers; the equivalence class of a set A is called its *degree of unsolvability* and denoted $\deg(A)$. We use letters a, b, ... to range over degrees of unsolvability. Given $a = \deg(A)$, $b = \deg(B)$ we take $a \leq b$ just in the case $A \leq_{\text{T}} B$, and $a < b$ if $a \leq b$ but $a \neq b$, i.e., if $A \leq_{\text{T}} B$ but not $B \leq_{\text{T}} A$. Note that $a = b$ iff $a \leq b$ & $b \leq a$.

Let $0 = \deg(N)$. Since N is recursive, we have $N \leq_{\text{T}} A$ for *any* set A, hence

(8) $0 \leq a$ for all degrees a.

It seems anomalous to call 0 a degree of unsolvability since for any A with $0 = \deg(A)$ we have $A \leq_{\text{T}} N$, hence A is recursive, i.e., is effectively decidable. However, 0 is a limiting case of the degrees:

(9) if $0 < \deg(A)$ then A is not recursive.

Let $0' = \deg(K)$. Then, by the lemma above, we have

(10) for each r.e. set A, $\deg(A) \leq 0'$.

As we observed in §13.2.2, every r.e. set A met in practice and which has been shown to be non-recursive, has been shown to be so by a chain of reductions leading to $K \leq_{\mathrm{m}} A$, and thus, of course, $K \leq_{\mathrm{T}} A$ in all such cases. Post (1944) raised the question whether this must be so in general. That is, he asked:

Post's Problem *Do there exist r.e. sets A with* $0 < \deg(A) < 0'$*?*

As (Post, 1944, pp. 289–290), puts it: "Our whole development largely centers on the single question whether there is, among these problems [for recursively enumerable non-recursive sets] a lower degree of unsolvability than that [of the highest degree $\deg(K)$], or whether they are all of the same degree of unsolvability". After crediting Turing (1939) for the basic idea of reducibility and for establishing in effect that for any set A there is one of a higher degree of unsolvability, Post goes on to say: "While [Turing's] theorem does not help us in our search for that lower degree of unsolvability, his formulation makes our problem precise. It remains a problem at the end of this paper. But on the way we obtain a number of special results, and towards the end obtain some idea of the difficulties of the general problem".

The 'special results' that Post refers to here concerned the existence of lower degrees of unsolvability with respect to the reducibility relations $\leq_{\mathrm{m}}$ and $\leq_{\mathrm{tt}}$. Thus he produced the existence of a non-recursive set S, which he called 'simple', such that K is not many–one reducible to S; however, $K \leq_{\mathrm{tt}} S$ (if one allows unbounded truth-tables). Then Post produced a non recursive r.e. set S^*, which he called 'hyper-simple', such that K is not truth-table reducible to S^*; however, $K \leq_{\mathrm{T}} S^*$, so Post asked whether there might not exist "hyper-hyper-simple" sets which evade this reduction. As the constructions became combinatorially more and more complicated, the difficulty of Post's problem became evident. Towards the end of his 1944 paper, Post said: "As a result we are left completely on the fence as to whether there exists a recursively enumerable [non-recursive] set of positive integers of absolutely lower degree of unsolvability than the complete set K, or whether, indeed, all recursively enumerable sets of positive integers with recursively unsolvable decision problems are absolutely of the same degree of unsolvability".

13.3.3 The solution of Post's problem and the flowering of degree theory.

The first results pushing toward a solution of Post's problem were obtained by Kleene and Post (1954). This also led to a basic bifurcation in the subject of degrees of unsolvability, already implicit in Post's remarks quoted above. Namely one can consider the relation $A \leq_T B$ without restriction on the way A, B may be defined. Post's problem had concerned the relation $\leq_T$ restricted to r.e. sets, but one may consider it for *any* sets A, B, one or both of which might not be r.e. As pointed out by Post, Turing's original construction (in his 1939 paper) in effect associated with each set A another set A' such that

(1) $\deg(A) < \deg(A')$,

by taking A' to be the diagonal halting problem K^A relativized to A, i.e.,

(2) $A' = K^A = \{x \mid M_x^A(x) \text{ is defined}\}$.

Thus K^A is obtained from a predicate (primitive) recursive in A by prefixing one numerical (existential) quantifier. In degree notation,

(3) $a < a'$, where $a' = \deg(K^A)$ for $a = \deg(A)$.

In particular,

(4) $0 < 0' < 0'' < \cdots$

But this is only a crude classification of the degrees of unsolvability of arithmetically defined sets. What Kleene and Post (1954) showed (among other things) is that between each of the inequalities $a < a'$ in (3) there are infinitely many other degrees, in fact, there is a subset D of $\{d \mid a < d < a'\}$ such that D is densely ordered by the $<$ relation. Naturally, if any d with $0 < d < 0'$ were the degree of an r.e. set this would be the solution of Post's problem. However, the Kleene–Post proof was not sufficiently effective to establish this, and their set D consists entirely of non-r.e. degrees.

While considerable effort was devoted by a number of logicians to Post's problem, in the dozen years following its publication there was no breakthrough until 1956 when the problem was solved, independently, by R. Friedberg and A.A. Muchnik (see the references in Rogers, 1967). At the time, Friedberg was a 20-year-old senior at Harvard University and Muchnik was hardly any older. Friedberg had learned of the problem while taking a course in recursion theory taught by Hartley Rogers of MIT. Friedberg and Muchnik's solution established the existence of *two r.e. sets A, B of incomparable degree*, i.e. for which

(5) (i) A, B are r.e., and (ii) neither $\deg(A) \leq \deg(B)$ nor $\deg(B) \leq \deg(A)$.

Thus, for $a = \deg(A)$ and $b = \deg(B)$ we have

(6) $0 < a < 0'$ and $0 < b < 0'$;

for if either one of a, b were equal to 0 or $0'$, then (5)(ii) would be false.

A special new technique, called the *priority method*, was introduced by Friedberg (and Muchnik) for the result (5). The sets A, B are constructed in stages and at each stage only a finite number of membership and non-membership relations are tentatively set down. Each stage is devoted to finding for specific i an n such that $c_A(n) \neq M_i^B(n)$ (and similarly for A, B reversed). If this is successful one puts n in A if $c_A(n) = 1$, otherwise it is out of A; and similarly for B. However, when we thus enlarge (the characteristic function of) A it may turn out that this affects the value assigned to $M_j^A(m)$ for some j, m at a previous stage. By assigning priorities to the actions, it is shown that at most a finite number of changes can take place for each i; this kind of argument is thus often called the *finite-injury method*. The argument for (5) is not long (it only took three pages to communicate it adequately in Friedberg (1957))[3] but its novelty, ingenuity, and the circumstances of its discovery were stunning, both to logicians and to a mass audience. (For example, the news was reported in *Time* magazine for March 19, 1956, p. 83). In the next few years Friedberg made several other interesting applications of the priority method, but after that left the field completely. The work of Friedberg and Muchnik opened the flood gates to the development of the *theory of degrees of unsolvability* (or simply *degree theory*, as it is often called) both for r.e. degrees and degrees of arbitrary sets. In a survey a decade later, (Simpson, 1977, p. 632) said that "...for many years now, degree theory has been one of the most technical and highly developed parts of mathematical logic. There are literally hundreds of papers in the literature, all devoted exclusively to degrees. The standard of originality in these papers is very high. Although certain ideas recur, the variety of methods employed is enormous".

The effect that this development had on recursion theory is evident from a comparison of the expositions of the subject in (Kleene, 1952, Part III) with that of Rogers (1967) or Odifreddi (1989), which are dominated by notions and results concerning the reductions $\leq_{\mathrm{m}}$, $\leq_{\mathrm{tt}}$ and $\leq_{\mathrm{T}}$ and the associated degrees. The first monograph to be devoted entirely to the subject of degree theory and which contained many important new contributions was Sacks (1963). (In particular, that extended the priority method to permit *infinite-injury* arguments.) More recently the books of Lerman (1983) and Soare (1987) have served to bring graduates and young researchers to the forefront of research, emphasizing, respectively, degrees of arbitrary sets and degrees of r.e. sets; the bibliography of the latter book contains

[3] See also the three-page proof in (Rogers, 1967, pp. 163–166). Regarding novelty, however, Rogers says (*loc. cit.*, p. 163) that "in their initial presentations, both Friedberg and Muchnik built on earlier ideas and results of Kleene and Post". As it turned out later, the priority arguments are not essential for the solution of Post's problem, owing to work of Kucera. See (Odifreddi, 1989, Ch. III) for this and other treatments more in the spirit of Post (1944); that text also gives interesting background on Post's problem, beginning with Post's own preliminary ideas on undecidability in the 1920s.

on the order of 600 entries. While the subject of recursion theory has also developed along other lines, some of which will be seen in the following, none matches degree theory for its level of difficulty and for the complexity of the arguments involved (as suggested by the quotation above from Simpson). It is for this reason that one sometimes hears of the periods of development of recursion theory being divided into 'pre-Post' and 'post-Post' or 'pre-Friedberg' and 'post-Friedberg' periods. On the face of it, the results of degree theory are irrelevant to computation theory, because they concern effectively unsolvable problems. However, they may be suggestive of analogous results for degrees of complexity of (effective) algorithms; we shall return to the possible connection in §13.6 below.

13.4 Uniform relative computability over the natural numbers

13.4.1 Relative computation procedures and partial recursive functionals

In Turing's conception of relative computability, the 'oracle' is queried for information about some fixed set. That is, given B, we try to find out for various A whether or not $A \leq_{\mathrm{T}} B$, i.e., whether we can find an i such that $A = M_i^B$. Put in terms of functions the question is, for given f, g, whether there exists i such that $f = M_i^g$.

Now, when dealing with the idea of *uniform relative computability* we shift attention from fixed f, g and possible computation instructions (coded by) i connecting them, to *fixed* i and the effect of varying g in M_i^g on $f = M_i^g$. That is, we fix a *relative computation procedure* and consider its effect on the functions which result when we vary the functions to which it applies. In general, such a procedure F will take us from several different functions $g_1, \ldots, g_m$ to a new function f in the form

(1) $f = F(g_1, \ldots, g_m)$.

Such procedures F are called *functionals*. Concrete examples of *effective functions* are provided by composition and minimalization, in the form:

(Comp) $f(x) = g_1\big(g_2(x), g_3(x)\big)$,
(Min) $f(x) = (\text{least } y)[g(x,y) = 0]$.

In (Comp), $f = F_1(g_1, g_2, g_3)$ and in (Min), $f = F_2(g)$. To begin with, these schemata were conceived of as applying to *total* functions g, g_1, g_2, g_3 on N and as leading to *total* functions f, provided for (Min) that $(\forall x)(\exists y) g(x,y) = 0$. In that schema, if we assume g is total but don't know whether there is always a y with $g(x,y) = 0$, we can conclude only that f is partially defined. Then the '$=$' sign in (Min) must be replaced by '$\simeq$', which means that if either side is defined, then so is the other, and they are equal. If we go further to allow partial functions to appear on the right-hand side of these equations then we must also replace '$=$' by '$\simeq$'

in (Comp) and in similar examples. Thus partial recursive functionals are allowed in general to operate on partial functions. When written in the form (1), the '=' sign is appropriate, since one obtains a well-determined partial function f from $(g_1,\ldots,g_m)$ by application of F. However, f itself might be nowhere defined. In general, then, we must write

(2) $f(x_1,\ldots,x_n) \simeq \big(F(g_1,\ldots,g_m)\big)(x_1,\ldots,x_n)$, for which we also write
(3) $f(x_1,\ldots,x_n) \simeq F(g_1,\ldots,g_m;x_1,\ldots,x_n)$.

The notion of partial recursive functional was introduced by Kleene in 1950. In a retrospective article he said that he had arrived at this "by considering Turing's computation by a machine having access to an oracle, but with the rules governing the machine ... fixed, *varying the oracle* so that she answers for one or another value of a ... function variable." (Kleene, 1981, p. 64). However, the direct treatment on either one of the machine approaches only works for functionals of *total* function arguments. For example, in the case of the register machine approach with (for simplicity) $n = m = 1$ in the above, the partial recursive functionals F are all those definable in the form

(4) $F(g) = M_i^g$

for any fixed i, so that $F(g;x) = M^g(x)$ for all g, x. In other words, one interprets the instruction set given by i, including rules of the form $r_j = g(r_k)$, to apply to *variable total* functions g. However, this does not work for *partial* function arguments g, because the computation will terminate once one hits any single place where $g(r_k)$ is undefined even though it could continue at other defined arguments. Instead, Kleene developed the basic theory of partial recursive functionals of partial function arguments in (Kleene, 1952, Ch. XII) via the Herbrand–Gödel equation calculus approach to recursion theory. Alternatively, as shown in (Rogers, 1967, pp. 146–149), an indirect explanation can be given on one of the machine approaches via the so-called enumeration and partial recursive operators.

13.4.2 The recursion theorems

Using any one of the approaches just explained, the following properties of partial recursive functionals in general are easily established. Again, for simplicity, we consider the case $n = m = 1$ and write $F(g;x)$ for $(F(g))(x)$.

Lemma *Suppose that F is a partial recursive functional of (possibly) partial function arguments.*

(i) (Monotonicity) *If g_1 is contained in g_2 then $F(g_1)$ is contained in $F(g_2)$.*

(ii) (Continuity) *If* $F(g;x) \simeq y$ *then for some finite* h *contained in* g *we have* $F(h;x) \simeq y$.

(iii) (Effectiveness) *If* g *is partial recursive then* $F(g)$ *is partial recursive.*

The fundamental result obtained by Kleene for these functionals is the following (*op. cit.*, p. 348):

The Recursion Theorem (functional form) *For any partial recursive functional* $F(g;x)$ *there is a least solution* f *to the equation* $f(x) = F(f;x)$*; moreover, that* f *is partial recursive.*

The proof simply takes $f =$ the union of the g_n, where g_0 is the empty function and $g_{n+1} = F(g_n)$. Then f is the least solution of the equation $f = F(f)$ by the monotonicity and continuity properties of F. To prove that f is partial recursive, one makes use of the effectiveness property.

The recursion theorem and another, index, form (already established by Kleene in 1938) have many applications in recursion theory. The reason for the importance of these results is that apparently they give the most natural effective versions of defining a function over the natural numbers recursively, i.e., in terms of itself. While the index form seems to be applied more often in practice, the functional form may be considered more fundamental. For, it is expressed in *intrinsic* terms, independent of any enumeration of the partial recursive functions. Moreover, it is of the same general character as definition by recursion in the wider set-theoretical setting.

13.4.3 Partial recursive functionals of finite type over the natural numbers

Notions of primitive, general and partial recursive functionals have been extended to various *finite type (f.t.) structures* $\mathcal{M} = \langle M_\tau \rangle$ over the natural numbers, where $M_0 = N$ and $M_{(\tau,\sigma)}$ consists of certain (possibly partial) operations f from M_τ to M_σ. In set theoretical terms one can define the structure HTF of *hereditarily total functionals of f.t.* by $M_{(\tau,\sigma)} =$ the set of *all* (total) $f : M_\tau \to M_\sigma$. Gödel in 1958 (see the Gödel (1990) collection) introduced a notion of primitive recursive functional which makes sense for objects in this type structure (but also for much narrower structures). Kleene (1959b) dealt with partial recursive functionals $F(f_1^{(n_1)}, \ldots, f_m^{(n_m)}) \simeq f^{(n)}$ where $f^{(n)}$ is an object of pure type n i.e., belongs to M_n, where those are defined by $M_{n+1} = M_{(n,0)}$. The type structure of hereditarily total functions can be reduced to that of pure types. There are also reasonable extensions of the notion of partial recursive functional to suitable structures of hereditarily continuous functions of f.t. (Kleene, 1959a, 1959b) and hereditarily monotonic partial functions of f.t. (Platek, 1966 and §13.5.3 below). It is not possible to explain all these notions without going into considerably more detail than we have space for here.[4] However this should begin to suggest that the idea of a partial recursive function(al) makes sense for a variety of *structures* $\mathscr{M}$ of objects which are not, themselves, effectively given. This constitutes the next chapter in the relativization of recursion theory.

13.5 Generalized recursion theory

13.5.1 Background and Overview

The development of recursion theory discussed up to this point has taken us from the early 1930s up to the late 1950s. The first part of this development, which took place up to roughly the mid-1940s, was devoted to foundational work, applications, and systematic organization of the subject. In that period, except for the introduction of Turing's concept of computation relative to an 'oracle', which remained untouched until 1944, recursion theory was conceived of in *absolute* terms. The second period of development saw a branching of the subject into distinctive distinctive subfields with increasing specialization and technical sophistication. Only two of these have been discussed above, namely *degree theory* and *the theory of recursive functionals*. A third area, which we shall not attempt to describe, concerns what is sometimes called *hierarchy theory* and, in particular, the extension of the arithmetical hierarchy to the *hyperarithmetical hierarchy* through all *recursive ordinals* (i.e., ordinals with a recursive order type); for this see Hinman (1978) and Sacks (1990). All of these concern *relativization* in one way or another, the last by transfinite iteration of certain 'jump' operations on sets. At the same time, the initial motivation to secure an 'absolute' concept of effective computability in order to establish the effective unsolvability of classical problems still maintained its force, with the eventual resolution (as has already been noted) of such outstanding questions as the word problem for groups and Hilbert's tenth problem. Still another use of both the basic theory and the theory of recursive functionals was in a more positive direction, namely to supply recursion-theoretic semantics for intuitionistic formal systems by means of so-called *recursive realizability interpretations* (for this see the survey article Troelstra (1977)).

During this same overall period, from the 1930s up to the late 1950s, mathematical logic as a whole was undergoing considerable development, following roughly the same pattern: in the pre-war period, by means of foundational and organizational work with basic applications, and in the post-war period through a split-up into more specialized and technically sophisticated research programs. But the field as a whole tended to be compartmentalized, with little interaction between the dif-

[4] See (Odifreddi, 1989, pp. 199–201) for a survey of various notions of recursion over higher-type structures.

ferent directions of work into what are still regarded as the main branches of logic: *set theory*, *model theory*, *recursion theory*, and *proof theory*.

One landmark event signaled the breakdown of this compartmentalization, namely a six-week long Institute in Symbolic Logic held in the summer of 1957 at Cornell University. This brought together leading research workers and their students from all the different fields of logic and encouraged a process of intercommunication and interaction which has continued unabatedly ever since. While each branch of mathematical logic still maintains a distinctive character and body of concerns, it is difficult to work in any one of them nowadays without using knowledge and techniques from one or more of the others.

Perhaps more than any other branch, recursion theory was to see an infusion of concepts, methods and examples from all the other branches which significantly affected its development in the following years. This was to transform its conceptual arena from the natural numbers (and related effectively enumerated structures such as word systems), in what has come to be called *ordinary recursion theory* (o.r.t.), to quite *general structures*, thus opening up the development of what is called *generalized recursion theory* (g.r.t.). This in turn has followed two lines: (i) the generalization of recursion theory to various *structures of sets and ordinals* and (ii) the generalization of recursion theory to (more or less) *arbitrary* ('*abstract*') *structures*.

The subject of g.r.t. is much more difficult to describe than the material discussed up to now in §§13.2–13.4, because of the welter of conceptual approaches, structures to which they are applied, and results obtained. Some initial impression of this variety can be obtained from the books Barwise (1975) and Fenstad (1980), the conference volumes Fenstad and Hinman (1974) and Fenstad et al. (1978), and the survey articles by Shore, Kechris and Moschovakis, Aczel, and Martin (in that order) in the *Handbook of Mathematical Logic* (Barwise, 1977). Last, but not least, to be mentioned is the survey and critical assessment of g.r.t. by Kreisel (1971), which represents the situation mid-stream. The book Sacks (1990) fills in much of the technical picture for recursion theory on sets and ordinals, though not for g.r.t. on arbitrary structures.

Given the complexity and heterogeneity of these developments, nothing short of a full survey and new critical assessment would do justice to g.r.t. My purpose in the following is just to give the reader *some* sense of how *some parts* of this have proceeded, as part of a picture of the further relativization of recursion theory, and in particular to emphasize the conceptual shift to a structural view of its subject matter.

13.5.2 Computability over sets and ordinals

Here I follow in part the survey Shore (1977), which gives a good introduction with historical background together with references that I shall not repeat (see also Kreisel 1971).

A notion of recursive function of ordinals was introduced by Takeuti in 1960 by means of schemata, where the schema for primitive recursion is expanded to all ordinals by taking the sup at limit ordinals x,

(1) $f(x) = \sup\{g(y) \mid y < x\}$.

Another generalization of recursion theory to ordinals was provided by Machover in 1961 using an extension of the Herbrand–Gödel equation calculus with certain infinitary rules of inference and by Lèvy in 1963 using an analogue of Turing machines; both Machover and Lèvy observed that one could work just as well with the ordinals less than a regular cardinal, since that set is closed under suprema (1). A still further refinement was made by Kripke in 1964 and Platek in 1966, who realized that a much wider class of ordinals, called the *admissible ordinals*, support a reasonable generalization of recursion theory. Kripke again used a form of the equation calculus, while Platek used both a definition by schemata and one by generalized computers. As described in (Barwise, 1975, p. 3), an ordinal α is called admissible if, for every α-(partial) recursive function f of ordinals, whenever $x < \alpha$ and $f(x)$ is defined, then $f(x) < \alpha$ where, moreover, f is α-(partial) recursive if its values can be computed by an "idealized computer capable of performing computation of less than α steps".

Meanwhile (1963–1965) Kreisel and Sacks had been developing recursion theory on the ordinal $\alpha = \omega_1^{\mathrm{CK}}$, the least non-recursive ordinal in the sense of Church and Kleene (the recursive analogue of the least uncountable ordinal ω_1); this turns out to be the first admissible ordinal greater than ω (the ordinal of the natural number structure $(N, <)$). Sacks sought to generalize results of degree theory from o.r.t. to recursion theory on ω_1^{CK}. For this it was necessary to have a suitable generalization of the relation $A \leq_{\mathrm{T}} B$ of relative computability for arbitrary sets A, B. The crucial ingredient was supplied by Kreisel in the form of a generalized notion of finiteness. Extended directly to arbitrary admissible ordinals α and subsets A of α, his proposal was to define:

(2) A is α-finite if A is α-recursive and bounded, i.e., if A is contained in β for some $\beta < \alpha$.

Now the Kreisel–Sacks definition of $A \leq_\alpha B$ (Turing reducibility generalized to recursion theory on an admissible ordinal α) is, roughly speaking, that every α-finite subset of A can be determined in an α-effective way from some α-finite subsets of B and its complement (to α), and similarly for every α-finite subset of the complement of A. Before long, Sacks and his students were extending one result after another from degree theory in o.r.t. to arbitrary admissible ordinals. In particular, in 1972, Sacks and Simpson established the analogue of the Friedberg–Muchnik solution to Post's problem: there exist α-r.e. sets A, B such that neither $A \leq_\alpha B$ nor $B \leq_\alpha A$. This makes use of an extension of the priority method to admissible ordinals, for which a full technical exposition is to be found in Sacks (1990).

Recursion theory on admissible ordinals also gave rise to a recursion theory on sets via the intimate relation between ordinals and constructible sets, in the sense of Gödel (see his 1940 monograph in the collection Gödel (1990)). Another form of recursion on sets, called *E-recursion*, that is distinct from admissible recursion theory, was introduced (independently) by Normann and by Moschovakis around 1978. A number of generalizations of degree theory have also been obtained for E-closed sets (also given an exposition in Sacks (1990)). Though there were many motivations for the generalization of recursion theory to ordinals and sets and these have been satisfied to a large extent by subsequent developments, the research program of generalized degree theory has been the direction which has been pursued the most vigorously and again with the most technically difficult results.

13.5.3 Computability over general structures

The idea of generalizing recursion theory to (more or less) arbitrary structures also began early in the 1960s. The article Kreisel (1971) provides a comprehensive source for the developments up to the time of that publication, and references cited there will not be repeated here. One of the first proposals was made by Fraïssé in 1961, in model-theoretic terms. A variety of subsequent proposals receiving special attention were those due to (among others) Lacombe, Montague, Moschovakis, Platek, and Friedman; we shall sketch only the latter two approaches here, in reverse historical order.

The work of Friedman (1971) generalized notions of Turing machines and register machines to arbitrary first-order structures

$$\mathscr{M} = \langle M_0, R_1, \ldots, R_l, g_1, \ldots, g_m \rangle$$

where now R_j are relations and the g_j are total functions on M_0. If the relation $x = y$ on M_0 is to be counted as computable then it must be included as one of the basic relations, but that is not assumed in general. In Friedman's generalization of register computability to an arbitrary structure $\mathscr{M}$, each register is empty or contains an element a of M_0. Then the actions specified by the instructions are of the form

(1) $r_i := g_j(r_{n_1}, r_{n_2}, \ldots)$ and
(2) if $R_j(r_{n_1}, r_{n_2}, \ldots)$ then go to I_k else to I_l.

The meaning of (1) is to replace the contents of r_i by $g_j(a_1, a_2, \ldots)$ where a_t is in the n_tth register; and that of (2) is to perform a conditional transfer with test $R_j(a_1, a_2, \ldots)$. Then a partial f is *register computable* over $\mathscr{M}$ if $f(a) = b$ just in the case that a computation with $r_1 = a$ as input terminates with $r_0 = b$ as output (and similarly for n-ary f). This generalizes the Shepherdson and Sturgis notion by taking $\mathscr{M}$ to be the structure $\langle N, R_1, 0, \mathrm{sc}, \mathrm{pd}\rangle$ where R_1 is the unary relation $\{0\}$, i.e., $R_1(x)$ iff $(x = 0)$; equivalently we may take $\mathscr{M}$ to be the structure $\mathscr{N} = \langle N, =_N, 0, \mathrm{sc}, \mathrm{pd}\rangle$. It also yields the relation $f \leq_{\mathrm{T}} g$ as a special instance since that holds just in the case when f is computable over $\langle N, =_N, 0, \mathrm{sc}, \mathrm{pd}, g\rangle$. Friedman's notion generalizes directly to many-sorted structures. Then he defines f to be register computable over $\mathscr{M}$ *with counting* if it is computable on the combined structure $(\mathscr{M}, \mathscr{N})$. Friedman also generalized computability by Turing machines to arbitrary $\mathscr{M}$, where the contents of a Turing tape cell may be empty or filled by an element of $\mathscr{M}$.

(Kreisel, 1971, p. 144) asked whether there is such a thing as "an extension of Church's Thesis to general (abstract) structures". In his discussion of this (*op. cit.*, pp. 175ff) he points out that "Evidently two elements are involved in Turing's analysis, ... the objects on which we operate, (and) the instructions or rules of computation". According to him, Turing's analysis requires a restriction on how the objects of computation may be presented to us, and what operations on them may be assumed. From this point of view, Turing computability on the structure $\langle N, \ldots, g\rangle$ where g is not recursive, is not a suitable structure for computation. Nor would a structure $\langle N, N_1, \ldots\rangle$ be admitted, for $N_1 = (N \to N) = \{g \mid g : N \to N\}$. Clearly the essence of Friedman's generalization of register computability is to give up any restriction on how the objects on which we operate are presented to us, but to maintain the form of the instructions or rules of computation. Shepherdson (1988) extended Gandy's principles for mechanisms to arbitrary structures, and it is there argued that Friedman's 'machines' lead to a general form of the Church–Turing Thesis. As has already been stated, we do not take issue here with such a position, one way or another.

However, it should be noted that not all structures for which we have a reasonable generalization of recursion theory fall under Friedman's definitions. In particular, recursion theory on admissible ordinals or sets beyond the natural numbers don't come out as special cases; the reason is that they embody essentially *infinitary operations* such as sup.

This brings us to Platek's generalization of recursion theory to arbitrary $\mathscr{M}$, carried out in his (unpublished) dissertation (Platek, 1966). Besides explicit functional definition (using the operations and relations of $\mathscr{M}$ as basic), this takes the recursion theorem

(3) $f = F(f)$

as its central means of definition. In order to make sense of this as providing the *least fixed point* $f = \mathrm{FP}(F)$, it must be at least assumed that F is a monotonic functional. But then the question arises, specifically which F are to be used: Platek's answer was that they must in turn be generated by the recursion theory over $\mathscr{M}$. He thus introduced a type structure HMF of *hereditary monotonic functionals over* $\mathscr{M} = \langle M_0, \ldots \rangle$. For this, a relation of inclusion is defined at each type, with f contained in g at type (σ, τ) if, for all x in M_σ, $f(x)$ is contained in $g(x)$ at type τ. Then $M_{(\sigma,\tau)}$ is taken to consist of all monotonic $f : M_\sigma \to M_\tau$ in this inclusion relation. Now for each F in $M_{(\rho,\rho)}$ where $\rho = (\sigma, \tau)$ there is a least f in M_ρ satisfying the fixed-point equation (3). Finally the operation $\mathrm{FP}_\rho : M_{(\rho,\rho)} \to M_\rho$ is itself in $M_{((\rho,\rho),\rho)}$. With each collection $\mathscr{F}$ of functionals in this type structure over $\mathscr{M}$ is associated the collection $\mathrm{Rec}(\mathscr{F})$ generated by explicit definition and all the fixed-point operators FP_ρ; the basic operations and relations of $\mathscr{M}$ are built into $\mathscr{F}$.

The type structure HTF(M_0) of hereditarily *total* functionals over M_0 can be extracted from HMF(M_0). In particular, when $M_0 = N$, Platek recaptured Kleene's (1959b) partial recursive functionals over N by taking $\mathscr{F} = \{0, \mathrm{sc}, \mathrm{pd}\}$, and Kleene's notion of (higher type) recursion in some particular functions or functionals $F_1, \ldots, F_m$ by taking $\mathscr{F} = \{0, \mathrm{sc}, \mathrm{pd}, F_1, \ldots, F_m\}$. In this way we can incorporate infinitary operations, e.g. the functional ${}^2E$ of Kleene, with

(4) ${}^2E(f) = 0$ if $(\exists x)(f(x) = 0)$ and ${}^2E(f) = 1$ otherwise, for $f : N \to N$.

(An important result of Kleene (1959b) was that the functions recursive in ${}^2E$ are exactly the hyperarithmetic functions.) But Platek also captured recursion on admissible ordinals and sets, by taking recursion in the functional Sup given by

(5) $\mathrm{Sup}(f,x) \simeq \sup\{\, f(y) \mid y < x \,\}$.

Is it necessary to go through all higher types in order to find the functions of type 1 in $\mathrm{Rec}(\mathscr{F})$? One of the principal results of Platek (1966) is that if every member of $\mathscr{F}$ is of type level ≤ 2 and f is in $\mathrm{Rec}(\mathscr{F})$ and of level ≤ 2, then we need only use the schemata for explicit definition and FP applied in type levels ≤ 2 in order to obtain f. The above examples with ${}^2E$ and Sup meet these conditions.

While Platek's approach is of impressive generality and builds on a natural idea (recursion as given by the FP operator), it does not cover all the cases one would want to include. In (Feferman, 1977, pp. 376–377), I discussed certain limitations of Platek's theory. In brief these are:

(i) It is assumed that there are pairing and projection functions on $\mathcal{M}$ in the basic $\mathcal{F}$, as well as distinct elements 0, 1 from M; thus M_0 contains an image of the natural numbers and the possibility of enumeration. It is preferable to separate out the natural numbers by a different basic sort if they are to be used at all.

(ii) The theory does not generalize relational notions of computability for which the paradigm is the Post–Smullyan approach (see Smullyan, 1961, Fitting, 1987).

(iii) Details of the extraction of the HTF type structure from the HMF type structure are very messy for types > 2, and this makes extraction of the general (Kleene, 1959b) notions very complicated.

Moving beyond the preceding, in the mid-1970s Moschovakis and I independently proposed getting around such defects by treating recursion in higher types as a special case of recursion on arbitrary structures, rather than as the means to define it. As I put it (Feferman, 1977, p. 373): "In contrast to Platek, higher-type structures are regarded here as just further examples which are to be *subjects* of the notions of g.r.t. rather than *tools* to explain the notions." My approach was sketched in Feferman (1977), but all the detailed work has been carried out by Moschovakis, first in collaboration with Kechris for the special case of Kleene recursion in higher types (see Kechris and Moschovakis, 1977), and then more generally in Moschovakis (1984), among other publications. Basically, the notions concern type-level-2 functionals F with arguments chosen from a collection $\mathcal{R}$ of type-level1 relations over a many-sorted structure $\mathcal{M} = \langle\langle M_k\rangle, \ldots\rangle$ where $\mathcal{R}$ is closed under unions of chains. In particular, $\mathcal{R}$ may be chosen to be all (sorted) partial functions over $\mathcal{M}$, or all (sorted) relations over $\mathcal{M}$. Then for any collection $\mathcal{F}$ of type level 2 monotonic functionals F over $\mathcal{R}$ one defines $\text{Rec}(\mathcal{F})$ to be the least collection of objects of type level ≤ 2 generated by explicit definition and the least fixed-point operators FP. This gives rise to a theory of great generality, although it is still not clear whether it covers *all* cases for which we have a reasonable generalization of recursion theory.[5]

There is one final significant step that Moschovakis took in his 1984 paper. This was to consider functionals F *operating across structures*. That is, given a class $\mathcal{K}$ of structures of the same similarity type one can give meaning to objects F defined over $\mathcal{K}$ such that for each $\mathcal{M} \in \mathcal{K}$, $F(\mathcal{M})$ is a functional $F_{\mathcal{M}}$ over $\mathcal{M}$ such that the functionals $F_{\mathcal{M}}$ all *act in the same way*. In other words, this provides

[5] In particular, Part II of Feferman (1977) was concerned with the question of whether the notion of partial recursive functional of *hereditarily total continuous objects* ('*countable functionals*') developed by Kleene and Kreisel in 1959 is covered by this theory. As far as I know, this is still open. However, a related and more extensive notion of recursion on the hereditarily *partial* continuous functionals was shown to be accounted for by recursive schemata in the above sense.

a notion of *uniform computability across structures*.[6] The significance of this for actual computability will emerge in the next section.

13.6 The role of notions of relative computability in actual computation

13.6.1 Computational practice and the theory of computation

The kinds of mechanisms we have in mind are high-speed, digital, general-purpose computers, from PCs to mainframes. For these, the aim of computational practice is to produce hardware and software that is reliable, efficient, flexible, and user-friendly. The aim of the theory of computation is to aid engineers in the design of hardware and software to meet these requirement by providing a body of concepts around which to organize experience and a body of results predicting correctness, efficiency, and versatility. Theory also serves to set limits to what is feasible and thus provides warning signals for when these limits are approached. The theory of computation employs logic and mathematics ranging from the most concrete combinatorial kind to the most abstract, algebraic, and topological kind. The following tells some 'stories' about the role that notions of relative computability play in computational practice; these sit somewhere in the middle between the two extremes of the theory of computation. Unlike the preceding sections of the chapter it is neither historical nor (for the most part) does it concern specific results.

The literature in theoretical computer science accepts the Church–Turing Thesis in principle, of course with the proviso that this must be supplemented by an assessment of time and space requirements for practice. Sometimes it is said that the notion of finite automaton must be substituted for that of Turing machine (or equivalent), to reflect the actual limitations on space. However, in practice memory (storage) is expandable, and the automaton model does not account for that. On the other hand, it is generally recognized that Turing machines themselves do not provide a realistic model of actual mechanisms since "... they are confined to specific data structures (the tapes) which have artificially high large access time (because in order to read a bit far away on a tape the respective head has to travel over all cells in between)" (Maas and Slaman, 1989, p. 80). Register machines instead provide a more realistic model of random access memory in practice (see *loc. cit.* and Aho et al. (1974)). Moreover, at least one style of programming is directly keyed to that mode, namely the ('von Neumann') imperative style, with assignment statements (e.g. as in Pascal). But the theory of computation must account for a number of other programming styles, such as functional programming or logic programming,

[6] Something like this was anticipated in a lecture for the Association for Symbolic Logic that I delivered in 1969 (see *JSL* **35** (1970), p. 179); regrettably, the material was never published though I circulated handwritten notes, 'Uniform inductive definitions and generalized recursion theory', at the time. See also (Kreisel, 1971, pp. 147–8).

which are less directly related to the nature of the hardware. For all these, and even for imperative style programming, the details of the underlying mechanism are largely considered to be irrelevant. Thus the question arises what the significance of the Church–Turing Thesis is for computational practice; on the face of it, the thesis seems to be a matter of basic creed which has nothing to do with day-to-day computational life. A four-fold response was provided to me by P. Odifreddi in correspondence, summarizing points in his introduction to the collection *Logic and Computer Science* of which he is the editor (Odifreddi, 1990):

(i) the notion of universal Turing machine is the idealization (and conceptual precursor) of modern *all-purpose computers*;
(ii) the enumeration theorem shows the *equivalence of programs and data*, basic for stored program machines;
(iii) the proof of Kleene's normal form theorem for partial recursive functions provides the theory underlying *interpreters*; and
(iv) various definitions of recursiveness provide the *computational core (and style) of different* programming languages (cf. for example Pascal, Prolog and LISP). In personal conversation, (and forthcoming work[7]), W. Sieg has further emphasized the constraints placed on actual computation by the theoretical analysis stemming from Turing (1936, 1937) and leading to Gandy's very general principles for mechanisms (Gandy, 1980): after all, what counts as computation in its everyday sense can't be completely arbitrary.

While I can hardly disagree with these points (and have already brought out (i) and (ii) in §13.2), it will be argued here that, nevertheless, notions of *relative* computability have a much greater significance for practice than those of *absolute* computability. The reason is very simple: as with all forms of technology, the requirements of efficiency, reliability, and usability force an organization of the devices and their control into *conceptual levels* and at each level into interconnected *components*. At the hardware level, one has a breakdown into such gross components as a central processing unit (CPU), memory locations, both read-only (ROM) and random access (RAM), a clock, and so on; then, for each of these, one has further refinements into subcomponents such as adders, down finally to the level of individual switches. At each level one depends on standard designs but always subject to improvements, so that if any one component is changed, the performance of the other components is not affected. Moreover, if the whole material basis of the technology is changed from, say, chips to fibre optics, the organization of components need not be modified. Nor is there a simple dichotomy between the hardware–software levels (or trichotomy, if one adds in the user). Rather, there is a stepwise

[7] See Sieg (2008).

ascent from hardware to software or, from the point of view of the programmer, descent from the programming language through a compiler or interpreter down to assembly language and, finally, to 'machine' language. And for the programmer there are, to begin with, the shifts in level from informally stated problems and tasks to their mathematical or symbolic formulation, down to a concrete program in one language or another, all well captured in the slogan of 'top-down design' (see for example Alagic and Arbib (1978), and Harel (1987)).

While notions of relative computability have some connection with the different conceptual levels of organizations in hardware and software, what is rather emphasized in the following is their significance, at a given level, for *modular organization*, i.e., *how things are packaged*, and *how they fit together.*

13.6.2 Built-in functions and black boxes

To become more specific, let us return to Turing's 'oracle' machines and the relation $f \leq_T g$. Actual computers have a variety of built-in functions g, whose values may be called at any point in a program. These are arithmetical operations on integers such as $+$, $-$, $*$, quot, rem; Boolean operations such as 'and', 'or', 'not'; operations from integers to Booleans such as lesseq; and sometimes also operations from and to (approximate) real numbers such as sqrt, sin, log, etc. As far as the programmer is concerned, each of these is given by a 'black box' – which is just another name for an 'oracle' – and a program to compute a function f from one or more of these $g_1, \ldots, g_m$ is really an algorithm for the computation of f relative to $g_1, \ldots, g_m$. Such an algorithm can thus build in commands to apply one of the g_j to arguments which arise in the course of the computation. Moreover, for certain purposes, measures of complexity can also be relativized to the black boxes, e.g. they might be assigned unit cost or even zero cost.[8]

13.6.3 Functional aspects of programming

These are both *implicit* and *explicit*. Examples of the former are provided by *flowchart analyses* for certain programming languages. Consider programs Π for a register machine. At any point in a computation the operation of such Π is determined by the state s of the contents of the various registers and the effect of Π is to change s to $\Pi(s)$. Thus Π may be considered as (determining) a function $\Pi : S \to S$ where S is the set of all states. Now, in a fragment of a flowchart program,

(1) $\to \Pi_1 \to \Pi_2 \to$

[8] Several people have suggested to me that *interactive computation* exemplifies Turing's 'oracle' in practice. While I agree that the comparison is apt, I don't see how to state the relationship more precisely.

indicates the composition $C(\Pi_1,\Pi_2)$, which has the effect

(2) $C(\Pi_1,\Pi_2)(s) \;=\; \Pi_2(\Pi_1(s))$.

The construction C here may be considered to be a functional on $S^S \times S^S$ where $S^S = \{\Pi \mid \Pi : S \to S\}$. As another example, conditional branching, whose flowchart follows Π_1 if R is true and Π_2 otherwise, where R is contained in S, gives rise to the functional

(3) $B_R(\Pi_1,\Pi_2)(s) \;=\;$ (if s is in R then $\Pi_1(s)$, else $\Pi_2(s)$).

Similarly, we may treat such program constructions as

(4) while R do Π, and
(5) do Π until R,

as functionals of Π and R (or of a program for R as a function from S to $\{T,F\}$). In these cases, the functionals may yield partial functions on states to states as values.

Examples of *explicit* functional operations are given by functional-style programming languages such as LISP and ML. In these we can form expressions involving functional recursion such as F defined by the following:

(6) $F(g,h;x) \;=\; [g(0) \text{ if } x=0 \text{ else } h(F(g,h;x-1),g(x))]$,

which has the solution $f = F(g,h)$ with

(7) $f(0) \;=\; g(0)$,
$\quad f(x') \;=\; h(f(x),g(x))$.

For example, $F(g,+,x)$ and $F(g,*,x)$ yield, respectively, the sum and product of terms $g(y)$ for $y \leq x$.

The use of higher-order functions permeates functional programming languages; see Reade (1989). They are generally based on some form of the untyped λ-calculus, though flexible ('polymorphic') systems of typing have also been imported (see *op. cit.* as well as Feferman (1990) for references). In these languages programs are represented by expressions, and operations on programs such as composition, conditional branching, iteration, etc. are represented by compound expressions. In the strictly typed λ-calculus there are rigid rules that govern the compounding of expressions and thus say exactly how the corresponding programs may be interconnected. In untyped calculi with polymorphic type-assignment systems (see for example Mitchell and Harper (1988)) such rules are considerably more flexible, permitting combinations forbidden in the strictly typed calculus but still providing for sensible interconnections of the corresponding programs. In terms of the theme of §13.6.1 above, these give systematic ways of representing the modular construction of software.

Research on type systems, logics, and semantics for functional programming languages is still being carried on vigorously by a number of authors (see the works cited above for further references).

13.6.4 Abstract data types

All programming languages deal with types of expressions, either internally within the syntax or externally in the semantics. One generally has such basic data types as integers, Booleans, and reals and then general type constructions, such as those for lists, arrays, stacks, queues, sets, trees, streams, etc. In functional programming lanuguages the concern is also with higher-order data types such as those for functions and functionals. Again, there are many approaches to dealing with these concepts and research is ongoing. The purpose here is to show how certain ways of looking at these connect with relativized recursion theory and, more specifically, with recursion theory over general structures.

From a semantical or external point of view, abstract data types are either specific structures $\mathcal{M}$ considered independently of the form of representation of their elements, in other words as *isomorphism types*, or as *collections* $\mathcal{K}$ *of structures* of a given similarity type. In either case, these structures may be prescribed by *axiomatic defining conditions* A, e.g., by equations or Horn clauses. In general such axiom systems are not categorical, unless supplemented by some second-order conditions, e.g., that $\mathcal{M}$ is the least structure satisfying A (or is an *initial structure* for A), or that $\mathcal{K}$ consists of all finite structures satisfying A, etc. In whatever way $\mathcal{M}$, respectively $\mathcal{K}$, is prescribed, one can give a semantics for programs on these structures using one of the generalizations of recursion theory mentioned or described in §13.5.3. For example, Tucker and Zucker (1988) considered various forms of schematic definability, while Moschovakis (1984) used the general form of inductive defining schemata. An interesting result from the latter (see p. 326) is that uniform global recursion on the class of finite structures with a linear ordering gives exactly the polynomial-time-computable relational queries for these structures (for which notion see Chandra and Harel (1980)).

Uniform global recursion provides a much more realistic picture of computing over finite data structures than the absolute computability picture, since finite data bases are constantly being updated. As examples, we may consider computations on weather data (given by finite samples from a continuous space) for weather prediction, or the status of communcation lines for routing in a telephone system, or airline reservation systems, and so on, with endless practical examples.

Limitation of space here does not allow me to go into the internal or syntactic representation of abstract data types. For some approaches see Mitchell and Plotkin (1984) and Feferman (1990).

13.6.5 Degrees of complexity

So far we have considered the significance for computation theory only of notions of relative computability other than those from the theory of degrees of unsolvability. But the latter would seem to provide a *prima facie* case not only of the application of notions, here to *complexity theory*, but also of methods and results. There is some dispute, though, as to whether the latter subject should be considered a part of computation theory or a part of recursion theory. Be that as it may, let us consider the situation as it appears at present. Here we rely on such sources as the venerable Garey and Johnson (1979) and the more up-to-date exposition Balcázar et al. (1988).

The theoretical basis for predicting the relative efficiency of algorithms lies in the assignment of time and space measures of complexity to algorithms. For example, one bounds how long it takes to compute a function $f(n)$ by a given algorithm, as a function of the size $|n|$ of input in binary notation. The crudest distinction puts tractable problems in the class that can be computed in $O(p(|n|))$ time for some algorithm and polynomial p, and intractable problems in the class that requires at least $O(2^{|n|})$ time for any algorithm. A function computable by an algorithm of the former kind is said to be *polynomial-time computable*, and the class of these is denoted by P. A set A of natural numbers (or the decision problem for membership in A) is in class P if c_A is in P. There are a number of decision problems that are not obviously in class P but lie in a class that is not as complicated as those requiring exponential time. This is denoted NP, for *nondeterministic polynomial-time computability*. Roughly speaking, these are problems for which one can check, for a given n, that n is in A, when in fact it does belong to A, by means of some certifying evidence which is itself verified to be such in P-time. For example, the problem of the satisfiability of a formula in classical propositional calculus is in the class NP. On the face of it, it is hard to decide whether such a formula is satisfiable since we must set up a truth table for it, and it contains n literals, which contains 2^n lines, each of which must be checked. But if a truth assignment s actually is one which makes the formula satisfiable, that fact can be checked in polynomial time. For problems A in the natural numbers the notion of NP-computability can be formulated as definability in the form

(1) x is in A if and only if $(\exists y)[|y| \leq p(|x|)\ \&\ R(x,y)]$

where p is a polynomial and R is in the class P.

The nondeterministic aspect of such problems lies in the fact that one may not have a feasible method of choosing in advance, given x in A, a y which quickly certifies that x is in A. The form (1) readily suggest an analogy:

(2) $NP \sim$ recursively enumerable, and

(3) $P \sim$ recursive.

Moreover, one has a concept of *polynomial-time reduction* of problems which is analogous to that of Turing reducibility, where $A \leq_{\mathrm{p}} B$ holds if, roughly speaking there is an algorithm which transforms any P-algorithm for B into one for A. Thus

(4) $(\leq_{\mathrm{p}}) \sim (\leq_{\mathrm{T}})$.

This immediately suggests the notion of NP-completeness analogous to that of Turing completeness in the class of r.e. sets: B is called NP-complete if every NP set A is P-reducible to B. There are, indeed, NP-complete problems; this was the major result of Cook (1971), who showed that the satisfiability problem for the propositional calculus is one such. Since then a number of other problems that arise naturally in practice have been shown to be NP-complete, including the 'travelling salesman' problem and the Hamiltonian path problem: see Garey and Johnson (1979). So far, so good: the parallels to degree theory are persuasive. One now comes to posing the analogue of the Church–Turing existence of effectively unsolvable problems:

(5) does $P = NP$?

No answer is yet known to this, though it is generally conjectured that $P \neq NP$. But here one has a break in the analogy. Namely, effective unsolvability in ordinary recursion theory relativizes, but the $P = NP$ question does not. That is, the halting problem H (or the diagonal halting problem K), which demonstrates the existence of r.e. but non-recursive sets, is such that, relative to any set A, we have

(6) H^A is not $\leq_{\mathrm{T}} A$.

Put in other terms, for any A, Rec^A is properly included in $\mathrm{R.E.}^A$. However, Baker, Gill and Solovay (1975) proved that

(7) there exist A such that $P^A = NP^A$, though there exist B such that $P^B \neq NP^B$.

If, as is generally conjectured, $P \neq NP$, it would be natural to investigate further the analogue to Post's problem:

(8) do there exist A in NP which are not in P and not NP complete?

Here, there is a positive answer (assuming $P \neq NP$) due to Ladner in 1975; see (Balcázar et al., 1988, p. 156). But it seems that none of the non-trivial techniques or results of degree theory has so far been of any use for this or any other result in complexity theory. Naturally, future developments may change that situation.

Finally, we should mention the development of *hierarchies* similar to the arithmetical, e.g. the *polynomial-time hierarchy* introduced by Meyer and Stockmeyer in 1973 and treated in (Balcázar et al., 1988, Ch. 8). However, many of the basic questions about this are open, such as whether the entire hierarchy goes beyond P or, alternatively, simply collapses.

13.6.6 Conclusion

Our final section has explored the question of the relevance of the mathematical theory of computability – in the guise of recursion theory – to the theory of computation, insofar as *that* is supposed to be a theory of computational practice. To avoid misunderstanding, I do not believe recursion theory is to be valued only if it can have such applications. Aside from the clear philosophical value of having a fundamental analysis of the notion of computation, there have been plenty of applications in logic and mathematics to justify its existence, if justification by external relevance is called for at all. But even in the most rarefied and recondite parts of the subject such as degree theory, persuasive intrinsic reasons can be given for the lines of development that have been taken and for the continued pursuit of internally driven problems (along with the feeling that "... we have to do it because it's *there*"). Be that as it may, the case has been made here that while notions of relativized (as compared to absolute) computability theory are essentially involved in actual hardware and software design, the bulk of the methods and results of recursion theory have so far proved to be irrelevant to practice. Whether and how these these disciplines might be brought closer together remains for the future to tell.

To conclude on a more positive but more speculative note, I think it can be argued that – whatever a 'fundamental' theory may tell us is the basic or underlying mechanism for given technological processes or systems – it is necessary for our design and use of such to think of them at various conceptual levels and with various modular forms of organization; that is, *we must think of them in structural terms*. The story of Turing's 'oracle' and its significance for actual computability is but one example among many of this characteristic *modus operandi* of human intelligence.

Postscript

It is now almost a quarter of a century since the original (Feferman, 1992) of this article appeared – a long time in our subject – but in my view the general perspective requires no revision.[9] However, there is much in the considerable subsequent literature that usefully amplifies on the topics dealt with here; I shall mention only

[9] I very much appreciate the invitation of Barry Cooper to contribute to the present volume in this way. Also, the reprinting here did provide a welcome opportunity to correct the treatment of partial recursive functionals at the end of §13.4.1.

a few items. First of all, one should look at the *Handbook of Computability Theory* (Griffor, 1999), especially the first three parts. Next, all Turing's articles on logic and the theory of computation may be found collected in Gandy and Yates (2001) with some useful introductory essays. More recently, the Turing Centennial in 2012 inspired a number of conferences and publications, most notably *Alan Turing. His Work and Impact* (Cooper and van Leeuwen, 2013) in which many of Turing's contributions across his wide field of interests are reproduced, along with a great variety of commentary. The collection of essays, *Computability. Turing, Gödel, Church, and Beyond* (Copeland et al., 2013) deserves special attention for historical and philosophical considerations. Finally, in the forthcoming article Feferman (2015), I have dealt in greater depth with computation and recursion on concrete and abstract structures and the question of generalizations of the Church–Turing Thesis.

References

A. Aho, J.E. Hopcroft, and J. Ullman 1974. *The Design and Analysis of Computer Algroithms*. Addison–Wesley.

S. Alagic and M.A. Arbib 1978. *The Design of Well-Structured and Correct Programs*. Springer Verlag.

T. Baker, J. Gill, and R. Solovay 1975. Relativisations of the P =? NP question. *SIAM J. Comput.*, 4(4):431–442.

J.L. Balcázar, J. Díaz, and J. Gabouró 1988. *Structural Complexity I*. Springer Verlag.

J. Barwise 1975. *Admissible Sets and Structures*. Springer Verlag.

J. Barwise, editor 1977. *Handbook of Mathematical Logic*. North Holland.

A.K. Chandra and D. Harel 1980. Computable queries for relational data bases. *J. Comput. System Sci.*, 21:156–178.

A. Church 1937. An unsolvable problem of elementary number theory. *Amer. J. Math.*, 58:345–363. Reprinted in Davis (1965).

S.A. Cook 1971. The complexity of theorem-proving procedures. In *Proceedings of the Third ACM Symposium on the Theory of Computing*, pp. 151–158, Shaker Heights, Ohio.

S. B. Cooper and J. van Leeuwen, editors 2013. *Alan Turing. His Work and Impact*. Elsevier.

B.J. Copeland, C.J. Posy, and O. Shagrir, editors 2013. *Computability. Turing, Gödel, Church, and Beyond*. MIT Press.

M. Davis 1965. *The Undecidable. Basic Papers on Undecidable Propositions, Unsolvable Problems and Computable Functions*. Raven Press.

M. Davis 1982. Why Gödel didn't have Church's thesis. *Information and Control*, 54:3–24.

S. Feferman 1977. Inductive schemata and recursively continuous functionals. In *Logic Colloquium '76*, pp. 373–392. North Holland.

S. Feferman 1998. *Turing in the land of* $O(z)$. In Herken (1988), pp. 113–147.

S. Feferman 1990. Polymorphic typed λ-calculi in a type-free axiomatic framework. In *Logic and Computation*, Contemporary Mathematics, volume 104, pp. 101–137. AMS.

S. Feferman 1991. Logics for termination and correctness of functional programs. In *Logic from Computer Science*, pp. 95–127. MSRI Publications, Springer Verlag.

S. Feferman 1992. Turing's 'oracle': from absolute to relative computability – and back. In *The Space of Mathematics*, J. Echeverria et al., editors, pp. 314–348. de Gruyter.

S. Feferman 2015. Theses for computation and recursion on concrete and abstract structures. To appear in *Turing's Revolution. The Impact of his Ideas about Computability*, G. Sommaruga and T. Strahm, editors.

J.E. Fenstad 1980. *General Recursion Theory: An Axiomatic Approach*. Springer Verlag.

J.E. Fenstad and P. Hinman, editors 1974. *Generalized Recursion Theory*. North Holland.

J.E. Fenstad, R. Gandy, and G. Sacks, editors 1978. *Generalized Recursion Theory II*. North Holland.

M. Fitting 1987. *Computability Theory, Semantics and Logic Programming*. Oxford University Press.

R.M. Friedberg 1957. Two recursively enumerable sets of incomparable degrees of unsolvability (solution of Post's problem 1944). *Proc. Nat. Acad. Sci.*, 43:236–238.

H. Friedman 1971. Algorithmic procedures, generalized Turing algorithms, and elementary recursion theory. In *Logic Colloquium '69*, pp. 361–389. North Holland.

R.O. Gandy 1980. Church's thesis and principles for mechanisms. In *The Kleene Symposium*, pp. 123–148. North Holland.

R.O. Gandy 1988. The confluence of ideas in 1936. In Herken (1988), pp. 55-111.

R.O. Gandy and C.E.M. Yates, editors 2001. *Collected Works of A.M. Turing. Mathematical Logic*. Elsevier.

M. Garey and D. Johnson 1979. *Computers and Intractability: A Guide to the Theory of NP-Completeness*. W.H. Freeman and Co..

K. Gödel 1986. *Collected Works Volume I: Publications 1926–1936*. Oxford University Press.

K. Gödel 1990. *Collected Works Volume II: Publications 1938–1974*. Oxford University Press.

E.R. Griffor, editor 1999. *Handbook of Computability Theory*. Elsevier.

D. Harel 1987. *Algorithmics: The Spirit of Computing*. Addison–Wesley.

R. Herken, editor 1988. *The Universal Turing Machine. A Half Century Survey.* Oxford University Press.

P. Hinman 1978. *Recursion-Theoretic Hierarchies*. Springer Verlag.

A. Hodges 1983. *Alan Turing: The Enigma.* Simon and Shuster.

A. Kechris and Y. Moschovakis 1977. Recursion in higher types. In: Barwise (1977), pp. 681–737.

S.C. Kleene 1938. On notation for ordinal numbers. *J. Symbolic Logic*, 3:150–155.

S.C. Kleene 1952. *Introduction to Metamathematics*. North Holland.

S.C. Kleene 1959a. Countable functionals. In *Constructivity in Mathematics*, pp. 81–100. North Holland.

S.C. Kleene 1959b. Recursive functionals and quantifiers of finite types I. *Trans. Amer. Math. Soc.*, 91:1–52.

S.C. Kleene 1981. Origins of recursive function theory. *Ann. History Comput.*, 3:52–67.

S.C. Kleene and E.L. Post 1954. The upper semi-lattice of degrees of unsolvability. *Ann. Math.*, 59:379–407.

G. Kreisel 1959. Interpretation of analysis by means of constructive functionals of finite types. In *Constructivity in Mathematics*, pp. 101–128. North Holland.

G. Kreisel 1971. Some reasons for generalizing recursion theory. In *Logic Colloquium '69*, pp. 139–198. North Holland.

M. Lerman 1983. *Degrees of Unsolvability*. Springer Verlag.

W. Maas and T. Slaman 1989. Some problems and results in the theory of actually computable functions. In *Logic Colloquium '88*, pp. 79–89. North Holland.

J. Mitchell and R. Harper 1988. The essence of ML. In *Proc. 15th ACM/POPL*, pp. 28–46.

J. Mitchell and G. Plotkin 1984. Abstract types have existential type. In *Proc. 12th ACM/POPL*, pp. 37–51.

Y. Moschovakis 1984. Abstract recursion as a foundation for the theory of algorithms. In *Computation and Proof Theory*, Lecture Notes in Maths. 1104, pp. 289–364. Springer Verlag.

P. G. Odifreddi 1989. *Classical Recursion Theory*. Elsevier.

P.G. Odifreddi, editor 1990. *Logic and Computer Science*. Academic Press.

R. Platek 1966. *Foundations of Recursion Theory*. PhD thesis, Stanford University.

E. Post 1944. Recursively enumerable sets of integers and their decision problems. *Bull. Amer. Math. Soc.*, 50:284–316.

C. Reade 1989. *Elements of Functional Programming*. Addison–Wesley.

H. Rogers 1967. *Theory of Recursive Functions and Effective Computability*. McGraw–Hill.

G.E. Sacks 1963. *Degrees of Unsolvability*. Annals of Mathematics Studies volume 55. Princeton University Press.

G.E. Sacks 1990. *Higher Recursion Theory*. Perspectives in Mathematical Logic. Springer Verlag.

J. Shepherdson 1988. Mechanisms for computing over arbitrary structures. In Herken (1988), pp. 581–601.

J. Shepherdson and H. Sturgis 1963. Computability of recursive functions. *J. ACM* 10:217–255.

R. Shore 1977. α-recursion theory. In Barwise (1977), pp. 653–680.

W. Sieg 2008. Church without dogma: xxioms for computability. In *New Computational Paradigms*, B. Löwe, A. Sorbi, S.B. Cooper, editors, pp. 139–152. Springer Verlag.

S. Simpson 1977. Degrees of unsolvability: a survey of results. In Barwise (1977), pp. 631–652.

R. Smullyan 1961. *Theory of Formal Systems*. Annals of Mathematics Studies volume 47. Princeton University Press.

R.I. Soare 1987. *Recursively Enumerable Sets and Degrees*. Springer Verlag.

G. Tamburrini 1987. *Reflections on Mechanism*. PhD thesis, Columbia University.

A.S. Troelstra 1977. *Aspects of constructive mathematics*. In Barwise (1977), pp. 973–1052.

J. Tucker and J. Zucker 1988. *Program Correctness over Abstract Data Types with Error-State Semantics*. CWI Monographs volume 6. Centre for Mathematics and Computer Science, Amsterdam.

A.M. Turing 1936. On computable numbers with an application to the Entscheidungs problem. *Proc. London Math. Soc.*, 42:230–265. Reprinted in Davis (1965) and Gandy and Yates (2001).

A.M. Turing 1937. A correction. *Proc. London Math. Soc.*, 43:544–546. Reprinted in Davis (1965) and Gandy and Yates (2001).

A.M. Turing 1939. Systems of logic based on ordinals. *Proc. London Math. Soc.*, 45:161–228. Reprinted in Davis (1965) and Gandy and Yates (2001).

14

Turing Transcendent: Beyond the Event Horizon

P.D. Welch

14.1 The beginning

Turing's seminal 1936 paper 'On computable numbers, with an application to the *Entscheidungsproblem*' is remarkable in several ways: firstly he had set out to solve Hilbert's decision problem of the title, with the definitive discussion of what a 'human computer' could do, and what would constitute an 'effective process', but secondly because it, well one cannot say 'almost in passing' because Turing was keen to put his observations on mechanical processes to work, but it was a second product of the paper that it lay also the foundations for the not yet even nascent theory of 'computer science'. The main point as Turing saw, was the 'universality' of his machine: there is a universal program or machine that can simulate any other. The story of the development of that science from Turing's work is retold elsewhere, and is not the aim or direction of this article. Nor are we going to trace the remarkable history of the theory of 'recursive functions' (now reclaimed as 'computable functions') thence until the present day. That story would involve not just the study of the Turing complexity of sets of numbers, and the computably enumerable sets, but later generalisations in 'generalised recursion theory' or 'higher type computability theory' where deep theoretical analyses of notions of *induction* and logical *definability theory* came into being in the late 1960s and 1970s. Such theories abstracted away from any machine model to notions of recursion or induction on abstract structures. These results are beautiful and very deep, and turn out to have intimate connections with mathematical logic and set theory.

Whilst this high road of increasing abstraction is very appealing intellectually, we have recently seen a flurry of research work drawing its inspiration from the original conception of Turing's machine model, and which tries to work with that, generalising what it could, in theory, compute when released from spatial and temporal boundaries. This article seeks to give some flavour of those ideas. Whilst it has to be said that a stretch of the imagination is required to send a computer off into a black hole, or to transcend ordinary time with its clock ticks of $0, 1, 2, \ldots$ into some transfinite stage 'ω' or even further beyond, thus 'transcending' the finite, this is a *mathematical* activity (or perhaps one in physics) where, as ever, we seek to further our knowledge of a theory, or of a model, its procedures, its limitations, by the process of generalising it. We try to relax current constraints purely 'to see what happens if ...'. It may be science fiction about Turing's machine, but it is a *mathematical* science fiction, and as such also has its boundaries. This kind of work perhaps argues less about the real, or even plausible, physical situation, but rather seeks to draw a containing line along the mathematico-logical boundaries to these kinds of speculations. It is, moreover, simply entertaining to think about these matters.

A computation is considered usually as a process where after a finite time we expect some output for an input. We are interested in the *result* of some sequence of manipulations of strings of symbols, perhaps (but not necessarily) coding numbers. This does not rule out continuous ongoing processes that machines can perform, when machines, or networks of machines, are designed to look after some system (such as the Web or other network), and in effect never finish their tasks, and are not turned off. When one observes closely, these machines are running processes that call for other 'subprocesses' which indeed do effect some computation and return some result. Such machines are administrating millions, if not more, computations of the kind to which I am alluding. Such a process can *interact* with its environment by asking for or receiving input following a query. Even the most complex of such interactive processes could be modelled by Turing's *o*-machines, or 'oracle-machines', which we shall take as the basic model for this article. For the purposes of this discussion such a machine has a single tape with a leftmost starting cell C_0 with further square cells to the right $C_1, C_2, \ldots$ We do not wish to bound in advance the machine's capabilities or requirements of space, so it is usual to allow *infinitely many* cells on the tape: $\langle C_i \mid i \in \mathbb{N} \rangle$. A read/write ('R/W') head moves back and forth a single cell at a time, reading *cell values* which we take from a simple alphabet of 0s and 1s. It has the capability of changing a read cell value, and then moving on. The *program* of such a machine is traditionally thought of as a sequence of tuples in a table, involving an initial *state* with each line of the table instructing the machine what to do when observing a particular symbol in a particular state, and to what state to change. There are only finitely many states, and hence only finitely many possible transitions from which to assemble a program. Programs themselves are finite lists of these transition instructions. How these instructions are written down is not important for this article, so we shall not go further into this. We shall note however that a single line, and thence a whole program, can be *coded* by a single number e. There are many ways of doing this: first assign code numbers to instructions, and then use prime powers to code finite lists of such numbers. The fundamental theorem of arithmetic is then used to decode from a natural number e the program (if any) that it codes up. This is itself a computable process – and this last fact is important for Turing's result that there is a *universal machine*. We shall refer to this list of programs as $\langle P_e \mid e \in \mathbb{N} \rangle$ (where we have harmlessly assumed here that every $e \in \mathbb{N}$ codes a program). An *oracle machine* allows for a second tape on which information is written – again a string, possibly infinite, of 0s and 1s, which can code further pieces of information. It is usual to think of the oracle tape as coding a subset A of the natural numbers. Again it is inessential how this information is coded as long as it is done in such a way that the machine can decode from the tape the information it requires. The two tapes are placed in parallel, one above the other, and the R/W head is able to read simultaneously from both tapes at once, whilst only being allowed to write to the principal tape. The instruction set is expanded to allow the R/W head to query of the oracle tape what is written in the particular cell of the oracle tape. In this way the machine has access to yes/no answers to queries of the form *Is* $n \in A$? We denote the set of programs with this possibility as $\langle P_e^Z \mid e \in \mathbb{N} \rangle$ for any potential oracle $Z \subseteq \mathbb{N}$.

We have stated this much detail already because we are going to think of these machines as having a finite nature (notwithstanding the possibly infinitely long oracle tape sequence); we shall later be allowing them to run possibly into the infinite, and so we should be clear in what follows as to what is finite about our machines and what is not. Note that we have already assumed some other-worldly features, since we cannot literally build a machine with infinitely many parts; we may choose to get around this, by saying that the machine has an arbitrarily large number of cells enough for any computation in hand, or else we add on more cells to the paper tape, as and when we need them. Turing machines (TMs) are really useful for *thinking about* the nature of computation: the usual kinds of coding and programs one sees in text books rapidly involve astronomical amounts of paper tape and stages in time when performed on a TM, and are way beyond real world physical feasibility in any case. However this is not the point, of course.

As Turing explained there is a *universal* such machine or program $\mathscr{U}$ that can emulate all others. Let us recall how this is done.

When the machine runs, at any time a finite amount of information, a 'snapshot', tells us exactly the state the machine is in. This consists of a pair $\langle q, r \rangle$. The first term, q, tells us the line number of the program table at which the machine is (and this tells us the current state of the machine, what it will do depending on the symbol read etc., and the state it will go to, and where it will move after the instruction is performed). The second term, the number r, codes two things: r_0, the number of the current cell the read/write head is hovering over; and r_1, a number coding the tape contents up to this point in time. (We assume the machine starts with a tape with cells C_i containing an initial sequence of 0s or 1s as input, and blanks thereafter running off infinitely far to the right say, and that the head writes only 0s and 1s; r_1 then must code up the current working tape's contents of a finite sequence of 0s and 1s. This also assumes that no blank cells intervene between a written 0 or 1 cell; with some care we can arrange our program to ensure this too.) A successful computation is then one that enters a *halting state*, and we deem the machine to have halted with whatever numerical binary output is on the working tape. (In other programs if we are only interested in a *yes/no* answer then we can ask that this be recorded as $1/0$ in C_0 say.)

The program $\mathscr{U}$ works by decoding the pair $\langle e, i \rangle$ where e is a code number of the program, and i is to be the binary input on the tape at the start (if any). The *oracle* as we have described it can be considered to be a bit-stream of 0s and 1s written to the oracle tape. Although this is not often stated, we do not wish to assume in advance a bound to the queries that a program on differing inputs can throw up, and it is considered quite usual to have an infinitely long such bit-stream. This again is an other-worldly feature.

In the computations that we are normally interested in, we consider only halting computations. Since this happens after a finite number of stages of time, only finitely many cells can have been written to, and only finitely many queries can have been made of the oracle. Thus, for this particular computation a purely finite machine, with finite-length tapes would have sufficed. However, for the next computation on this program, but with a different input, perhaps more, perhaps less tape and more or fewer queries would be needed. So we allow arbitrary lengths of tapes.

An infinite sequence is thought of as a *function* $f : \mathbb{N} \longrightarrow \{0,1\}$ with $f(k)$ the value at the kth place. If a program P_e on input k halts with the (binary) value m on the output tape, we denote this by $P_e(k){\downarrow}m$. We may think of the program P_e as computing a function $f : \mathbb{N} \longrightarrow \mathbb{N}$ given by $f(k) = m$ iff $P_e(k)\downarrow m$, and such a function is deemed (TM-)*computable*. We allow for the possibility that the function may not be defined for every input k (this means that if it is not defined for k, but we insist on running the machine on input k, then the computation $P_e(k)$ will run for ever). Again, by means of coding pairs of integers by integers such a function can be identified with a subset of $\mathbb{N}$ and in turn any subset $X \subseteq \mathbb{N}$ can be identified with its characteristic function $c_X : \mathbb{N} \longrightarrow 2 = \{0,1\}$ via $k \in X \leftrightarrow c_X(k) = 1$. An infinite strings of 0s and 1s then can be thought of as such a characteristic function; the set of all such strings is denoted $2^{\mathbb{N}}$.

The rightly famous *halting problem* (in one version) asks: if $\mathscr{U}$ is given $\langle e, 0 \rangle$ will it halt? (We abbreviate this as $\mathscr{U}(\langle e, 0 \rangle){\downarrow}?$) Turing showed by a so-called diagonalisation argument, that there is no machine, i.e. program, which will itself give $0/1$ output answering this question on input e for every e. We can see in a naive fashion the problem: we may run $\mathscr{U}(\langle e,0\rangle)$ simulating $P_e(0)$ and, after finitely many stages, if the latter simulation halts, we can arrange for a 1 to be in C_0 and then the overall computation $\mathscr{U}(\langle e,0\rangle)$ halts. This process is finite, perhaps $P_e(0)$ takes N steps itself, and indeed we can code up the whole successful *course of computation* $\langle q_0,r_0\rangle, \langle q_1,r_1\rangle,\ldots,\langle q_N,r_N\rangle$ by a single integer $y=y(e)$ (where $\langle q_i,r_i\rangle$ is the 'snapshot' at the ith stage). This y then contains all the information needed to reconstruct the whole sequence of computations. However if $P_e(0)\uparrow$ (meaning $P_e(0)$ never halts) then clearly there is no such y. Since $\mathscr{U}$ can do these simulations in a uniform manner (meaning the way it does this is not dependent on any special features of the program coded by e), we could think of $\mathscr{U}(\langle e,0\rangle)$ as calculating for us such a $y=y(e)$, but again *only if it exists*. Otherwise $\mathscr{U}(\langle e,0\rangle)$ will run forever. Such ideas are behind Kleene's T-theorem which formalised these ideas of Turing in a single summary statement (here somewhat simplified just for one-place functions).

Theorem 14.1.1 (Kleene's T-theorem) (i) *There is a computable predicate T of three variables such that for any program index e:*

> $P_e(k)\downarrow n \leftrightarrow$ *there is a minimal $y\in\mathbb{N}$ such that $T(e,k,y)$; and moreover such a y codes the course of computation witnessing that $P_e(k)\downarrow n$.*

(ii) *Moreover, given any e, there is an algorithm for determining an e' with the property that for any $k\in\mathbb{N}$ we have*

$$P_e(k)\downarrow n \leftrightarrow P_{e'}(k)\downarrow y \text{ with } T(e,k,y).$$

Kleene used (the full version of) this as a basis for establishing a series of fundamental results on computable functions: the *enumeration theorem* the *S-n-m* theorem, and the *recursion theorems*.

The computation of $\mathscr{U}(\langle e,0\rangle)$ is such that we can regard it as *searching for* a y that does the job. We could arrange for $\mathscr{U}(\langle e,0\rangle)$ to simply inspect each $y=0,1,2,\ldots$ in turn until it finds the right one (running off to do tests each time to see if y indeed does code the right snapshots). The program $\mathscr{U}$ is thus performing an *existential search*: it searches for an existent y that does the job. Such existential searches thus typify the limits of the capability of the Turing machine model: we can always get a program/machine to look for a number which has some simple properties, such as being the code of a course of computation, and if successful output it. What has been slipped in here is the phrase 'simple properties'. This has to be clarified or else we may end up classifying the halting property of $P_e(0)$ as 'simple'. What is required of 'simple' is that it be a property Φ that is *decidable* by

Figure 14.1 S.C. Kleene 1909–1994

a Turing machine in *finite time*, i.e. a routine may be run to check $\Phi(n)$ which will assuredly halt with a yes/no answer.

Examples of simple: a prime number; a power of 3; a code of a program state; a code of a program; a course of computation of some $P_e(k)$; being a solution to a priorly given polynomial.

Examples of non-simple: a number e such that $P_e(0)\downarrow$; a code of a number pair $\langle e,n\rangle$ such that $P_e(n)\downarrow$; a number e such that $P_e(k)\downarrow$ for infinitely many k ...

The point to note about simple properties is that they are all *local*: a computation involves only numbers *local or near* to the given number (typically often only smaller numbers). The process may involve a *search* through some numbers, but it will be a *bounded search*: we shall know in advance how far to look. In terms of logical expressions a 'simple property' turns out to be precisely those that can be written out by a number-theoretic formula which involves only 'bounded quantifiers'. (These are called by logicians 'Δ_0 properties'.) The non-simple examples given involve *unbounded searching* or equivalently *unbounded quantifiers*: to justify $P_e(0)\downarrow$ we must be prepared for a search through all possible course of computation numbers y. We ask for something to exist: it is an unbounded existential quantifier '$\exists y[\cdots]$' and is thus an *unbounded existential search*. To justify the negation, i.e. $P_e(0)\uparrow$, we have to search through all numbers to check that something does not happen. This is an unbounded universal statement:

$$\text{`}\forall y \textit{ not } [\cdots]\text{'}.$$

Logicians classify these statements as '$\exists$' (also called 'Σ_1') and '$\forall$' (also called 'Π_1') respectively. The last statement that "there exist infinitely many k such that ..." is more complex still: it is '$\forall\exists$' or Π_2: $\forall n \exists k[n < k \& \cdots]$. In general a formula of arithmetic may have more quantifiers still, and they are classified as Σ_n which means n alternations of quantifiers starting with a $\exists$, and then a matrix $[\cdots]$ which contains no unbounded quantifiers. Similarly we define 'Π_n', which are then the negations of Σ_n. Sets of numbers defined by such formulae are, in the language of arithmetic, appropriately called 'arithmetic'.

Another point to note is that if a property Φ has both an existential form, $\exists n \Psi_0(n)$, and a universal form, $\forall m \Psi_1(m)$, then it is decidable: we may start searching simultaneously for an n so that $\Psi_0(n)$ and for an m so that $\neg\Psi_1(m)$, now two existential searches: if Φ holds the former will be successful after a finite stage; if $\neg\Phi$ then the latter, again after a finite stage.

Existential properties are also called semi-decidable: what we have just argued is that if both Φ and $\neg\Phi$ are semi-decidable, then in fact Φ (and so also $\neg\Phi$) is decidable. Thus we can effectively decide both by the use of TMs.

14.2 Limit decidable

We have seen that an existential search can be effected on a TM. We may thus have a function $f(k) = n$ which is definable by a Σ_1-property $\Phi(k,n)$, computed by a machine. To compute $f(k)$ the machine tests increasing values of m to see if $\Phi(k,m)$ and outputs n when (and if) it finds it. If $\text{dom}(f) = \mathbb{N}$ then this will always halt, but if $f(k)$ is undefined then on input k this search continues forever. The following model is almost the next best thing.

Suppose we have a property $\Phi(n)$ of numbers. We say that $\Phi(n)$ is *decidable-in-the-limit* if there is a computable function $f : \mathbb{N} \times \mathbb{N} \longrightarrow \{0,1\}$ such that:

$$\begin{aligned} \Phi(n) &\longleftrightarrow \text{Lim}_{y\to\infty} f(n,y) = 1, \\ \neg\Phi(n) &\longleftrightarrow \text{Lim}_{y\to\infty} f(n,y) = 0. \end{aligned} \qquad (*)$$

The 'Limit' notation above means that if, e.g., $\Phi(n)$ holds then the value of $f(n,y)$ as y increases will at some point eventually be 1 *and will stay at* 1 for all larger values of y. The form of $(*)$ says that *either* the value will be 1 from some point on, *or* it will be 0. If only we could stand back at the 'end of time' after all the finite steps, and look back and see to which value $f(n)$ has settled down, then we should

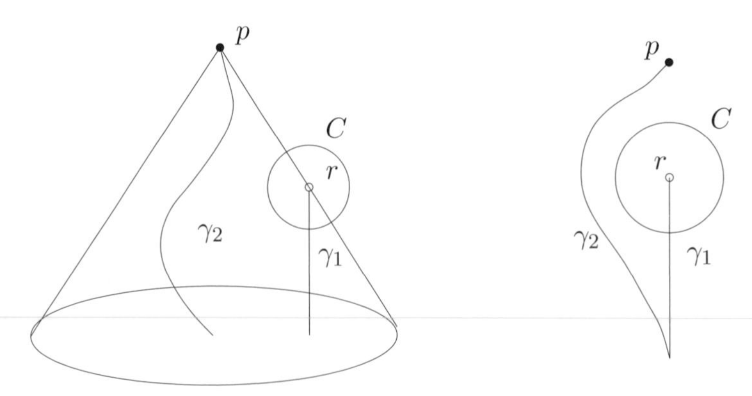

Figure 14.2 MH spacetime.

know if $\Phi(n)$ holds! Eventually because of the properties of f given above, 'in the limit' we should have the right answer on the tape.

We see the essential feature that both the expression formulating the property, and its negation, are Σ_2: they are of the form $\exists y_0 \forall y > y_0 \cdots$ meaning that 'something holds from some point onwards'. Moreover it can be shown that any property with this essential feature can in turn be expressed also in the form $(*)$ above. However, we also have to have the capacity to go to the end of the stages $0, 1, 2, \ldots$ in time to check our output tape. We cannot know at any finite stage of time whether we have arrived at the final answer. Nor, it can be shown, is there any way to define a computable function that given n will tell us how long to wait! We truly do have to wait through all time to know that we have the right answer on the tape. How can we do this?

14.3 Malament–Hogarth spacetimes

Various authors (Etesi-Neméti, Hogarth, Pitowski) have pointed out at various times that Einstein's equations for general relativity (GR) allow for spacetimes where in the causal past of one observer, call her $\mathcal{O}_p$, there may lie a path of infinite proper time length for a second observer $\mathcal{O}_r$. Such spacetimes have been dubbed 'Malament–Hogarth spacetimes' (MH). Such a situation is pictured in Figure 14.2 on the left.

This is non-contradictory since $\mathcal{O}_r$ is heading towards a singularity at r along the path γ_1. One can imagine such a situation as arising in the following way. One takes a regular flat Mostowski spacetime, where outside a compact region C the curvature is flat. Then one smoothly allows the curvature to go off towards infinity as r is approached. The point r is then removed from the manifold; but the whole path γ_1 remains in the causal past of $\mathcal{O}_p$ although now, by the calculations of GR it has infinite proper time length. Ideally then, one can send a TM along the path γ_1 checking say some universal $\forall$ property, such as Goldbach's conjecture (which says that for all even numbers greater than 2, they are the sum of two primes) or that one's favourite mathematical theory is *consistent* – for example PA, the axioms due to Peano for arithmetic; or ZFC, the axioms of Zermelo–Fraenkel set theory (the *consistency* statement is again a 'for all' one: it says for all finite sequences of formulae making up a proper proof from the axioms, if the final concluding sentence is s say, then s is not a contradiction, such as '$0 = 1$'). If the property fails, this is because some counterexample to Goldbach's conjecture has been found: namely an even number that fails to be a sum of two primes, or, a proof of '$0 = 1$' from the PA axioms. The TM then sends a signal to $\mathcal{O}_p$. If a signal is received, then $\mathcal{O}_p$ knows that the Goldbach conjecture is false, or that PA is inconsistent. Theoretically (and summarised diagrammatically on the right of Figure 14.2) then, the observer $\mathcal{O}_p$ (travelling along γ_2) could herself set the TM running along γ_1, go out to lunch, and then if no signal is received at point p when she returns, say an hour later, then she knows that the universal statement is true since no counterexample has been reported as found.

Now of course in many respects this situation is wonderfully fantastic: the TM that is to be sent down the path γ_1 now absolutely has to have an arbitrarily long tape, since it is going to have to check Goldbach's conjecture for arbitrarily large even numbers, and representing these on the tape alone requires arbitrarily long sequences of 1s even before any calculation takes place. Since we cannot literally get hold of an infinite tape, we have to tell a story about the machine, or the generations of workers we send with the machine, extending the tape as it, or they, go along etc. Others have pointed out physical difficulties of another kind, concerning the arbitrarily large blue-shifts in the signals that the human observer receives (as for her the time of the TM becomes arbitrarily speeded up, and hence the wavelengths of the signal she receives could be arbitrarily small), or the increasing energies needed for the signals to be distinguished from background radiation.

Nevertheless certian spacetimes that have been studied earlier and quite independently of this issue have the MH property: Gödel's famous model where there are time loops, anti-de Sitter spacetime, and Reissner–Nordström spacetimes. Etesi and Neméti have been proponents of Kerr's solution to the GR equations in the form of a rotating black hole – for which there is observational evidence (see Figure 14.3). In such a model the singularity is not in the form of a point or centre, but is a ring around the black hole. In their arrangement the observer $\mathcal{O}_q$ descends into the ring, whilst the TM orbits the black hole! They have arguments to get around the blue-shift problem mentioned above (for example by the computer calculating what the blue-shift might be from the human observer's end and sending a rocket

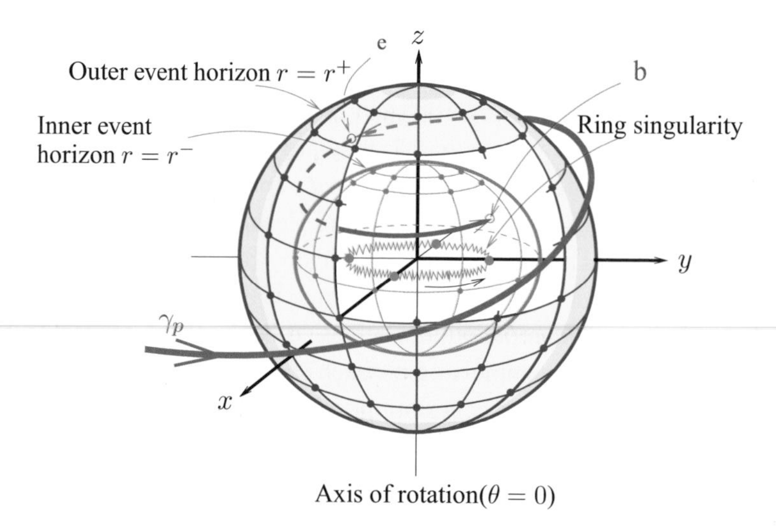

Figure 14.3 A slowly rotating (Kerr) black hole has two event horizons and a ring-shaped singularity. The ring singularity is inside the inner horizon $r = r^-$ in the 'equatorial' plane of axes x, y. With grateful acknowledgements to G. David and I. Neméti 'Relativistic computers and the Turing barrier', *App. Math. and Comp.*, **178**, 2006.

in the reverse direction to counter its effects!) and for some of the following difficulties. The spacetimes listed all have varying causal properties but, in common, as Hogarth showed, they all violate a property called 'global hyperbolicity', which in essence says that the future can be determined by a full account of the initial conditions over a complete hypersurface – a so-called 'Cauchy surface'. Thus an assumption of global hyperbolicity rules out any spacetime as being MH.

Besides this, other hypotheses, variants on Penrose's *cosmic censorship hypothesis* which rules out 'naked' singularities (i.e. singularities that are not 'hidden' by an event horizon such as the points r on the curves γ_1), also fail. These are reasonable hypotheses about causation in spacetimes, in general, but which have not been deduced from the GR equations. Nevertheless they encapsulate properties that one might want any GR solution to have, and although the matter is far from settled, many physicists endorse one or other of them.

However, let us cavalierly brush all these grumbles aside and see what *logically* comes out.

14.4 Infinite ordinals: beyond arithmetical

To discuss sets of numbers that are more complex than arithmetic (and later, computations that are longer than those given by natural number stages alone), we must invoke Cantor's theory of ordinal numbers. We shall not need this whole theory, but

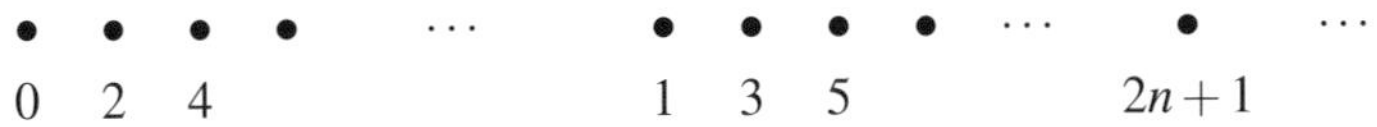

• • • • ... • • • • ... • ...

0 2 4 1 3 5 $2n+1$

An ordering $\prec$ of the natural numbers in order type $\omega+\omega$

• • • • ... • • • ... • • • ... • • • ... • • • • ...

0 1 2 ω $\omega+1$ $\omega+\omega$ $\omega.3$ ω^{ω} $\omega^{\omega}+3$

Figure 14.4 Some ordinals.

we do need the notion of a *countable wellordering*. (A set is called *countable* if it can be put into one-to-one correspondence either with a natural number $k \in \mathbb{N}$ or with all of $\mathbb{N}$.) Some infinite wellorderings are pictured in Figure 14.4.

Figure 14.4 gives pictorial representations of these ordinals, in the same way that four dots in a row can represent an ordering of four. Cantor showed how we can have arithmetical operations on ordinals that extend the usual addition, multiplication, etc. that we have on the finite numbers. That won't concern us much here, but two things will count: (a) the pictures of these transfinite ordinals and (b) the fact that countable ordinals can be represented by characteristic functions of subsets of $\mathbb{N}$.

To illustrate (b), consider $\omega+\omega$. (In set theory the class of finite numbers is given different names depending on the role it is playing at any moment: by '$\mathbb{N}$' we mean the *set* of all the usual finite ordinal numbers starting with 0; on the other hand 'ω' whilst being strictly the same set, emphasises the *ordering* role of the set of numbers. We think of an ordinal number as the set of its predecessors. This makes $3=\{0,1,2\}$; $n=\{0,1,\ldots,n-1\}$ and $\omega=\{0,1,\ldots\}$, then $\omega+1=\{0,1,\ldots\omega\}$. The ordering we are considering is then that on the set $\omega+\omega=\{0,1,\ldots,\omega,\omega+1,\ldots\}$.)

We can represent pictorially that ordering, as at the top of Figure 14.4, by putting all the even numbers before the odd numbers to get something of 'double $\mathbb{N}$ length'. However we need a mathematical definition of this picture, something that a computer can work with, so we define mathematically a new ordering $\prec$ on $\mathbb{N}$ by $i \prec j \Longleftrightarrow i,j$ *are either both even or both odd and* $i<j$ *or* i *is even and* j *odd.* Then if we draw the picture of $\prec$ it looks like the first example above. Now consider the function $\pi(n,m)=2^n\cdot(2m+1)-1$. This is a one-to-one mapping, a injection, from pairs from $\mathbb{N}$ into $\mathbb{N}$. We can use this to code orderings or indeed any subsets of $\mathbb{N}\times\mathbb{N}$: we let $E_{\prec}\subseteq\mathbb{N}$ be the set of k of the form $\pi(i,j)$ where $i\prec j$. This $E_{\prec}$ codes exactly the ordering $\prec$. Conversely given any $F\subseteq\mathbb{N}$ we can look at $\prec_F =_{\mathrm{df}} \pi^{-1}\text{``}F=\{(i,j)|\pi(i,j)\in F\}$. Then it may or may not be the case that $\prec_F$ is an ordering at all. But to be a *linear ordering* it is an $\forall\exists$ question about F:

First: F is a *transitive* ordering iff $\forall p,q,r(p\prec_F q \& q\prec_F r \Rightarrow p\prec_F r)$ iff $\forall k_0\in F\forall k_1\in F\forall p,q,r\exists k_2[\pi(p,q)=k_0\wedge\pi(q,r)=k_1\longrightarrow\pi(p,r)=k_2\in F]$.

Second: two more clauses that express that $\prec$ is *connected* and a *strict* ordering ($n \not\prec n$) have to be added to this to ensure that F codes a *linear* ordering of a subset of $\mathbb{N}$. This makes sure that all the elements in the picture are lined up. It was Cantor who suggested that we extend the quintessential feature of the natural number system: that it is *well-ordered*; this means for a linear ordering $\prec_F$ that if $X \subseteq \text{Field}(F)$ and X is non-empty, then there is an $\prec_F$-*least* element k_0 in X. Thus: there is no $p \in \text{Field}(F)$ with $p \prec k_0$. All the pictures above have this property. But the property is a special one: many natural orderings are ill-ordered: the negative integers with their usual ordering are ill-ordered, because they themselves have no least element; the usual ordering of the rational numbers in $[0,1]$ is ill-founded, as the set $\{\frac{1}{n+1} \mid n \in \mathbb{N}\}$ has no least element. Notice also that being well-ordered has to be expressed by a quantification over all subsets X of $\mathbb{N}$ since it cannot be expressed by number quantifiers $\forall n, \exists m$, etc. In fact many sets F that code well-orderings are rather simple; the set coding the ordinal $\omega + \omega$ of evens followed by odds is actually computable: we can write a program P_e such that on input k it applies π^{-1} to k, so yielding a pair (n,m), and checks whether $n \prec m$ in that ordering. However, many code-sets of wellorderings are highly complex; indeed the set $\{e \mid P_e$ computes a wellordering$\}$ in the above manner is itself a highly non-computable set of program indices.

As well as defining wellorderings we can drop the requirement that the orderings be linear. We allow pictures of *trees* (Figure 14.5) where we draw the tree *downwards* from its root (thus we have $2 \prec 5$ in the tree). We say that $F \subseteq \mathbb{N}$ codes a tree, if we drop the connectedness requirement from being a linear ordering. It is then a *well-founded tree* if it is both a tree and, now, for any $X \subseteq \text{Field}(F)$, we still require if it is non-empty that there is some element $k_0 \in X$ that is $\prec_F$-*minimal*. (Now of course there may be many such $\prec_F$-minimal elements of X because the tree splits in many ways. Indeed any node may have *infinitely many* elements immediately below it. Notice that any such countable well-founded tree can be considered (and is often called) a *finite path tree* because any path starting anywhere in the tree and proceeding only downwards is finite in length. (It cannot be infinite because that path would constitute a subset $X \subseteq \text{Field}(F)$ without any minimal element.)

Now notice that no Turing machine, given say X as an oracle, can answer the question of whether X codes a linear ordering (we have seen that already requires a two quantifier condition to be answered). Nor can we say given a program number e, whether we can run another program to decide whether P_e computes the characteristic function of a linear order: this would require a machine to answer infinitely many questions about P_e's behaviour which no machine can do. Because this is a Π_2 question, even if we allow computation-in-the-limit it is still impossible, as this latter kind of computation only allows us to answer questions which can be simultaneously expressed as Σ_2 *and* as Π_2; questions that are Π_2 expressible it cannot

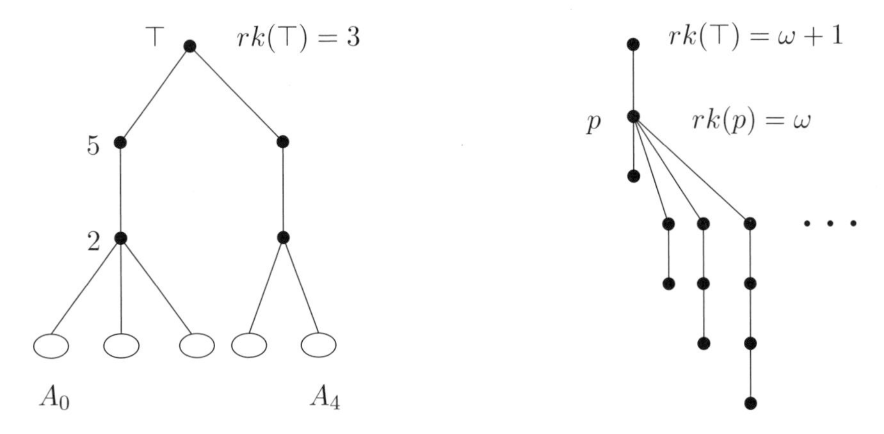

Figure 14.5 Finite-path trees.

handle. We have to go to a stage at which we can look back at the previous sequence of all stages and come to a judgement: we need to be standing at time $\omega+1$, at some node or point beyond all times $0, 1, 2, \ldots, k, \ldots$ How can we do this?

Hogarth (2004) suggested putting diagrammatic elements of his spacetime continua (the right of Figure 14.4) together to be able to answer any arithmetic question. If one were to draw a schematic diagram of such a question, such a tree would be of the kind on the left of Figure 14.5.

The subquestions that need to be answered by such an array of machines, in an array of spacetime regions, can be represented as a special kind of finite path tree, where we allow infinite splitting below nodes, but *every path must have a fixed bound*, k say, on its length, corresponding to the number of quantifiers in the question.

We say that a finite-path tree T has a *(finite) rank* k if, assigning the leaves rank 0 and for any node $p \in T$, we define $\mathrm{rk}(p) =$ least number $k >$ all $\mathrm{rk}(q)$ for $q \prec_T p$. If $\top$ is the topmost node, then we set $\mathrm{rk}(T) = \mathrm{rk}(\top)$. Then $\mathrm{rk}(T)$ is always finite in the case of trees associated to arithmetic-deciding bits of spacetime. But what if $\mathrm{rk}(\top) = \omega$? Or something larger? Initially the reader may be surprised that a tree can have infinite rank but only finite-length paths. However, see the right-hand diagram of Figure 14.5. Such trees can be used to build sets B of complexity far beyond the arithmetical and hence properties ('being a member of B') of numbers that are more complex than arithmetical. The first such sets B are *relatively* simple because we can describe a *computable protocol* for their construction: we take a *computable* finite path tree, and at the minimal bottom-most nodes we attach some simple sets A_i $(i \in \mathbb{N})$, say computable or arithmetically defined ones. Then we inductively proceed up the tree, attaching to every node $p \in T$ a set A_p. If p is terminal we have already done this; if the node immediately above p, p' say, only has p below it, we set $A_{p'}$ to be the complement of A_p; otherwise $A_{p'}$ is the union of all the A_q, with q immediately below p'. In this way a set gets attached to the topmost node $\top$, and we set $B = A_\top$.

The countable collection of sets defined in this way are called the *hyperarithmetic sets*. Note two things: first, that they include all the arithmetic sets (we indicated that these are obtained by using just trees of finitely bounded path length) and thus are far from computable sets. Second, although we emphasised that we were taking computable trees, with computable assignments of arithmetic sets to the leaves thereby obtaining one of a potentially countable set of possible computable descriptions of such B, mathematically we could have dropped these computability requirements and enlarged our class by considering *all* possible finite path trees and *all* possible assignments of arithmetic sets to leaves. One can argue that as opposed to the hyperarithmetic sets, which form a countable collection, there are then uncountably many such sets so created – this much wider class is called the class of *Borel* sets, and they are arranged in a hierarchy of complexity depending on the minimal ordinal rank of a constructing tree.

14.5 Returning to MH spacetimes

Now we can see that we could extend the argument of Hogarth deciding membership of arithmetic sets, to membership of hyperarithmetic sets, or even potentially of any particular Borel set B given the right spacetime manifold. We need a region of spacetime with a tree structure of SAD components reflecting the tree structure of B's construction. We use the Hogarth-type 'arithmetic trees' as starting component regions for the leaves at the bottom; these components must then be considered as within a hypothetical region that itself reflects the finite path tree constructing B with those arithmetic sets at the leaf positions.

Indeed, with care, and if we were able, we could enumerate all possible computable protocols $e_0, e_1, \ldots$ enumerating the hyperarithmetic sets, and one could hypothetically find a region of spacetime that would accommodate all such trees simultaneously. One could have an overall Command TM, that when submitted a pair (e,n) representing the query "*Is* $n \in B_e$?" where B_e is the eth hyperarithmetic set, would send n to the machine at the entry point to the eth region. In short, we should have a *hyperarithmetically deciding region*, HAD. The reader may be forgiven at this point for thinking that we have departed from reality by now (if not some while ago). For example, the enumeration of the computable protocols cannot be a computable process for one thing (we could get a contradiction from assuming it were, by a diagonal argument). Nor is it even arithmetic. We also have the *recognition problem* of even seeing when or where there is a SAD region in our spacetime, let alone these exotic varieties. However, the point is to push the envelope of what is *logically* conceivable with MH spacetimes, granted the physical assumption that a single SAD is consistently available. We are not discussing the physics or (even potentially) the physical realisability.

However we can establish a purely logico-mathematical limit to even to these very theoretical, very speculative, theoretical computations in MH spacetimes. Given an MH spacetime (perhaps our universe after all is such a one), could there be sufficiently many regions that any Borel question could be answered? The answer (Welch, 2008) is no: we do hit a limit, a *universal constant* of the spacetime manifold $\mathscr{M}$:

Theorem 14.5.1 *Given an* MH *spacetime manifold* $\mathscr{M}$ *there is a countable ordinal* $w(\mathscr{M})$ *such that no Borel question of rank* $w(\mathscr{M})$ *can be 'answered' by a region contained in* $\mathscr{M}$.

14.6 The $\aleph_0$-mind

What the above forcefully illustrates is that we have to have some way of coping with infinitistic arguments concerning infinite sets and processes: we can only think about and work with a hyperarithmetic set if we can in some way comprehend the totality of its description (which, although given by a number index *e* say, has to be unpacked into the characteristic function of an infinite finite path tree). If only our minds were capable of this! The phrase '*the* $\aleph_0$*-mind*' has been used to describe a mind capable of assimilating and manipulating countable amounts of numerical information in one step. One way of thinking further about this is to imagine a Turing machine where we may survey the whole infinite tape at one glance, and ask of an oracle Z "Is the *whole of the work-tape's* contents an element of Z?" Thus Z is now to be thought of as not just a set of integers but a set of infinite strings of 0s and 1s. Depending on an answer, or our program, we then in a single movement effect some computable process on the whole tape (such as flipping every other bit, or moving and storing the string somewhere). Now the notion of computation is radically changed: we are computing not on natural numbers but on infinite (but countable) objects, the infinite strings on the tape.

In a series of papers in the late 1950s and early 1960s, S.C. Kleene developed an equational calculus for such processes and even higher generalisations. Kleene was a pioneering researcher in Princeton in the early days of computability theory working with Church and Gödel in the 1930s.[1] He developed an equational calculus for presenting the 'generalised recursive functions' essentially due in one form to Gödel. This class of functions Turing showed to be precisely the class

[1] This early history is well surveyed in detail in Soare (2009).

computed by TMs. The calculus is a sort of axiomatic listing of clauses as to how such functions from $\mathbb{N}^k$ to $\mathbb{N}^n$ (for any k,n, say) could be built up in an inductive fashion. Numbers are thought of as being of *type 0*, namely at the bottom of a hierarchy. A function $f : \mathbb{N} \to 2$ is a *type 1* object. A *type 2* object would be a function $F : \mathbb{N}^{(2^{\mathbb{N}})} \to 2$, and so on. Kleene axiomatised *higher type recursion theory* by extending his original axioms for the generalised recursive, or Turing computable, functions in a natural way to axioms for 'type-n recursion'. This subject became a significant area of research in the 1960s and early 1970s, with people trying to explore his definitions, ironing out the quirks or looking at possible alternatives in the axioms. This bottom-up approach rapidly becomes rather complicated (as are his original equational sets). In the late 1960s, work of Moschovakis (later joined by Kechris, Harrington, and Normann amongst others) began to build on the work of Kleene, Spector and Gandy, to give a definition of a generalised recursion theory which emphasised *logical definability* rather than *mechanical* definability (or perhaps better put, *computably derived* notion of definability of Kleene's equations). This was a very flexible approach and rapidly connections were made with the theory of inductive definitions and descriptive set theory (the latter seeks to prove results concerning, not all sets of numbers, but sets as classified by their descriptions, i.e. logical definitions; this has become a central area of modern set theory.) Indeed Moschovakis was able to define notions of 'inductive' and 'hyperarithmetic' not just for the natural number system but for *any* structure satisfying a modicum of coding possibilities. (Such structures he called *acceptable*. For example such a structure should have a way of pairing off elements of the domain M: there should be in the structure $\mathscr{M}$ a function $\pi : \mathscr{M} \times \mathscr{M} \to \mathscr{M}$.) Whilst such a view can be said to have given what must be considered the definitive theory of inductive definability and the associated recursion theories, these leave the machine or mechanical approach or intuition far behind.

Let us return to the $\aleph_0$-mind. Kleene showed that his equational calculus allowed a notion of what is now called 'Kleene computation', involving, as we have said, sets of numbers (or equivalently the infinite strings of 0s and 1s that make up their characteristic functions). Thus a type-1 recursion can be thought of as computed by the thought-experiment machine where, as we have suggested, we manipulate the whole string and ask the oracle questions about the whole string in single steps. He showed that in this case, what corresponded to the computable or decidable sets of integers, would now be the hyperarithmetic sets of integers and subsets in $2^{\mathbb{N}}$; what corresponded to the computably enumerable sets of integers would be the subsets of $2^{\mathbb{N}}$ of a complexity no more than that of the wellorderings of $\mathbb{N}$. (In descriptive terms of analysis, Kleene's 'computable enumerable' sets are Π^1_1 (also called 'co-analytic') sets, and all such sets can be computed (in the usual sense) from WO, the set of such codes. (Actually Kleene allowed elements of $2^{\mathbb{N}}$

as constants or parameters in these computations, thus in effect allowing not just computable finite path trees into one of his computations, but any such; the upshot is that for Kleene the ground computable sets were the Borel sets, not just the hyperarithmetic ones.) An essential feature of Kleene computations is that a particular computation could require an infinite string of subcomputations $s_0, s_1, \ldots$ to be completed to deliver up some information for the main computation to continue. (We see this for example in the description of the hyperarithmetic sets above, when the tree is infinitely branching: to know whether n gets into a set $A_{p'}$ at such a node p' we must look at all the possible sets A_q at nodes immediately below $p' \ldots$). A successful Kleene computation is then best thought of as having itself a *well-founded computation tree* structure; diagrammatically this would show the structure of all these subcomputations. We say well-founded, because we do not want an infinite string of sub-routine calls to endlessly go down an infinite path, as such a computation would never terminate.

The $\aleph_0$-mind that is trying to keep an eye on all these subcomputations, indeed must do so over the whole tree at once. At a particular node the process at that node must wait for information to come back up from lower nodes; it would be possible to 'linearise' this whole process, fulfilling in a linear order such computation calls. But, as indicated, some such calls would have to wait for infinitely many steps below the node to be completed. Such a linear order would have to be a wellordering of length, well what? ... some countable ordinal, as the overall tree is a countable finite-path tree. So maybe we should in any case analyse linear, but transfinite, computations? To do this we have to cross a different kind of event horizon: that of ω, the first infinite ordinal.

14.7 Infinite-time Turing machines

Infinite-time Turing machines (ITTMs) are just such an attractive model of computation, invented by J. Hamkins and J. Kidder in Berkeley in the 1990s (Hamkins and Lewis, 2000). Their motivation was really to find a definition of how one could allow a Turing machine to run transfinitely. This would mean a well-ordered sequence of computations with successor computation steps given by a standard Turing program. The machine was to be equipped otherwise with the usual Turing machine capabilities. Thus the *length* of a terminating computation would be some countable ordinal, now possibly infinite. Really all that has to be added is a description of what happens at limit ordinal times. Thus, for example, we may run the machine through finite steps $0, 1, 2, \ldots$ If this is a halting computation in the usual Turing sense, then for some $k \in \mathbb{N}$ the machine stops. All well and good. But if not? It will run throughout all the natural number stages. What happens at stage ω?

We avoid all considerations of 'Thomson lamp'-like difficulties (a body of literature in the philosophy of science that asks what happens after a light has been switched on and off infinitely often) by simply declaring by mathematical fiat that a cell C_k has the value at time ω say (which we shall denote $C_k(\omega)$), given in the following way: if there was a stage $n < \omega$ such that at all later stages m, with $n < m < \omega$, the value $C_k(m)$ is unchanged, then we set that value to be the value at time ω, $C_k(\omega)$. Otherwise we reset it to zero: $C_k(\omega) := 0$. An equivalent way to put this is to say the value at stage ω is 0, unless from some time before ω it was 1 and that 1 remained in the cell thereafter. This specifies what the cell values $C_k(\omega)$ are for all k. We have to specify where the R/W head got to! The first case to consider is that as $k \to \omega$ there was a particular cell C_n, say, that it returned to visit for unboundedly many stages $k < \omega$. In that case, for the least such n, call it n_0, we'll put the R/W head over this C_{n_0} at time ω. In the other case, the head has crept off down the tape to infinity: the 'liminf' of the cell numbers the head visits is not finite but is ω itself: for every cell C_n there is some time $t(n)$ such that for all later times t with $t(n) < t < \omega$ the head is reading a cell beyond C_n on the tape; i.e., it never revisits a cell C_m for any $m \leq n$ for $t > t(n)$. In that case we replace the head back on C_0 – the tape start at time ω, Finally we have to specify the next instruction state on the machine's program. This program is entirely a standard TM program, so it is a finite list $I_0, \ldots, I_N$ say. Hence if at time t we denote the instruction about to be performed as $I_{q(t)}$, then there must be a first (in our list) instruction state I_q on the list that was performed unboundedly often below ω: in other words, we have that this q is the liminf of the $q(t)$ for times $t < \omega$.

			R/W							
Input:	1	1	0	1	1	0	0	0	0	...
Scratch:	0	1	1	1	1	1	1	0	0	...
Output:	1	0	0	0	1	1	0	1	0	...

Figure 14.6 A three-tape infinite time Turing machine.

One attribute of these definitions is that they have the effect that the head (in the first case above) is thus placed, and the 'next instruction' is thus declared to be so that the machine is at the beginning of the outermost loop of any subroutine in the program that was called unboundedly often in time below ω. (Actually this is somewhat unimportant in what follows; indeed Hamkins and Kidder did not use this rule: they declared that the head would always return at limit stages to C_0 and the machine would enter a 'special limit state' added to the list of normal

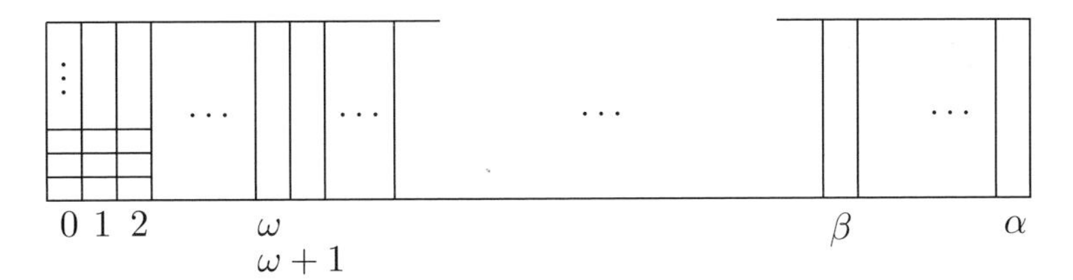

Figure 14.7 A course of computation of length α.

Turing states. They also used 'limsups' rather than 'liminf' considerations, but it is easy to argue that all these variations are minor and do not alter the classes of functions or sets computable by such machines.) What we have described at stage ω is now declared to hold true similarly for any limit ordinal $\lambda = \omega+\omega, \omega^2, \omega^\omega \ldots$ by replacing ω by λ in the above. The rules stay exactly the same.

Lastly, for conceptual ease, it is useful to think of the machine as having three tapes: for input, for scratch work, and for output, placed in parallel one above the other. The head reads simultaneously one cell from each tape but can only write to the scratch and output tapes. This again (with a minor caveat) has no difference in computational power from a single-tape machine.

Now we are in business! A number of things immediately occur to us: we have the possibility that a computation may halt not just at a finite time $t < \omega$ but at some transfinite ordinal time $\omega \leq t$ (we shall see an example of this below). However, some computations still will never halt. As we have set it up, such computations will continue throughout the eternity of all the ordinal numbers ON. However, it is intuitive to see (and can be proven) that any non-halting computation will essentially start looping at some *countable* ordinal. Let us say a 'snapshot' of the machine at a time t is a complete list of all the cell contents $\langle C_k(t) \mid k \in \mathbb{N}\rangle$, together with $q(t)$ and $r(t)$. In order for the machine to be permanently looping there must be a sequence of countable ordinals $\alpha_0 < \alpha_1 < \cdots < \alpha_n \cdots < \alpha_\omega$ where α_ω is the 'limit' of the α_n (thus any ordinal $\beta < \alpha_\omega$ is $< \alpha_n$ for some n), and where the snapshots at the times $t = \alpha_n$ are identical for all $n \leq \omega$. At this stage α_ω we know that thereafter the computation is doomed to cycle endlessly through the same snapshot configurations. (Figure 14.7 illustrates a course of computation of length α. One should think of each vertical section as a picture of the infinite three-tape sequence as time increases to the right.)

The other immediate possibility is that we can put an infinite string of 0s and 1s on the input tape at the start, and the machine has enough time to read the whole input tape and act on it. Likewise if the machine halts, there may be an infinite string of information on the output tape. In such cases we may think of the machine computing a function $F : 2^{\mathbb{N}} \to 2^{\mathbb{N}}$. We are thus definitely in the realm of computing on type-1 objects.

However, before we leap into this new world, there are plenty of questions to consider first about functions $f : \mathbb{N} \to \mathbb{N}$ (or perhaps $f : \mathbb{N} \to 2^{\mathbb{N}}$) where we may think of input n as being a string of $(n+1)$ 1s followed by 0s, or even just functions on 0 input? Actually, in terms of ordinal time, how long do such computations go on for. What kind of strings are on the output tape after the machine halts? More generally what kinds of reals can appear at *any* time? One can ask equivalents to the halting problem: what is $h =_{\mathrm{df}} \{e \in \mathbb{N} \mid P_e(0)\downarrow\}$? Suppose we write $P_e(x)\downarrow y$ for '*Programme indexed by e on input x halts with* $y \in 2^{\mathbb{N}}$ *on its output tape*'. We shall call such y '*writable-from-x*'. If $x \in \mathbb{N}$ we shall just say '*writable*'.

Question *What exactly are the writable sequences* $y \in 2^{\mathbb{N}}$*?*

Let us call a countable ordinal α *writable* if some writable $y \in 2^{\mathbb{N}}$ is a code for α.

Question *What exactly are the writable ordinals?*

This at first glance looks mysterious. Call an ordinal *clockable* if it is precisely the length of a halting computation $P_e(0)$ for some e.

Question *What are the clockable ordinals? Is every clockable ordinal writable?*

Question *What are the decidable, or the semi-decidable, predicates?*

These questions make sense for computations $P_e(x)$ where we consider priming the input tape with an input sequence x. We have then this higher type computational behaviour of the model to investigate. Again we may ask what are the decidable, or the semi-decidable predicates. More generally:

Question *What are the relations to the earlier Kleene recursion?*

Question *How do these ITTM-defined relations fit in with the later abstract recursion theories?*

Theorem 14.7.1 (ITTM T-theorem) (i) *There is an ITTM-computable predicate* T *of three variables such that for any program index e and for any* $x \in \mathbb{N} \cup 2^{\mathbb{N}}$*:*

> $P_e(x)\downarrow z \leftrightarrow$ *there is a* $y \in 2^{\mathbb{N}}$ *such that* $T(e,x,y)$*, and moreover y codes the course of computation witnessing that* $P_e(x)\downarrow z$.

(ii) *Moreover given any e, there is an (ordinary) algorithm for determining an* e' *with the property that for any* $x \in \mathbb{N} \cup 2^{\mathbb{N}}$ *we have*

> $P_e(x)\downarrow z \leftrightarrow P_{e'}(x)\downarrow y$ *with* $T(e,x,y)$.

Looking at the above we see the immediate difference between this and Kleene's original theorem: the presence of the y as now not an integer, but as an element of $2^{\mathbb{N}}$. However, this has to be so: even if $x \in \mathbb{N}$ or is zero, a course of computation is now in general a transfinite sequence of snapshots, each of which is essentially an element of $2^{\mathbb{N}}$. These must then all be amalgamated and coded together into a single $y \in 2^{\mathbb{N}}$. In part (ii), note that in order for there to be any chance of a program $P_{e'}(x)$ outputting such a y, we *must* have that y contains within it a code of the *length* of the halting computation, $P_e(0)$, say (if $x = 0$). For this to happen we need as a minimum that the ordinal length of any halting computation can itself be output as the result of another computation: that is, any 'clockable ordinal' must also be 'writable'. Fortunately this turns out to be the case and the theorem holds.

After the model's introduction, subsequent work gave us the answers to all these questions, and we now have a pretty exact idea of what the ITTM machine can do, at least in the case of integer input (see the survey article Welch (2009)).

It is at least pertinent to ask *What is this good for?* Clearly it has little to do with concrete everyday desk-top computation, so we are really working with a mathematical idealisation where we can compute *as if* we could transcend the finite integers. We started with a pure conceptualisation out of sheer curiosity, and ran with it to see where it goes. We do not expect it to directly affect everyday life. However, that much is true of most pure mathematical endeavours (at least initially). Here we do have a mathematical model that is an excellent test-bed, so to speak, for other models that compute discretely using fixed steps but also allow some infinitary action to take place. So, we can *simulate* the action of sending ordinary TMs through sequences of SAD or HAD regions of spacetime on an ITTM. In this respect, ITTMs are similar to TMs. Turing showed that many forms of recursion could be simulated on his machines; ITTMs too can mimick any form of (finite) computation on integers. Thus far ITTMs can simulate other forms of infinitary computational machine-models on integers which have been independently defined.

Just as the halting problem h_0 for TMs in a precise senses delineates the boundaries of what TMs can calculate, so we know now exactly what kind of set h is presented by the halting problem set for ITTMs. We thus have a similar boundary mapped. It turns out that this boundary lies far beyond that of Kleene recursion. The decidable predicates of reals in Kleene's sense are precisely the hyperarithmetic sets, but ITTMs can decide whether a $y \in 2^{\mathbb{N}}$ codes a wellordering, and this is beyond the capabilities of Kleene recursion. Although there is a sharply delineated boundary for the decidable predicates of ITTMs it is not the case that they could answer membership questions about *any* set built up from an *arbitrary* finite path tree description *unless* that description is coded into the input string. However if the protocol for building B is so coded on the input string, then for any number k the ITTM can answer if $k \in B$. It is just that there are only countably many programs whilst there are uncountably many such finite-path trees, and hence constructing protocols: the ITTM has only the ordinary Turing programs $\langle P_e \mid e \in \mathbb{N} \rangle$ to work with. Without the extra description as input, there are only just so many finite-path trees that can be defined by the ITTMs alone; but the rank of those path trees (or equivalently the wellorderings that are writable by an ITTM alone without input) give us the level at which the machines can calculate 'for themselves' so to speak. And that indicator is well beyond Kleene recursion and the hyperarithmetic sets.

14.8 Register machines and other generalisations

After the publication of the ITTM model, and with the realisation that this tied in with earlier work on generalised recursion theory, people started to look for ways to generalise it. Indeed, as many remarked at the time, the beautiful simplicity of the model, which extends the action of TMs into the transfinite, could easily have been invented decades earlier.

There are two possible approaches to generalisation: (i) try changing the properties of ITTMs or (ii) look at other models of computation from the standard everyday, finite arena, and let them loose into the transfinite realm.

We consider (i) first (although this is not in historical order). The over-arching feature of the ITTM is of course what it does at limit stages. One could (and should) reasonably argue that the whole of the capabilities of the machine have been hard-wired into the limit rules. Are there potential 'hypermachines' which exceed the ITTM's capabilities by strengthening the limit rules in some way? The rule as given above is essentially an $\exists\forall$- or Σ_2-rule. The value of a cell is 1 if and only if $\exists t < \lambda \forall s (t < s < \lambda \to C_k(s) = 1)$. Here the quantifiers on the right-hand side have this pattern. Could there be a Σ_3-rule or a $\exists\forall\exists$-rule? A Σ_n-rule for larger n? The presumption is that if such were sensibly defined, then a 'Σ_3-hyper-ITTM' could use the contents in previous stages in more subtle ways than simple liminf, in order to come up with a value at a limit stage λ.

It turns out that such Σ_n-hypermachines are definable for each $n \in \mathbb{N}$ and that they do indeed have the expected properties. In fact any real number whose existence can be proven outright in mathematical analysis can be 'computably written' in the sense above by some such machine in some Σ_n-class, with some such program on zero input. It has to be said that the Σ_n-limit rules are increasingly complex as n increases, and even for $n = 3$ the intuitive rationale of them is not obvious: instead of taking a 'liminf' along *all* ordinal times $t < \lambda$, one takes an unbounded subset of them which reflects some stable informational content. If anything the machine has said during the course of computation before λ about some real appearing on its tapes before stage α has also been said prior to α then α is said to be a point of such stable informational content. To calculate the value of $C_k(\lambda)$ we take a liminf then for those $\alpha < \lambda$ whose ordinals of stable informational content are the same as those of λ which are below α. To define rules for Σ_4 or higher one proceeds in an inductive manner. However, even for Σ_4, the rule becomes rather complex, as indeed is inevitable, as this reflects the extra quantifier and such rules seem to move away from any machine-like intuition.

Another possibility for ITTMs, given that we have relaxed the time element in computations, is to relax the space element. Why not have tapes with cells that are ordered corresponding to some ordinal α? Or indeed why not go the whole hog and imagine a tape with a cell C_β for every ordinal number β? We stay with the original Turing programs but now have the possibility that the R/W head may escape from the finite numbered cells to, first of all, C_ω. We redesign the R/W rules so that at a limit stage its position is now the liminf of previous stages, even if that liminf is ω or some larger limit ordinal. We prevent the head moving back one cell, if it is at a limit cell and the program instructs it so to move (which it cannot do, since by definition a limit numbered cell has no immediate predecessor); then instead by *fiat* it gets sent back to C_0. Otherwise all is as before. Such machines now operate not just with infinite sequences in $2^{\mathbb{N}}$, but actually with potentially ON length such sequences of 0s and 1s. Whilst we will not go into detail here, such sequences potentially could code any set at all (assuming the Axiom of Choice), and one finds that such machines can output (a code of) any set X that occurs in Gödel's hierarchy of constructible sets L, with which he proved the consistency of the Axiom of Choice and the (Generalised) Continuum Hypothesis along with the other axioms of set theory. Such 'ON-ITTMs' thus give another presentation of this L-hierarchy.

If we restrict the lengths of such tapes to suitably closed limit ordinals α, called 'admissible ordinals', then one obtains a presentation of the sets which occurred in theories of 'α-recursion theory' in the early and middle 1970s. (The theoretical match-up here is not precise, because such recursion theories are involved very much with Σ_1 or extensionally defined processes whereas these machines are producing output by using a Σ_2-rule.)

We now consider possibility (ii), of generalising other machines from the finite arena. The immediate candidate is the notion of *register machine* (RM), due to Shepherdson–Sturgis and Minsky. Such a machine has a finite number of registers $R_0, \ldots, R_n$ to hold natural numbers. It also has an instruction set: to set to zero, or to increment by 1, a register; to transfer the contents of register R_i to R_j, say (any $i, j \leq n$); and to *jump* from instruction I_p to I_q if a particular register content is equal to 0. It is a long-established fact that the RM model computes exactly the same functions as the TM model. Indeed a *universal* RM program can be written which requires surprisingly few registers. We may define a transfinite behaviour for such machines, by the same liminf rule for the 'next instruction' at limit stages. Now, for each register R_k, we may set the value of $R_k(\lambda)$ at some limit time λ to be the $\liminf_{\alpha \to \lambda} R_k(\alpha)$ of the previous values *unless* that liminf is ω, in which case we re-set the register value to 0. (In this way we ensure that only natural numbers occupy registers.) This completely specifies the infinite time register machine (ITRM). Several surprises are in store: in contradistinction to the finite case RMs, there is no universal ITRM or program. This is essentially because one can show that, as we increase the number of registers, the power of the machines strictly increases. There is also a marked difference between the power of the ITRMs (taken together) and that of the ITTMs, which are, by a long way, considerably stronger. *Proof theorists* who try to measure the logical strength of various axiomatic theories have analyses of most mathematical systems that occur in mathematics or mathematical physics; however, the strength of the ITTMs outstrips our current proof-theoretical knowledge. ITRMs outrun Kleene recursion (just) since there is a fixed Turing program that when run on an ITRM (with a fixed, but small, number of registers) with an oracle $Z \in 2^{\mathbb{N}}$ can determine whether Z codes a wellordering. Indeed we have an exact measure of the theoretical strength of the class of the ITRMs taken as a whole in terms of a classical theorem of mathematical analysis: the latter states that any *closed* subset, C, of the real line $\mathbb{R}$ is the union of two parts; the first is a countable set of 'isolated' points (meaning that any such point has a neighbourhood which has no other points of C in it); the second part is a 'perfect' set of points, meaning that for every point p in the perfect part, the opposite happens: *every* neigbourhood of p contains an element of C. This theorem can be derived from the statement: *every ITRM with any number of registers, and any oracle Z, either halts or goes into a loop*. And the converse also holds true: from the theorem on closed sets we can deduce the assertion of the ITRM behaviour. We thus have two statements seemingly remote from one another actually being equivalent in the sense that each is provable from the other. (Strictly speaking we should also be specifying a weaker *base theory* in which we make these comparisons.)

Having defined RMs containing numbers, we could allow registers to contain ordinals – as these just generalise natural numbers. Now we no longer need to reset register contents if they get out of hand: if the liminf of a register content becomes infinite, or reaches a limit ordinal, then that ordinal is placed in the register at that limit stage. We have to specify an action if we try to decrease a register by 1 when it contains a limit ordinal; let us say that in this case the register content is set to zero. It can be shown that, remarkably, even with a small number of fixed registers such a machine can correctly decide the truth value of statements about any sets from Gödel's constructible hierarchy L of sets. That this works relies on the fact that any constructible set occurs at some least level L_α of that hierarchy, and each level is itself well-ordered. We may thus label each constructible set with a finite sequence of ordinals $\langle \alpha, \ldots \rangle$; any formula about a finite sequence of such sets is then transformed into one about a finite sequence of ordinals, which may be coded up via pairing functions and submitted to a register machine.

Figure 14.8 The pilot ACE computer.

14.9 Conclusions

What we have not done here is mention other discrete but potentially transfinite processes: automata theorists for example have also models with automata recognising words consisting of infinite sequences of 0s and 1s. We we have not touched on these. There are other models of computation: the Blum–Shub–Smale model of computation on algebraic rings – with one example being the ring of reals $\mathbb{R}$. Such a model also considers computations that halt in finite time where now a real number (which, we have already said, is also an infinitistic object, as in general it requires an infinite expansion of decimal or binary digits to specify) is treated as a single object of computation. Can these models also be remastered into a transfinite format? Lastly, there are many infinite mathematical processes, such as occur in ergodic theory or the theory of group actions. If we leave the machine model behind perhaps such mathematical processes, which are usually considered as limits of ω many natural-number steps, can be extended into the transfinite beyond ω?

As an example one may consider the action of an n-register ITRM as acting on an n-dimensional torus lattice of positive integer points. This gives it a quasi-dynamic feel, and questions asked about functions acting on points of such a torus can be translated into ITRM questions and vice versa. The results on the analytical strength of the ITRMs are then equivalent statements about the functions on the torus. Some of these no doubt will turn out to be rather trivial or uninteresting extensions, but this may not be the case for them all; this is indeed speculative, but there may be some fascinating mathematics to uncover.

In conclusion, then, the above examples illustrate the richness of Turing's original conception: one that goes far beyond his original purposes. We may think in these 'mechanistic' terms in ways that he did not intend but which his model, and his example, has inspired.

References

J.D. Hamkins and A. Lewis (2000). Infinite time Turing machines. *J. Symb. Logic*, 65(2):567–604.

M. Hogarth (2004). Deciding arithmetic using *SAD* computers. *British J. Phil. Sci.*, 55:681–691.

R.I. Soare (2009) Turing oracle machines, online computing, and three displacements in computability theory. *Ann. Pure Appl. Logic*, 160:368–399.

P.D. Welch (2008). Turing unbound: the extent of computation in Malament–Hogarth spacetimes. *British J. Phil. Sci.*, 59(4):768–780.

P.D. Welch (2009). Characteristics of discrete transfinite Turing machine models: halting times, stabilization times, and normal form theorems. *Theor. Comp. Sci.*, 410:426–442.

15

On Attempting to Model the Mathematical Mind

Roger Penrose

Abstract

In his important 1939 paper, Alan Turing introduced novel notions such as ordinal logics and oracle machines. These could be interpreted as possible ingredients of an approach to model human mathematical understanding in a way that goes beyond the conventional ideas of formal systems of axioms and rules of procedure. A hope appears to have been that in this way one might circumvent the limitations to formal reasoning that are revealed by Gödel's incompleteness theorems. In line with such aims, an idea of a *cautious* oracle device is here introduced (differing, in intention, from related ideas put forward by others previously), which is supposed to give accurate answers to mathematical questions whenever it claims to have an answer, but which may sometimes confess to being unable to provide an answer and sometimes continues trying indefinitely without success. Despite such devices seeming to be somewhat closer to human mathematical capabilities than appears to be provided by a standard Turing machine, or Turing oracle machine, they are still limited by being subject to a Gödel-type diagonalization argument. Although leaving open the question of what actual physical processes might underlie human mathematical insight, these arguments appear to indicate a significant constraint on any such hypothetical process.

15.1 Turing's ordinal logics

In early September 1955, I attended a lecture given by Max Newman[1] on the topic of *ordinal logic*. I found the lecture to be one of the most fascinating that I ever

Published in *The Once and Future Turing*, edited by S. Barry Cooper & Andrew Hodges. Published by Cambridge University Press

[1] Max Newman had been a close friend and strong supporter of Alan Turing since their pre-war Cambridge days. Max had also been a good friend of both my parents and, as it would eventually turn out, he would become my stepfather after my father died, nearly 20 years after this talk was given.

attended. Alan Turing had died only a little over a year earlier and this talk was dedicated to him, being essentially based on Turing's 1939 paper on this topic. Newman also started his lecture by providing, as Turing had done in his paper, an introduction to Church's λ-calculus. It has been said that the somewhat limited initial impact that Turing's 1939 paper had on the mathematical community at that time may have been partly due to his phrasing the paper in terms of the λ-calculus, which is hard to employ in an explicit way and makes the reading difficult. Nonetheless, one of the things that did strike me particularly about Newman's lecture was the extraordinary economy of concept exhibited by Church's calculus.[2] No doubt this had impressed Turing also, as he could have presented his ideas in other ways, using his own approach based more explicitly on his notion of a 'Turing machine',[3] which much more directly encapsulates the intuitive notion of *computability*.

The basic idea that Turing made use of in his 1939 paper was that if a particular formal system $\mathscr{F}$ is to be trusted as sound, then the extended system, to which the Gödel sentence $G(\mathscr{F})$ (according to either Gödel's first or second incompleteness theorem) is adjoined as a true statement, must also be trusted as sound. If the ordinal number α is associated with the system $\mathscr{F}$ then the ordinal $\alpha+1$ is to be associated with the extended system. This is fairly straightforward so far, but the idea extends also to *limit* ordinals; so, for example, if the process is continued indefinitely to obtain formal systems associated with successive ordinals α, $\alpha+1$, $\alpha+2$, $\alpha+3,\ldots$ then the entire process can be systematized, to provide us with a still trusted system encompassing the whole lot, and is to be associated with the limit ordinal $\alpha+\omega$. Then the original process is continued again to provide stronger and stronger trusted systems associated with $\alpha+\omega+1$, $\alpha+\omega+2$, etc. This kind of procedure continues for any computable ordinal. It begins to encounter difficulties and ambiguities, however, as the procedure of systemization is found to need new formulations at later stages. By the very nature of what is being attempted, the procedure cannot be systematized as a whole. Moreover, there are related problems that arise, such as the knotty issue of how to tell whether or not two computable ordinals are actually identical.[4]

Nevertheless, I believe that Turing's ideas of ordinal logic, as described by Newman in his lecture, had a considerable influence on my later thinking. As Solomon Feferman has pointed out, in his fine description of Turing's notion of ordinal

[2] I was sufficiently impressed that I tried to persuade Christopher Strachey, when I worked for him briefly at National Development Research Council in 1956, that the λ-calculus might have some value in computer operation (see Penrose, 2000). I was surprised to find, much later, that this kind of idea had become an important ingredient of the programming language LISP.

[3] See Turing (1937a). Church was Turing's research supervisor (Hodges, 1983), which may have influenced his choice.

[4] Turing (1939), Feferman (1962, 1988).

logic,[5] Turing had been interested in the question of what kind of mathematical system might appropriately describe the capacity of human understanding in ascertaining the truth of mathematical propositions. He well appreciated how Gödel's incompleteness theorems demonstrate the human capability to transcend the scope of any particular formal mathematical system whose trustworthiness had already been accepted, and he was interested to see how far this particular human insight could be carried. This was in the years preceding Turing's wartime activities at Bletchley Park, and his own views as to the ultimate nature of human mathematical insight and understanding still appeared to be somewhat fluid and much less constrained to a 'computationalist' perspective than they appear to have become later.[6] His experiences at Bletchley Park brought him into direct contact with the burgeoning of electronic technology that was to become so vital to the physical manifestation of his ideas, and the extraordinary potential power of this technology was then just beginning to make itself evident. Turing's subsequent computationalist perspective on how the mind might operate perhaps became more plausible to him as a consequence of this experience.

In the present commemorative article, I try to build upon Turing's earlier line of thinking, as represented in his 1939 paper, that (idealized) human understanding of mathematical truth might be represented in some way that could transcend the potential capabilities inherent in his own notion of a *Turing* machine.[7] In various earlier writings,[8] I have tried to express the view that Gödel's incompleteness theorems indeed show that human mathematical understanding is something beyond that which can be modelled in terms of strict Turing machine actions and that this tells us something profound about the very nature of the physical processes that underlie conscious brain action. I shall briefly summarize these ideas in Section 15.3 (and see Section 15.8).

15.2 Mathematical trust

The basic philosophical issue is a simple, albeit somewhat subtle, one, clearly well appreciated by Turing himself, although I still find it surprising how many other people still appear to miss the point.[9] The point concerns the nature of mathematical belief, in relation to the idea of formal proof. One aspect of a completely *formal* demonstration of a mathematical proposition P is that all the steps of such a formal argument can be checked entirely mechanically (i.e. computationally), with no further mathematical understanding of the meanings of any of the symbols being

[5] Feferman (1988); see also Hodges (1983, 1988).
[6] Turing (1950)
[7] Turing (1937)
[8] See Penrose (1989, 1994, 1996, 2011a,b).
[9] For example, see M. Detlefsen's response following my article Penrose (2011b).

involved. In order that such a demonstration be considered to constitute an actual *proof* of P, however, it is necessary not only that the rules $\mathscr{R}$ of the formal system be strictly obeyed – which is what the computational checking is intended to achieve – but also that the rules themselves be such that strict adherence to them does indeed provide an argument establishing the mathematical truth of P. In order for us to trust the argument, we must have trust in the *soundness* of $\mathscr{R}$ in relation to the meanings of its symbols. In particular, we must believe that $\mathscr{R}$ is consistent for, if it were not, then both P and its negation would be derivable by the rules $\mathscr{R}$, regardless of whether or not P is actually true. We need this trust in order to *use* $\mathscr{R}$ in the way in which it is intended to be used, so as to establish the truth of P. But Gödel shows us how we can use this very *same* trust to establish propositions that are *beyond* the scope of $\mathscr{R}$. In the case of Gödel's second incompleteness theorem, it is the particular aspect of this trust that tells us to trust $\mathscr{R}$'s *consistency* (since otherwise $\mathscr{R}$ is useless) and that transports our trust in $\mathscr{R}$ to a trust in the Gödel proposition $G(\mathscr{R})$, where it is $G(\mathscr{R})$ that asserts of $\mathscr{R}$s consistency, translated into a proposition of the type that the proof procedures $\mathscr{R}$ are designed to address directly, whereas we note that in this case these very proof procedures are themselves *incapable* of directly establishing the actual truth of $G(\mathscr{R})$!

Thus we see that whereas the *rules* of $\mathscr{R}$ are *not* sufficient in themselves to allow us to transfer our trust in $\mathscr{R}$ to a trust in a particular proposition P, where $P = G(\mathscr{R})$, Gödel's arguments nevertheless do tell us that our understanding of the *meanings* of the concepts underlying the symbols that are being manipulated do indeed allow us to transfer our trust in $\mathscr{R}$ to a trust in P. This is precisely the type of reasoning that Turing himself used when developing his ideas of ordinal logics, as indicated in Section 15.1, in which one's trust in the system associated with an ordinal number α is transferred to one's trust in the system associated with the ordinal $\alpha+1$. In the case of passing to a limit ordinal β we need to be sure that the systematization has actually been carried out correctly, so that our combined trust in *all* the systems associated with ordinals smaller than β can be carried over to a trust in the one associated with β. As pointed out at the end of Section 15.1, there are various difficulties involved in considering these issues for higher ordinals but in the arguments that I shall be providing here, I am not particularly concerned with the 'ordinal' aspects of Turing's 1939 paper, but with another idea that he introduced in this paper, namely that of an 'oracle' machine. I come to this in Sections 15.4 and 15.5.

Another point that needs to be made here is that some specific proposition $G(\mathscr{R})$ that Gödel actually constructs will depend on the particular *Gödel numbering* that happens to be adopted for the symbols that are being employed. There is inevitably some considerable choice in how one actually specifies such a Gödel numbering in full detail. Gödel's original procedures, involving prime factorization of the number involved, are very complicated to use, and some kind of 'alphabetical ordering' is much simpler. Although the particular choice of Gödel numbering that is adopted affects the particular 'Gödel sentence' (e.g. $G(\mathscr{R})$) that one arrives at, this does not affect the overall scope of the system obtained in moving from α to $\alpha + 1$ in Turing's ordinal-logic procedure.[10] In fact, in what follows, I shall not be at all concerned with the details of how the Gödel numbering is to be carried out. All that I require is the *possibility* of carrying out an explicit Gödel numbering according to a strictly *recursive* procedure (i.e. a computable one, with the reverse operation being also computable) – and, moreover, a *trusted* procedure. In what follows, I shall argue for the possibility of doing this by simply by depending upon an unfettered use of "Church's thesis" (in Turing's form[11]). The Gödel numberings that I shall be concerned with here will, accordingly, not actually be made explicit at all; their explicit form would be irrelevant to the arguments being presented.

15.3 Physical processes underlying mathematical understanding?

I have long found it to be a very mysterious fact that whereas there need be *nothing at all sophisticated* about any specific procedure that may be adopted in the Gödel-numbering process, it is nevertheless, the very fact that such a procedure *is* being used that enables us to broaden our trust in a deeply subtle way, from that underlying the given system of proof rules $\mathscr{R}$, to encompass, in addition, a trust in an actual *mathematical proposition* – of the kind that $\mathscr{R}$ is designed to address but happens not to be able to establish directly – namely Gödel sentences such as $G(\mathscr{R})$! What the procedure of Gödel numbering allows us to achieve is to be able to stand back from the formal structure that we are trying to employ and to apply the fruits of our conscious understanding to that very structure itself.

My own perspective on what may be achieved by understanding and insight – in some kind of idealized form – is that this is *not* something that is constrained by the ordinary notions of computation and that we shall indeed need something more. Whereas I am enough of a 'materialist' to believe that whatever mathematicians are capable of achieving by their deliberations must be the result of the functioning of their material brains, operating strictly in accordance with physical laws, I defer from the common view of an entirely *computational* temporal evolution that the picture presented by current fundamental physics seems to imply. Our picture of a computational deterministic universe is rooted in the deterministic and

[10] See Feferman (1962).

[11] It should be made clear that I am referring here to the *original* (Turing) interpretation of 'Church's thesis', i.e. to the fact that any process that is 'obviously computable', in the sense of being a totally mechanical process, can in fact be carried out using Church's λ-calculus – although I do not explicitly refer to the operations of Church's calculus here, relying, instead on the equivalent (see Turing 1937a) notion of what can be carried out according to the procedures encompassed by a *Turing machine*. I explicitly do *not* mean that I take the view that *physical* processes are necessarily of this nature (contrast Deutsch, 1985).

continuous evolution equations of current classical physics, and also in the continuous and deterministic dynamics of the unitary evolution of quantum mechanics, as exhibited in Schrödinger's dynamical evolution.[12] As I have expressed in many places elsewhere,[13] I believe that present-day quantum mechanics is incomplete (and even inconsistent), with its two contrasting evolution processes of continuous Schrödinger dynamics on the one hand, and of the discontinuous 'quantum jumping' that is involved in quantum measurement process, on the other. I contend that, despite its enormous and currently uncontradicted success, present-day quantum mechanics must be only an approximate and provisional description of reality. In this I am in basic agreement with at least four of the pioneers of the subject (Einstein, Schrödinger, de Broglie, and Dirac), but additionally I regard the missing physics that would be needed to complete our present-day quantum picture to be necessarily *non-computational* in nature.

The form of this (in my opinion) needed extension of current quantum theory would have to combine (and develop) foundational ideas from both current quantum mechanics and Einstein's general theory of relativity,[14] and the needed changes would become manifest when sufficient displacements of mass are involved in a quantum superposition, the lifetime of such a superposition being limited by such foundational considerations. Moreover, in order that this might be able to result in significant influences in the way that a conscious human brain might operate – so that some form of non-computational behaviour might be the result – it would be necessary that significantly large-scale quantum coherence can occur within relevant brain structures, so that a level can be reached at which the differences between the putative non-computational ingredients of the missing 'New Theory' and the current quantum/classical expected behaviour might be achieved.

One of the strongest forms of criticism that these ideas have encountered have come, indeed, from seemingly cogent arguments that the 'warm and messy' brain is highly uncongenial to the significant maintenance of large-scale quantum coherence.[15] The conventional picture of a brain operation entirely controlled by the firing of neurons is indeed in conflict with such a role for large-scale quantum coherence. However, there is the definite possibility that neuron firing by no means presents a full picture and represents mainly the broad level of control by the brain, whereas it is nerve transmission that provides both the input of information into the brain and the output that leads to control of the body. It has been argued[16] that

[12] There are, however, some distinct subtleties involved when we try to apply the notions of 'computability', in the strict Turing sense, to continuous systems evolving according to partial differential equations. See, for example, Penrose (1989).

[13] See, for example, Penrose (1994, 2011c).

[14] See, for example, Penrose (2011c, 2014).

[15] See Tegmark (2000); compare Hagan et al. (2002).

[16] See Hameroff and Penrose (1996, 2014).

there is a deeper level of activity which has an important influence on the strength of synapses and that *neuronal microtubules* are fundamentally important at this deeper level of control.

This article is not concerned with such matters of neuro-anatomy, but it is worth pointing out that some remarkable recent experimental research, by Anirban Bandyopadhyay and his colleagues,[17] appears to have demonstrated effects (such as 'ballistic conductance') indicative of the actual presence of the kind of large-scale quantum-coherent activity in body-temperature living (pig-brain) neuronal microtubules that would be an essential pre-requisite of such a picture.

15.4 Π-sentences

The sentences $G(\mathscr{R})$ that arise in Gödel's first and second incompleteness theorems are mathematical assertions of a particularly simple logical form, namely what are called Π_1-*sentences*, these being statements of the form "the computation C never terminates". In the case of the second incompleteness theorem, C would be the computation which seeks, among the theorems of the system, both some proposition P and its negation $\neg P$. Many of the best-known theorems and conjectures in number theory also have the form of Π_1-sentences, such as the famous Fermat's last theorem (proved by Andrew Wiles in 1994), Lagrange's 1770 theorem that every number[18] is the sum of four square numbers, and the still unproved Goldbach conjecture that every even number greater than 2 is the sum of two primes.

The conundrum raised by the inevitable Gödel incompleteness of (sufficiently extensive) formal proof procedures – and, apparently, of the inevitable incompleteness of modelling the ideal notion of mathematical 'trust' in terms of a *single* knowable Turing-computable system, as described in Section 15.2 – leads us to examine some of Turing's other ideas of how one might extend the notion of computability. As noted above, the central issue raised by Gödel incompleteness can be phrased entirely in terms of Π_1-sentences. This leads us to consider the possibility that some kind of *device* capable of ascertaining the truth of such sentences might conceivably allow us to model what the idealized human mind may be, in principle, capable of achieving. Such a hypothetical device is described by Turing's idea of an *oracle*-machine, which involves a component referred to as an *oracle*, able to ascertain the truth of certain types of mathematical assertion. Since the most immediate type of mathematical assertion that arises from Gödel incompleteness has this form of the ascertaining of the truth of a Π_1-sentence, the oracle that is often considered is one which supplies the valid response T (for *true*) or F (for false) when presented with (the Gödel number of) any Π_1-sentence.

[17] See Sahu *et al.* (2013a,b), Hameroff and Penrose (2014).

[18] In this article a *number*, without further qualification, will be a *natural number*: 0, 1, 2, 3, ...

In this article we shall be concerned with the somewhat more general notion of a Π_n*-sentence*, where n can be any natural number. We can consider such sentences that depend, in an explicit Turing-computable way, on a finite number, k, of natural-number variables $x_1, x_2, x_3, \ldots, x_k$. A Π_0-sentence is to be simply an explicit *recursive* propositional function – with values T for *true* and F for *false* – of its variables. That is, for a Π_0-sentence in $x_1, x_2, x_3, \ldots, x_k$, there is an explicit terminating Turing computation, in terms of these natural-number variables, that concludes either T or F for each set of values of $x_1, x_2, x_3, \ldots, x_k$. Then a Π_n-sentence P, for $n > 0$, depending on $x_1, x_2, x_3, \ldots, x_k$, is defined, inductively, as having the form

$$P(x_1, x_2, x_3, \ldots, x_k) = \nexists x Q(x_1, x_2, x_3, \ldots, x_k, x)$$

where Q is some Π_{n-1}-sentence depending, computably, on $k+1$ natural-number variables. Of course, the computable dependence on k variables can always be reduced to a computable dependence on a single variable, e.g. by repeated application of the standard process whereby the two natural numbers v and w can be encoded as the single number u defined by

$$u = \frac{(v+w)^2 + 3v + w}{2}$$

(so that, in the reverse direction, we obtain $v = u - t$ and $w = r - v$, where $t = r(r+1)/2$ is the largest triangular number that is no larger than u, where $r \geq 0$).

It may be noted that there are well-known theorems and conjectures that have the form of Π_2-sentences, the most familiar such conjecture is perhaps the unproved assertion that there are infinitely many prime pairs, i.e. prime numbers p, where $p+2$ is also a prime (since this has the form of an assertion: "there does not exist a prime p for which $p+2$ is also prime and for which there does not exist a prime q with $q > p$, for which both q and $q+2$ are prime"). In fact, Π_2-sentences need not be hard to prove, since Euclid's result that there are infinitely many primes also has the same logical form of a Π_2-sentence.

The notation '$\twoheadrightarrow$' will be used with a Π_n-sentence to mean 'takes the truth value', so $P \twoheadrightarrow \mathsf{T}$ means that P is true and $P \twoheadrightarrow \mathsf{F}$ means that P is false. It may also be noted that any Π_n-sentence is also a Π_{n+1}-sentence. This follows immediately by induction from the fact that any Π_0-sentence P is also a Π_1-sentence since, defining f to be $f(x) = 0$ if $P \twoheadrightarrow \mathsf{F}$ and $f(x) = 1$ if $P \twoheadrightarrow \mathsf{T}$, then the Π_1-sentence $\nexists x(f + x = 0)$ is trivially equivalent to the Π_0-sentence P.

I use the terminology 'Π-sentence' for a Π_n-sentence for which the value of n is not specified. We can make a Gödel numbering of all Π-sentences dependent on a single free variable x: we list all 1-variable Π_n-sentences, for all $n = 0, 1, 2, 3, \ldots$, taken together, in the same list. This will be referred to as the Π-list Gödel numbering. It may be noted that the Π-sentences that have no x-dependence are also included in this Π-list Gödel numbering, as those that simply are independent of x.

15.5 Cautious oracles

I wish to consider a generalization of the Turing machine concept which, although very much built upon Turing's 'oracle' idea, is somewhat more flexible than the oracle notion that Turing put forward in his 1939 paper. The particular concept of a 'cautious oracle' that I am putting forward[19] is aimed at being rather closer to something that could represent the capabilities of an idealized human mathematician than Turing's oracle would be. In putting this forward, I am following the line of thinking, alluded to in Section 15.3, that we do in fact need something beyond what can be achieved by a straightforward Turing machine if we are to have any real chance of modelling the idealized capacity of human mathematical understanding (even allowing that such a machine might need to be bounded in some way in total capacity, accuracy, and allowed time of operation).

Turing's original concept of oracle was some kind of 'black box' which could answer mathematical questions of specific types (e.g. ascertaining the truth of Π_1-sentences that are put to it; in fact, Turing in his 1939 paper was particularly concerned also with oracles able to address Π_1-sentences and Π_2-sentences.[20]) Here I explore a notion[21] of *cautious-oracle device*[22] intended to model a tentative possible view of how, in idealization, mathematical understanding and insight might effectively operate. The "device" is, in effect, a Turing machine with an additional component that I shall refer to as a ϖ-oracle (where the Greek letter "ϖ" is being used, this being a non-standard version of "π" that in some British university courses is often pronounced "pomega"), which can be consulted at any stage of the device's operation. In what follows, I shall insist that the ϖ-oracle never makes mistakes in its pronouncements but that it may sometimes admit to being unable to answer the question put to it and that it may sometimes continue to ponder indefinitely without coming to any conclusion at all.

For definiteness, I shall require that our ϖ-oracle is capable of addressing mathematical assertions that have the specific form of Π-sentences. As far as I can see,

[19] Ideas of this general nature, although generally with a different purpose in mind, have been studied for many years. See Cooper (1990, also 2004).

[20] See Feferman (1988).

[21] See Kleene (1952) for the somewhat related notion of partial recursive function

[22] I prefer to use the word 'device' here rather than 'machine', since to my way of thinking, it is useful, in philosophical argument, to reserve the word 'machine' for something entirely under *computational* control (perhaps with additional genuinely random components). It might, nevertheless, conceivably be possible to construct such a 'device' in a laboratory, provided that some non-computational physics is incorporated into its operation, perhaps in accordance with the ideas of Section 15.3. Some other ides for a non-computational input (not very plausible, in my own opinion) have been proposed by Copeland and Proudfoot (2000) and by Copeland (2002).

the arguments that follow would actually apply equally well to cautious oracles able to assess other types of mathematical assertions provided that these assertions are appropriately well defined[23] and capable of being Gödel-listed. But it is helpful, as here, to have the family of propositions that ϖ is able to assess made completely explicit, as is indeed the case with Π-sentences.

If S is a Π-sentence, then the assessment that ϖ makes when presented with S will be written ϖS, or equivalently $\varpi(s)$, where s is the Gödel number of S in the Π-list. The possible responses that ϖ is allowed to make are:

$\varpi S \twoheadrightarrow \mathsf{T}$	for 'ϖ concludes that S is true'
$\varpi S \twoheadrightarrow \mathsf{F}$	for 'ϖ concludes that S is false'
$\varpi S \twoheadrightarrow ?$	for 'ϖ confesses to being unable to assess S'
$\varpi S \rightsquigarrow \infty$	means ϖ loops trying to assess ϖ, failing to come to a conclusion.

Here 'loops' refers to a continuing action that never terminates. It is required that each of the two responses $\varpi S \twoheadrightarrow \mathsf{T}$ or $\varpi S \twoheadrightarrow \mathsf{F}$, whenever it occurs, is necessarily *truthful* and is trusted to be truthful if ϖ is considered to be a trusted cautious oracle.

These alternatives apply also to the final result of the action of a device $\mathscr{D}_\varpi$ that acts as the kind of Turing 'oracle machine' that is appropriate here, and about which I shall be more explicit in Section 15.6. The symbol $\mathscr{D}$ here refers to the 'Turing-computational' structure of the device, into which alternative possible *particular* choices of cautious oracle might be inserted. If this cautious oracle is indeed the particular one that I am denoting by ϖ, then the resulting oracle device is denoted by $\mathscr{D}_\varpi$. We are interested in the action of $\mathscr{D}_\varpi$ on various types of entity, where these might be Π-sentences, but which could be other types of mathematical object, the only requirement being that the kind of entity being acted upon by $\mathscr{D}_\varpi$ has a *Gödel numbering*. We may express the action of $\mathscr{D}_\varpi$ as directly operating upon the Gödel number g of the entity that we are interested in, where $\mathscr{D}_\varpi$ is to provide a truthful answer $\mathscr{D}_\varpi(g)$ to some specific true/false question about that entity whenever it finds itself to be capable of supplying an answer. As with the cautious oracle itself,

[23] We would get into difficulties, however, if the ϖ-oracle were allowed *directly* to assess the soundness of a $\mathscr{D}_\varpi$-action (the kind of action described in Section 15.6). The computational structure $\mathscr{D}$ can be satisfactorily specified by the Δ-list Gödel numbering of Section 15.6 but not the cautious oracle ϖ itself. Otherwise ϖ could enter in a circular way, and the argument of Section 15.7 could then give rise to a contradiction.

we are allowed the following possible responses:

$\mathscr{D}_{\varpi}(g) \twoheadrightarrow \mathsf{T}$	if $\mathscr{D}_{\varpi}$ reaches the conclusion 'true'
$\mathscr{D}_{\varpi}(g) \twoheadrightarrow \mathsf{F}$	if $\mathscr{D}_{\varpi}$ reaches the conclusion 'false'
$\mathscr{D}_{\varpi}(g) \twoheadrightarrow \mathbf{?}$	if $\mathscr{D}_{\varpi}$ concludes it is unable to reach a definite answer
$\mathscr{D}_{\varpi}(g) \rightsquigarrow \infty$	means $\mathscr{D}_{\varpi}$ loops, continuing indefinitely to try to find an answer.

Again, 'loops' refers to a continuing action that never terminates, there being no implication that there is anything manifestly repetitive in this unending activity. For a *trustworthy* $\mathscr{D}$, it is required that each of the two responses $\mathscr{D}_{\varpi}(g) \twoheadrightarrow \mathsf{T}$ or $\mathscr{D}_{\varpi}(g) \twoheadrightarrow \mathsf{F}$, whenever it occurs, is necessarily *truthful* and is trusted to be truthful if ϖ is considered to be a trusted cautious oracle.

A possible subtlety arises here, in that the 'trustworthy' nature of $\mathscr{D}$ is intended to refer only to $\mathscr{D}$'s computational structure with regard to the intended interpretation of the Gödel numbering of the objects that it addresses. Thus our belief in the trustworthiness of $\mathscr{D}$ is required irrespective of the *particular* choice of cautious oracle ϖ, our trust in $\mathscr{D}_{\varpi}$ being dependent on ϖ only with regard to our trust in ϖ's truthfulness.

15.6 The operation of cautious-oracle devices

The idea of an 'oracle machine' as developed from Turing's original concept is much studied in current literature,[24] often in the context where one is not so much concerned with the 'truth' of the assertions derivable by such a device but more with the scope of what is accessible by their use and the issue of whether one such device might be more or less powerful than another or whether they might, in some appropriate sense, be 'Turing equivalent' to one another.[25] In this general context, a Turing-oracle machine would be a standard Turing machine, which we can take to have a two-way infinite read/write tape with discrete spaces for marks (one space at each integer location) and which, at any particular stage, has a finite number of marks (**1**), the rest being blank (**0**), but where there would be an additional 'read-only' tape, this tape being allowed to have a *fixed* set of marks (**1**) and the rest being blank (**0**), but where the number of **1**s is not restricted to be finite, as it would be the case with a standard Turing-machine read/write tape. The positions on the read-only tape are (as for the read/write tape) taken to be labelled by integers, where there must be a specified starting position on the tape '0'. The specific locations of the **1**s would encode the answers to all the questions that might be posed to the oracle.

[24] See Cooper (1990, also 2004).
[25] See Cooper (1990, 2004) for information on such issues.

A *cautious-oracle* device $\mathscr{D}_{\varpi}$ acting on some number x is taken to operate in essentially the same way, but there are some important complicating issues arising from the fact that ϖ will often provide no clear answer to the question posed to it. It will be useful to think of the action of $\mathscr{D}_{\varpi}$ as consisting of a sequence of 'steps' – as the device reads, or marks its read/write tape, or consults its cautious oracle ϖ. However, we shall find that because of the lack of clear conclusions that ϖ may come up with, we shall be led to consider what has to be thought of as vastly parallel alternative actions going on simultaneously, as ϖ perhaps continues to 'ponder' about some question posed to it, being unable, at that stage, to commit to one of its allowed conclusions $\twoheadrightarrow \mathsf{T}$, $\twoheadrightarrow \mathsf{F}$, or $\twoheadrightarrow$ **?**. It will, indeed, be important that we do not require that ϖ respond immediately when it is presented with (the Gödel number of) some Π-sentence S, so it is allowed to ponder for as many steps as is needed. Only if it continues to ponder indefinitely do we ultimately express this as $\varpi S \rightsquigarrow \infty$. When, at some stage of the operation of $\mathscr{D}_{\varpi}(x)$, ϖ is indeed continuing to ponder, then the notation

$$\varpi S \rightsquigarrow \cdots$$

will be used.

It is necessary that such 'parallel' actions be allowed to take place since otherwise, whenever a $\varpi S \rightsquigarrow \infty$ response were to be the outcome of ϖ, the device would be frozen in its action, waiting forever, and therefore providing us with $\mathscr{D}_{\varpi}(x) \rightsquigarrow \infty$. However, this may not necessarily be the outcome we require of $\mathscr{D}_{\varpi}(x)$ in such circumstances (since in some cases the truth or falsity of S might turn out to be irrelevant to the question that $\mathscr{D}_{\varpi}(x)$ is designed to address, and the fact that ϖ can make no headway with this particular S should not be allowed to prevent $\mathscr{D}_{\varpi}(x)$ from continuing). This is dealt with in a cautious oracle device simply by running both responses $\varpi S \twoheadrightarrow \mathsf{T}$ and $\varpi S \twoheadrightarrow \mathsf{F}$ simultaneously in parallel whenever $\varpi S \rightsquigarrow \cdots$ is first encountered in an oracle consultation, and both these simultaneous branches are continued, as long as the pondering process $\varpi S \rightsquigarrow \cdots$ continues. A similar situation arises when $\varpi S \twoheadrightarrow$ **?**: again we must treat this by running $\varpi S \twoheadrightarrow \mathsf{T}$ and $\varpi S \twoheadrightarrow \mathsf{F}$ in parallel.

Of course, although the direct action of a Turing machine is very much a 'serial' rather than a parallel process, we can perfectly well simulate parallel actions by suitably interleaving them in a serial way and keeping track of everything that is going on with appropriate 'markers'. For ease of description, however, I shall phrase these 'simultaneous' actions *as though* they are indeed being carried out simultaneously, noting that each single 'parallel step' of such a parallel action might encompass vast numbers of 'serial steps' of the interleaved Turing-machine action of $\mathscr{D}$.

We must bear in mind that during such a 'parallel step' there can be a 'pondering' process ($\rightsquigarrow \cdots$) taking place within ϖ; indeed there may well be several such pondering processes going on at once because there may be a later consultation of ϖ while, in an earlier such consultation, ϖ is still in the process of pondering. I am taking the view that ϖ is perfectly capable of interleaving its ponderings concerning each of several different questions that might have been posed to it at various stages, just as it would be if there were several copies of the cautious oracle ϖ all doing their pondering simultaneously but where the total 'serial time' taken by all these operations would be the result of all the "parallel times" added together. This serial-time picture arises if we take the view that there is, in fact, only one instance of the cautious oracle ϖ present in the device, which is sharing its time between all the different Π-sentences being presented to it.

When does all this branching activity come to an end? Well, some of the (normally vast number of) branches will be eliminated whenever a $\rightsquigarrow \cdots$ process becomes $\twoheadrightarrow \mathsf{T}$ or $\twoheadrightarrow \mathsf{F}$. Moreover, if at any stage of the operation *all* branches have terminated (irrespective of the possible continuing presence of any markers that indicate ϖ-pondering) then the ultimate result (**?**, T, or F) of each branch must be examined. If it happens that all are T, irrespective of continuing pondering markers, then we must finally conclude that $\mathscr{D}_\varpi(x) \twoheadrightarrow \mathsf{T}$. If all are F, again irrespective of pondering markers, then we finally conclude that $\mathscr{D}_\varpi(x) \twoheadrightarrow \mathsf{F}$. If neither of these holds (either because *both* T and F appear among different branches or because **?** remains) then the process of awaiting the resolution of incomplete ϖ-ponderings must continue. If (and only if) all branches finally terminate, and that any continuing pondering has ceased to have influence on the ultimate branch outcomes *and* there is either a **?** remaining among the alternative conclusions or both T and F appear among the conclusions, then we must finally conclude that $\mathscr{D}_\varpi(x) \twoheadrightarrow$ **?**. In many situations, however, the entire process will continue indefinitely (which can happen for many different kinds of reason) in which case the 'conclusion' of this entire action is $\mathscr{D}_\varpi(x) \rightsquigarrow \infty$.

I imagine that it would not be a particularly simple task to program all this in terms of explicit Turing machine actions. Nevertheless, it would in principle be a straightforward and completely explicit computational procedure to work out the action of $\mathscr{D}_\varpi(x)$ once we know the Turing-machine structure of $\mathscr{D}$ and the value of x, and we are given the responses of ϖ ($\twoheadrightarrow \mathsf{T}$, F, **?**, or $\rightsquigarrow \cdots$) when they are asked for. Moreover, one could certainly provide a Gödel listing of all the possible such $\mathscr{D}$-actions. As with the Π-listing given by the Gödel numbers for Π-sentences (see the end of Section 15.4), this will provide us with a list of all $\mathscr{D}$-operations acting on a single natural number x. This listing will be referred to as the Δ-*list Gödel numbering*. No attempt is made to provide any sort of 'Gödel listing' of the different possible cautious oracles, however. It is hard to see that this would be at all possible, in fact. Fortunately it is not needed for the argument provided in the next section.

15.7 A Gödel-type theorem for cautious-oracle devices

It may seem remarkable, in view of the general and flexible nature of the notion of cautious oracle, that a Gödel-type diagonal argument can be provided which allows us to extend our trust in any given cautious-oracle device to something lying outside its scope, just as it is with conventional formal systems. In fact, something very similar to the standard Gödel–Turing argument will work. Let us see how this comes about. Suppose we have a trusted cautious oracle ϖ. Then we can make the following definition.

Definition A non-termination assessor $\mathscr{A}_\varpi$ is a ϖ-device whose job is to try to ascertain whether ϖ-device actions fail to terminate. More specifically, we require of $\mathscr{A}_\varpi$ that when it is presented with the Δ-list Gödel number d of an x-dependent cautious oracle action $\mathscr{D}$, and with any particular value of x, it has the property that it can address $\mathscr{D}_\varpi(x)$'s action so that *if* $\mathscr{A}_\varpi$'s action provides the result $\twoheadrightarrow \mathsf{T}$, it has truthfully ascertained that $\mathscr{D}_\varpi(x)$ fails to terminate:

$$\mathscr{A}_\varpi(d,x) \twoheadrightarrow \mathsf{T} \quad \textit{implies} \quad \mathscr{D}_\varpi(x) \rightsquigarrow \infty.$$

What will be shown is that, given any such a trusted non-termination assessor, we can always construct a particular x-dependent ϖ-action which we trust does not in fact terminate, but where $\mathscr{A}_\varpi$ is incapable of ascertaining this fact. To achieve this, we first modify $\mathscr{A}_\varpi$ slightly, to obtain another non-termination assessor $\mathscr{E}_\varpi$ which is almost identical with $\mathscr{A}_\varpi$ in that it has the same property as $\mathscr{A}_\varpi$ as stated above:

$$\mathscr{E}_\varpi(d,x) \twoheadrightarrow \mathsf{T} \quad \textit{implies} \quad \mathscr{D}_\varpi(x) \rightsquigarrow \infty,$$

but where we also require that $\mathscr{E}_\varpi$ is designed so that the alternative terminating responses $\twoheadrightarrow \mathsf{F}$ and $\twoheadrightarrow$ **?** are never obtained. This is easily achieved by adopting the simple expedient of putting $\mathscr{E}_\varpi$'s process into a deliberate loop ($\rightsquigarrow \infty$) as soon as one of $\twoheadrightarrow \mathsf{F}$ or $\twoheadrightarrow$ **?** is obtained by $\mathscr{A}_\varpi$, but where $\mathscr{E}_\varpi$ is otherwise identical to $\mathscr{A}_\varpi$. As with $\mathscr{A}_\varpi$, the response $\twoheadrightarrow \mathsf{T}$ of $\mathscr{E}_\varpi$ is taken to be sound and trusted for all d and x for which $\mathscr{E}_\varpi(d,x)$ produces a definite response, although for many values of d and x we may find that $\mathscr{E}_\varpi(d,x)$ results in $\rightsquigarrow \infty$ and so provides no definite response.

Theorem *Given any trusted cautious oracle ϖ and trusted non-termination assessor $\mathscr{A}_\varpi$, we can construct a specific $\mathscr{D}_\varpi$-action which is trusted not to terminate but for which $\mathscr{A}_\varpi$ fails to establish this fact.*

Proof Put $x = d$ in the above $\mathscr{E}_\varpi(d,x)$ relation to obtain

$$\mathscr{E}_\varpi(d,d) \twoheadrightarrow \mathsf{T} \quad \text{implies} \quad \mathscr{D}_\varpi(d) \rightsquigarrow \infty.$$

Now $\mathscr{E}_\varpi(x,x)$ is a ϖ-device action that operates on just one parameter x, so it must appear in the Δ-list Gödel numbering at, say, the value h, whence

$$\mathscr{E}_\varpi(x,x) = \mathscr{H}_\varpi(x),$$

where $\mathscr{H}_\varpi(x)$ is that particular 1-parameter ϖ-action whose Δ-list Gödel number is h. Now putting $x = h$ we find

$$\mathscr{E}_\varpi(h,h) = \mathscr{H}_\varpi(h),$$

so that taking the particular ϖ-device $\mathscr{D}_\varpi$ for which $d = h$, we have $\mathscr{D}_\varpi = \mathscr{H}_\varpi$ and

$$\mathscr{E}_\varpi(h,h) \twoheadrightarrow \mathsf{T} \quad \text{implies} \quad \mathscr{H}_\varpi(h) \rightsquigarrow \infty,$$

so that, by the equality above,

$$\mathscr{H}_\varpi(h) \twoheadrightarrow \mathsf{T} \quad \text{implies} \quad \mathscr{H}_\varpi(h) \rightsquigarrow \infty.$$

Hence we derive (and thereby trust),

$$\mathscr{H}_\varpi(h) \rightsquigarrow \infty,$$

since the alternative responses or $\twoheadrightarrow \mathsf{F}$ or $\twoheadrightarrow$ **?** are excluded by definition. Moreover, since $\mathscr{E}_\varpi(h,h) = \mathscr{H}_\varpi(h)$, our non-termination assessor $\mathscr{E}_\varpi$ is unable to establish this fact. By the definition of $\mathscr{E}_\varpi$ in relation to $\mathscr{A}_\varpi$ we see that $\mathscr{A}_\varpi$ is also incapable of establishing the non-termination of $\mathscr{H}_\varpi(h)$. □

15.8 Physical implications?

What are we to deduce, from the theorem of Section 15.7, about the physical underpinnings of mathematical understanding? Before attempting to address this issue, I should first briefly refer to the three basic counter-arguments that are most commonly levelled against the type of argument that I, and others,[26] have put forward against an entirely computationalist model of the conscious mind, as based on Gödel–Turing-type theorems:

(1) *Errors argument*: Human mathematicians make errors, so rigorous Gödel-type arguments do not apply.
(2) *Extreme complication argument*: The algorithms governing human mathematical understanding are so vastly complicated that their Gödel statements are completely beyond reach.

[26] See Nagel and Newman (1958), Lucas (1961), Penrose (1989, 1994, 2011a).

(3) *Ignorance of the algorithm argument*: We do not know the algorithmic process underlying our mathematical understanding, so we cannot construct its Gödel statement.

I have tried to argue elsewhere[27] that (1), (2), and (3) do not invalidate the conclusion that our conscious understandings cannot be entirely the product of computational actions, and it is not my purpose to repeat such arguments here. In essence, I do not believe that (1), (2), or (3) really address the *ideal* reasonings for which mathematicians strive, these being mathematical notions which are basically profound and precise rather than complicated and often erroneous. In practice, individual mathematicians may frequently fall significantly short of these ideals, but the ideals are what these strivings are measured against and therefore have a reality of their own beyond what we actually achieve in practice. In essence, it is these *ideals* of reasoning that my arguments are concerned with rather than the operations of particular mathematicians' minds.

In this article I have explicitly been concerned, indeed, with *idealized* mathematical thought, and I have been trying to address a supposition that it might be possible to model such idealized thought processes in terms of ϖ-devices. I would contend, however, that the theorem of Section 15.7 provides evidence that any such device would necessarily fall short of what can actually be achieved by such idealized thought. I would further contend that the mere fact that we are capable of appreciating such ideals and of striving for them is evidence that actually our conscious minds do *not* fully operate in the way suggested by ϖ-devices.

All this notwithstanding, as I have tried to indicate in Section 15.3, I believe that such arguments as I have presented here may offer a few clues as to what might actually be operating in the physical process that underlie the workings of our conscious minds. I have tried to argue previously[28] that there must be a role not only for essentially quantum processes in our conscious thinking, but also for something that necessarily lies *beyond* the quantum theory that we employ today, and which addresses the mysterious boundary that lies between the quantum and classical worlds; it is in this boundary that an undiscovered non-computational physics must be having a central role to play. Do the arguments of this article have anything to say about the nature of this putative non-computational physics? Just possibly so; for it seems to me to be conceivable that there is something in the vastly parallel processes that appear to be necessarily part of the detailed operation of a ϖ-device, as described in Section 15.6, that bears some comparison with the vast quantum superpositions that are the trademark of quantum computation.[29] Yet something important is indeed still missing, and it may well be that the very limi-

[27] See Penrose (1994, 1996).
[28] See Penrose (1994, 2001c).
[29] See Deutsch (1985).

tations of present-day quantum mechanics[30] have something to do with this. Mere unrestricted parallelism is not enough and something beyond this seems to be required, purely from the physical arguments that suggest fundamental limitations on present-day quantum mechanics. Perhaps this issue has some relation to the fundamental limitations of ϖ-devices, as illustrated by the theorem of Section 15.7.

References

Cooper, S.B. (1990). Enumeration reducibility, nondeterministic computations and relative computability of partial functions. In *Recursion Theory Week, Oberwolfach 1989*, K. Ambos-Spies, G. Müller, G. E. Sacks (eds.), Springer-Verlag, pp. 57–110.

Cooper, S.B. (2004). *Computability Theory*. Chapman and Hall.

Copeland, B.J. (2002). Accelerating Turing Machines. *Minds and Machines* **12**(2), 281–300.

Copeland, B.J. and Proudfoot, D. (2000). What Turing did after he invented the universal Turing machine. *J. Logic, Language and Information* **9**(4), 491–509.

Deutsch, D. (1985). Quantum theory, the Church–Turing principle and the universal quantum computer. *Proc. Roy. Soc. Lond. A* **400**, 97–117.

Feferman, S. (1962). Transfinite recursive progressions of axiomatic theories. *J. Symb. Log.* **27**, 259–316.

Feferman, S. (1988). Turing in the land of $O(z)$. In *The Universal Turing Machine: A Half-Century Survey*, R. Herken (ed.), Kammerer and Unverzagt.

Hagan, S., Hameroff, S. and Tuszynski, J. (2002). Quantum computation in brain microtubules? Decoherence and biological feasibility. *Phys. Rev. E* **65**, 061901.

Hameroff, S.R. and Penrose, R. (1996). Conscious events as orchestrated space–time selections. *J. Consciousness Studies* **3**, 36–63.

Hameroff, S. and Penrose, R. (2014). Consciousness in the universe: a review of the 'Orch OR' theory. *Phys. Life Rev.* **11** (1), 39–78 (also 104–112).

Hodges, A.P. (1983). *Alan Turing: The Enigma*, Burnett Books and Hutchinson; Simon and Schuster.

Hodges, A.P. (1988). Alan Turing and the Turing Machine. In *The Universal Turing Machine: A Half-Century Survey*, R. Herken (ed.), Kammerer and Unverzagt.

Kleene, S.C. (1952). *Introduction to Metamathematics*. North-Holland.

Lucas, J.R. (1961). Minds, machines and Gödel. *Philosophy* **36**, 120–124; reprinted in Alan Ross Anderson (1964), *Minds and Machines*, Prentice–Hall.

Nagel, E. and Newman, J.R. (1958). *Gödel's Proof*, Routledge and Kegan Paul.

Penrose, R. (1989). *The Emperor's New Mind: Concerning Computers, Minds, and the Laws of Physics*, Oxford University Press.

Penrose, R. (1994). *Shadows of the Mind: An Approach to the Missing Science of Consciousness*, Oxford University Press.

Penrose, R. (1996). Beyond the doubting of a shadow. *Psyche* **2**(23), 89–129. Also available at `http:psyche.cs.monash.edu.au/psyche-index-v2_1.html`.

[30] See Penrose (2011c).

Penrose, R. (2000). Reminiscences of Christopher Strachey. *Higher-Order Symb. Comp.* **13**, 83–84.

Penrose, R. (2011a). Gödel, the mind, and the laws of physics. In *Kurt Gödel and the Foundations of Mathematics: Horizons of Truth*, M. Baaz, C.H. Papadimitriou, H.W. Putnam, D.S. Scott, and C.L. Harper, Jr. (eds.), Cambridge University Press.

Penrose, R. (2011b). Mathematics, the mind, and the laws of physics. In *Meaning in Mathematics*, John Polkinghorne (ed.), Oxford University Press.

Penrose, R. (2011c). Uncertainty in quantum mechanics: faith or fantasy? *Phil. Trans. Roy. Soc. A* **369**, 4864–4890.

Penrose, R. (2014). On the gravitization of quantum mechanics 1: quantum state reduction. *Found. Phys.* **44**, 557–575.

Sahu S, Ghosh S, Ghosh B, Aswani K, Hirata K, Fujita D, and Bandyopadhyay, A. (2013a). Atomic water channel controlling remarkable properties of a single brain microtubule: correlating single protein to its supramolecular assembly. *Biosens. Bioelectron* **47**, 141–148.

Sahu S, Ghosh S, Hirata K, Fujita D, and Bandyopadhyay, A. (2013b). Multi-level memory switching properties of a single brain microtubule. *Appl. Phys. Lett.*, **102**, 123701.

Tegmark, M. (2000). Importance of quantum coherence in brain processes. *Phys. Rev. E* **61**,4194–4206.

Turing, A.M. (1937a). On computable numbers, with an application to the Entscheidungsproblem. *Proc. Lond. Math. Soc. (ser. 2)* **42**, 230–265; a correction, **43**, 544–546.

Turing, A.M. (1937b). Computability and λ-definability. *J. Symb. Log.* **2**, 153–163.

Turing, A.M. (1939). Systems of logic based on ordinals. *Proc. Lond. Math. Soc.* **45**(2), 161–228.

Turing, A.M. (1950). Computing machinery and intelligence. *Mind* **59**, 236.

Afterword

The world is open for fresh ideas about fundamentals of mind, matter, information, space and time. Turing made just such a plunge in 1936, undeterred by being young and new to the field. Compared with the carefully delineated and trained trajectory expected of modern research students, his début seems amazing. It was extraordinary even then, being without precursor papers or collaborators. But, of course, the young Turing was not alone: Einstein and Eddington, von Neumann and Russell, had spoken volumes to his receptive mind, and he had worked hard through Cambridge mathematics. He was justifiably proud of his breakthrough as a 24-year-old, and correspondingly interested in spotting new youthful talent. His highly unconventional running with young Garner had a parallel: a wish to hand on scientific inspiration to a young person, perhaps in a way quite outside the standard academic framework. This Alan Turing indeed achieved, to some extent, but maybe he has still more to do.

These essays themselves, with their unusual connections and incomplete conclusions, might stimulate new minds with new thoughts. Miguel Walsh, winner of the 2014 Ramanujan Prize, explains how he learned about outstanding problems in mathematics from the Internet, while otherwise isolated in Argentina[1]. We are reminded of how Srinivasa Ramanujan himself emerged from isolation, in the very India where Alan Turing was conceived, through the old mechanism of print and post offices. In the last century, the universal machine has greatly accelerated and amplified such global interaction. Who knows! The history of science depends on the strangest encounters of human brains. The once and future Turing both take their life from the unpredictable and sometimes highly inconvenient magic of human thought.

[1] `http://www.ictp.it/about-ictp/media-centre/news/2014/6/2014ramanujanprize.aspx.`